Imaging from Spaceborne and Airborne SARs, Calibration, and Applications

SAR Remote Sensing

A SERIES

Series Editor
Jong-Sen Lee

Imaging from Spaceborne and Airborne SARs, Calibration, and Applications, *Masanobu Shimada*

ADDITIONAL VOLUMES in PREPARATION

Imaging from Spaceborne and Airborne SARs, Calibration, and Applications

Masanobu Shimada

CRC Press
Taylor & Francis Group
Boca Raton London New York

CRC Press is an imprint of the
Taylor & Francis Group, an **informa** business

CRC Press
Taylor & Francis Group
6000 Broken Sound Parkway NW, Suite 300
Boca Raton, FL 33487-2742

CRC Press is an imprint of Taylor & Francis Group, an Informa business

No claim to original U.S. Government works

International Standard Book Number-13: 978-1-138-19705-3 (Paperback)

Library of Congress Cataloging-in-Publication Data

Names: Shimada, Masanobu, author.
Title: Imaging from spaceborne and airborne SARs, calibration, and applications / Masanobu Shimada.
Description: Boca Raton, FL : Taylor & Francis, 2018.
| Series: SAR remote sensing
| Includes bibliographical references.
Identifiers: LCCN 2018030229 | ISBN 9781138197053 (hardback : alk. paper)
Subjects: LCSH: Synthetic aperture radar.
| Environmental sciences--Remote sensing.
Classification: LCC TK6592.S95 S55 2018 | DDC 621.3848/5--dc23LC record available at
https://lccn.loc.gov/2018030229

Visit the Taylor & Francis Web site at
http://www.taylorandfrancis.com

and the CRC Press Web site at
http://www.crcpress.com

Contents

Preface

I turned 61 years old in March 2016. Time flies by very fast. I now have started my second research life at Tokyo Denki University after retiring from the Japan Aerospace Exploration Agency (JAXA) in 2015. Following my interest in rockets and airplanes during my childhood, I studied aerodynamics at Kyoto University where I received my bachelor's and master's degrees before starting my career at the Japanese Space Agency (National Space Development Agency of Japan, NASDA) in 1979, miraculously one year after the launch of Seasat (although I did not know its importance at the time). Departing from my original field of interest, however, I was assigned to be involved in the microwave remote sensor group, where I later would develop a microwave scatterometer, conducting research on radar backscattering and wind vectors over the ocean. Unfortunately, budget cuts stopped further scatterometer research, and I joined the team developing the receiving station at NASDA's Earth Observation Center.

While there, I learned a lot about remote sensing from the space—such as how to know the position of a satellite from the rotating Earth, how to capture a very fast moving target at the ground station (antenna), how the rotating Earth could be observed from space, how a map could be generated from satellite-based observation, how the geometric and radiometric accuracy of measurements are defined, and much more. It was the end of 1980s, when work on the first Japanese spaceborne SAR (JERS-1) and SAR processing system was ongoing. Development of the SAR processor was strongly connected with that of the main frame, and Japanese six mainframe companies started competitive research on SAR processing. My own SAR research began in the early 1990s at NASA's Jet Propulsion Laboratory (JPL) with estimation of the SIR-B antenna elevation pattern using Amazon Rainforest images obtained by SIR-B. This research allowed me to learn about SAR imaging, the relationship between an SAR image and small scatterers, and about speckle noise distributions. At that time (even more so than today), JPL was the mecca of SAR studies and home base for several quite famous researchers. JPL offered a fantastic research environment supplemented by softball games and mountain hiking, in addition to open discussions on state-of-the-art research.

I was stimulated significantly by these circumstances. At the age of 35, I began the start of my later research life. Since returning to Japan my research has focused on the coding SAR processor on a Macintosh computer with a limited memory and a hard disk. I continue to keep the faith in order to understand the phenomena of simulation work over coding on the computer. This follows the doctrine of my two bosses at JPL, John Curlander and Anthony Freeman. John once told me, "Understanding digital signal processing is quite difficult. Simulation allows you a deeper understanding, and coding for simulation is the best and shortest way to reach that understanding." Being stimulated by these words, I have continued to conduct a life as a researcher for more than 25 years since then. My SAR processor, Sigma-SAR, became JAXA's standard processor for JERS-1, ALOS and the research processor for JERS-1, ALOS, ALOS-2, Pi-SAR-L/L2.

Of course, my own research covers the precise imaging of the spaceborne and airborne SAR, where calibration (including polarimetry) sits at the center and deformation detection, forest classification, and deforestation are the application research that surrounds it. We have created a global deforestation map and produced a processing system for emergency disaster mitigation. It is my great pleasure to work closely with the SAR system—a liner system evolving in technology and variously related to elements of Earth's cultures. In the 1990s, microwave sensors (synthetic aperture radar) were little utilized. Nowadays, their use has expanded so much, it makes me happy to see the ongoing development of the next generation of SARs.

This book was written to summarize my SAR research during my time with JAXA and during my second life at this university. It summarizes the SAR product generation stream algorithm and application from end-to-end. I had been interested in writing a book and, following my old friend Jong-Sen Lee's invitation, I decided to write this one. It will be my great honor to have you read it. If you find an erroneous description, please let me know.

Masanobu Shimada, Dr., Professor
Tokyo Denki University, School of Science and Engineering
April 12, 2016

Epilogue

When I began working with JAXA in the early 1980s, the study of SARs was not my first area of specialty. However, since the late 1980s, my second JAXA career—SAR research—has fascinated and challenged me from both an engineering and a scientific perspective in being addition to being a pleasure to study. Differing from my university major, aerodynamics and fluid dynamics, remote sensing dynamically compounds electromagnetic engineering, information theory, sampling theory, and stochastic processes, with the Newtonian equation as a base, but it also balances well with my original field of study. SAR is now a well-developed system for monitoring our changing Earth. All the challenges of satellite remote sensing have struggled with learning about the Earth's natural systems. On a satellite, SAR is one of the most precise systems from an engineering point of view, while the natural earth is a different type of system, balancing human activities with biophysical and solid Earth elements. The adaptation of SARs for monitoring the dynamic Earth was the one of the goals when the earth observation system was begun in the 1970s. Through this book, I have reviewed the general points of SAR through descriptions of theories and applications. I have extensively described the basics of SARs so that younger students or researchers can better understand them and perhaps consider being future SAR scientists.

While I have enjoyed writing this book, probably I will update or improve the contents in the future, so I am happy to receive comments from readers.

Acknowledgments

In writing this book I would like to specially thank the following colleagues. First, I would like to express my sincere gratitude to all my colleagues at Japan Aerospace Exploration agency (JAXA), especially those at the Earth Observation Research Center (EORC): Drs. Takeo Tadono, Takeshi Motooka, Masato Ohki, Masato Hayashi, Rajesh Thapa, Nicolas Longepe, Ryo Natsuaki, Hiroto Nagai, Shinichi Suzuki, and many more. I would like to express my sincere thanks to Dr. Manabu Watanabe and Christian Koyama of Tokyo Denki University for their great collaboration with me. In addition, I would like to express sincere thanks to all the members of Japan's Remote Sensing Technology Center (RESTEC), especially, Dr. Osamu Isoguchi, Mrs. Hideyuki Noguchi, Takahiro Ohtaki, Atsushi Sugano, and Ms. Keiko Ishii. I would also like to express great thanks to Prof. Yoshio Yamaguchi of Niigata University, who always impressed and supported me with his polarimetric research, Dr. Jon Sen Lee who provided me with the opportunity to write this book as well as the scientific impressions through his life's work, Professor Emeritus Wolfgang Boerner of Illinois University, who always supported my research career and stimulated my research on SAR. I would also like to express my sincere thanks to Dr. Anthony Freeman, who taught me about SAR calibration and how an SAR image is formed theoretically. Finally, I would like to express enormous thanks to my wife, Mrs. Sanae Shimada, who enriched my life not only from an SAR point of view but also in terms of quality of human life.

Author

Masanobu Shimada (M'97–SM'04–F'11) received his BS and MS degrees in aeronautical engineering from Kyoto University in 1977 and 1979, and his PhD degree in electrical engineering from the University of Tokyo in 1999. He joined the National Space Development Agency of Japan (NASDA, formerly Japan Aerospace Exploration Agency, JAXA) in 1979 and conducted research projects for 34 years. During that period, his three greatest achievements were in the areas of (1) sensor developments (i.e., Ku-band Scatterometer, Polarimetric Interferometric Airborne L-band SAR 1 and 2), (2) operational algorithm developments in SAR imaging, interferometry, polarimetry, calibration, mosaicking, and applications, and (3) leading endeavors such as JERS-1 SAR calibration and validation (1992–1998), the JERS-1 science project (global rainforest and boreal forest mapping project and SAR interferometry for detecting deformation distribution), the ALOS science project and PALSAR CALVAL, and the Kyoto and Carbon Initiative project using the time series PALSAR/PALSAR2 mosaics. Among his achievements, those with the most impact were the deformation detection of the Hanshinn-Awaji-Earthquake using the JERS-1 SAR interferometry, generation of the world's first global SAR mosaics depicting annual deforestation change, and experimental near-real-time deforestation monitoring using the ALOS/PALSAR ScanSAR.

His current research interests include high-resolution imaging for spaceborne and airborne SARs (PALSAR-2 and Pi-SAR-L2), calibration and validation, SAR applications including polarimetric SAR interferometry, and MTI using UAVSAR interferometry. Since April 1, 2015, he has been a professor at Tokyo Denki University, an invited researcher at JAXA, and a guest professor of Yamaguchi University, and, beginning in 2018, a lecturer at the Nara Women University.

1 Introduction

1.1 BACKGROUND

Global warming is one of the serious issues that humans face, and they should take action for their future survival. Global warming occurs due to interactions between humans and nature and has accelerated since the Industrial Revolution of the late 18th century. During the last several decades, thematic understanding of the Earth's environment has increased our knowledge about nature's diversity, warning us about the impact of man's actions on our way of life. Since the early 1970s (and particularly with the development of the Landsat program in 1972), the use of remote sensing technologies based on space- and aircraft has been responsible for much of our understanding about the Earth. The Paris Agreement at the 21st Conference of the Parties to the UN Framework Convention on Climate Change (COP21) implies that the world has decided to take action in fighting the gradual increase of carbon emissions. To this end, remote sensing using spaceborne and/or airborne equipment plays a very important role in observing and measuring the Earth's surface components: forest, ocean, sea-ice, natural disasters, and so on.

Within the various remote sensors, synthetic aperture radar (SAR) has the simplest but strongest functions capable of measuring the backscattering properties of targets (on the Earth), and the distance between the target and the sensor, with high resolution and high accuracy. Knowing and measuring the backscattering properties of a target allows that target's identification and helps in the interpretation of that target on the Earth.

Global environmental changes, such as warming, ozone depletion, natural disasters (floods and hurricanes), and earthquakes, require a total understanding of the Earth's system. The Earth is a very complex thermodynamic ecosystem, balancing solar energy, radiation, and the inhabitants' energy consumption. If the energy interactions among the elements (i.e., forests, desert, water, land, ice, and snow) are expressed accurately, we may obtain knowledge of some key factors for describing the mechanisms of those environmental changes, which may result in our developing ideas about how to prevent or delay them. These phenomena occur almost randomly on different spatial and time scales (several kilometers to several thousand kilometers and several seconds to several years). Earth-orbiting satellites carrying the appropriate sensors may detect these phenomena because of their large spatial coverage and frequent availability. Active microwave instruments using relatively low-frequency signals (i.e., several hundred megahertz to several gigahertz) are vital for observing the Earth because those signals are self-generated, do not rely on sunlight, and can directly carry the Earth's surface information without being affected by the atmosphere.

SAR is an active microwave instrument that performs high-resolution observation under almost all weather conditions. The measurements are made using backscattered signals from targets on Earth in response to transmission codes. The optimal correlation of these signals with the ideal SAR receiving signal achieves high-resolution imaging of several meters. These correlated signals (or simply SAR images) contain information on targets, wave-propagation media, the distance between the SAR and the target, and the SAR characteristics. If the image is well modeled for these targets, the SAR observation can retrieve the target itself. However, there are two difficulties. The first is how well the SAR characteristics are eliminated from the SAR images and how accurately the signal power or the backscattering coefficient of the target is determined. The second is how accurately the backscattering model can be built for each target. The first difficulty is the so-called SAR calibration. Because an erroneous calibration of SAR images causes an incorrect understanding of the targets, the SAR calibration is very important.

1.2   SHORT HISTORY OF THE SAR

High-resolution imaging by the two-dimensional correlation of a scattered signal can be performed optically or digitally. The SAR imaging principle was first developed by Wiley (1965) in the 1950s and was verified by the end of the 1950s, using the experimental SAR instruments and optical image processor (Ulaby et al. 1982). Digital SAR processing and its application evolved after an effective computing approach was proposed by Wu (1976). Wu's approach reduced the heavy mathematical load by employing the Fast Fourier Transform (FFT) for the range and azimuth correlation processing (this method is called the range-Doppler method). SAR imaging algorithms were developed in many ways, such as range-Doppler, the seismic method (Caffirio et al. 1991), and chirp scaling (Raney et al. 1994). All these algorithms have reduced possible distortion and have yielded higher throughput.

Seasat was the first satellite to have spaceborne SAR. Developed by NASA's Jet Propulsion Laboratory (Jordan 1980), it was launched into an 800-km polar orbit in 1978 with other microwave instruments as a "proof of concept" mission. This SAR featured a high resolution of 30 m (in azimuth, four looks) and 10 m (in slant range), L-band and HH polarization (horizontal transmit and horizontal receive), a fixed off-nadir angle of 23 degrees, an imaging width of 100 km, a passive array antenna that was 12 m by 2 m, a transmission power of 1,000 W, an onboard data formatter that enabled on-ground digital data generation, and repeat-pass interferometry, which was arranged after the launch. Although the Seasat mission was terminated three months after the launch due to a power unit failure, most of the SAR missions succeeded. The huge amount of data acquired has advanced SAR technologies not only in sensor development but also in digital signal processing, data interpretation, and in Earth sciences. Following Seasat, the shuttle imaging radar SIR-A in 1982 and SIR-B in 1984 (Cimino et al. 1986) promoted radar science. The SARs onboard the European resource-sensing satellites ERS-1 (1990) and ERS-2 (1995) (Joyce et al. 1984), the Japanese Earth Resources Satellite JERS-1 (1992) (Nemoto et al. 1991; Yoneyama et al. 1989), and Radarsat (1995) (Raney et al. 1991; Ahmed et al. 1990) are being used more operationally at different frequencies, polarizations, and incidence angles. To increase the measurement consistency (same time, same place) at different radar parameters, SIR-C/X-SAR (1994) (Jordan et al. 1991; Huneycutt 1989) was launched on a later space shuttle mission. The data acquired in the two ten-day missions covered 19% of the Earth surface (100 million km^2) within ± 60 degrees of latitude in the L, C, and X bands with full polarizations and wide incidence angles. Almaz, the Russian SAR system, operated at the S band. Moon surface observation by Apollo 17 and observation of Venus by the Magellan SAR (Kwok and Johnson 1989; Ford et al. 1989) are planetary applications of SAR technology (Way and Smith 1991).

Compared to the 1990s when the SAR system was developed and launched with more operational intentions, the 21st century has seen more advanced SAR system functions and enhancements. L-band SARs, such as ALOS/ALOS-2, which were launched in 2006 and 2014, respectively, are used for operationally monitoring the Earth's surface, utilizing the L band's signal penetration capabilities for detecting deformation of the land surface and for forest observations. C-band SARs—such as Envisat, Sentinel-1, and RADARSAT-2—were also launched and used for operational monitoring of the Earth. X-band SARs, such as TerraSAR-X, TanDEM-X, and COSMO-SkyMed, were launched in the polar orbit for monitoring the high-frequency SARs. The SARs in the 2000s and 2010s have been significantly evolved in fully adopting enhanced digital technologies (i.e., distributed transmit-receive modules, the systematic driving synchronized with the relative position between the satellite and the Earth's surface) and the satellite technology on orbit maintenance. Digital processing method was the state of the art technology to counter with the relevant issues raised during the development of the SAR system so far, although the system deging and manufacturing has become significantly complicated.

1.3   THE SCOPE OF THIS BOOK

This book covers a variety of SAR topics, especially data processing and utilization. Topics covered in this volume's 15 chapters include SAR imaging theory (strip mode and ScanSAR [scanning synthetic aperture radar] mode), radiometric models (including saturation and speckles), geometric expressions (including ortho and slope corrections), interferometric SAR processing for digital terrain model (DTM) generation and surface deformation detection, corrections for the troposphere, large-scale SAR imaging and SAR mosaicking, and application research (i.e., forest observation and disaster observation).

This chapter provides a general introduction to SARs. Chapter 2 offers a general description of SAR hardware. Chapter 3 outlines SAR imaging basics, and Chapter 4 introduces a radiometric model. Chapter 5 covers ScanSAR imaging, Chapter 6 polarimetric calibration, Chapter 7 antenna pattern calibration, Chapter 8 geometric and ortho-rectification, and Chapter 9 covers the calibration of the SAR image. In Chapter 10, SAR imaging for a moving target is discussed, and Chapter 11 covers large-scale strip processing and mosaicking. Chapter 12 discusses interferometric SAR processing, and Chapter 13 addresses varied and irregular SAR images. Chapter 14 outlines the application introduction sampled for forest observation and landslide detection, and Chapter 15 covers forest-non-forest map generation, conclusions and acknowledgments.

REFERENCES

Ahmed, S., Warren, H. R., Symonds, M. D., and Cox, R. P., 1990, "The Radarsat System," *IEEE T. Geosci. Remote*, Vol. 28, No. 4, pp. 598–602.

Caffirio C., Prati, C., and Rocca, F., 1991, "SAR Data Focusing Using Seismic Migration Techniques," *IEEE T. Aero Elec. Sys.*, Vol. 27, No. 2, pp. 194–207.

Cimino, J. B., Elachi, C., and Settle, M., 1986, "SIR-B the Second Shuttle Imaging Radar Experiment," *IEEE T. Geosci. Remote*, Vol. GRS-24, No. 4, pp. 445–452.

Ford, J. P., Blom, R. G., Crisp, J. A., Elachi, C., Farr, T. G., Saunders, R. S., Theilig, E. E., Wall, S. D., and Yewell, S. B., 1989, "Spaceborne Radar Observations, A Guide for Magellan Radar-Image Analysis," NASA, Jet Propulsion Laboratory, California Institute of Technology, Pasadena, CA.

Huneycutt, B. L., 1989, "Spaceborne Imaging Radar-C Instrument," *IEEE T. Geosci. Remote*, Vol. 27, No. 2, pp. 164–169.

Jordan, R. L., 1980, "The SEASAT A Synthetic Aperture Radar System," *IEEE J. Oceanic Eng.*, Vol. OE-5, No. 2, pp. 154–163.

Jordan R. L., Huneycutt, B. L., and Werner, M., 1991, "The SIR-C/X-SAR Synthetic Aperture Radar System," *Proc. IEEE*, Vol. 79, No. 6, pp. 827–838.

Joyce H., Cox, R. P., and Sawyer, F. G., 1984, "The Active Microwave Instrumentation for ERS-1," in *Proc. IGARSS 84 Symp.*, Strasbourg, France, ESP SP-215, pp. 835–840.

Kwok R. and Johnson, W. T. K., 1989, "Block Adaptive Quantization of Magellan SAR Data," *IEEE T. Geosci. Remote*, Vol. 27, No. 4, pp. 375–383.

Nemoto Y., Nishino, H., Ono, M., Mizutamari, H., Nishikawa, K., and Tanaka, K., 1991, "Japanese Earth Resources Satellite-1 Synthetic Aperture Radar," *Proc. IEEE*, Vol. 79, No. 6, pp. 800–809.

Raney K., Luscombe, A. P., Langham, E. J., and Ahmed, S., 1991, "Radarsat," *Proc. IEEE*, Vol. 79, No. 6, pp. 839–849.

Raney R. K., Runge, H., Bamler, R., Cunning, I. G., and Wong, F. H., 1994, "Precision SAR Processing Using Chirp Scaling," *IEEE T. Geosci. Remote*, Vol. 32, No. 4, pp. 786–799.

Ulaby, F., Moore, R., and Fung, A., 1982, *Microwave Remote Sensing: Active and Passive, Volume 1, Fundamentals and Radiometry*, Addison-Wesley, Boston, MA, p. 8.

Way, J., and Smith, E. A., 1991, "The Evolution of Synthetic Aperture Radar Systems and their Progression to the EOS SAR," *IEEE T. Geosci. Remote*, Vol. 29, No. 6, pp. 962–985.

Wiley, C. A., "Pulsed Doppler Radar Methods and Apparatus," U.S. Patent 3,196,436. Field August 13, 1954, patented July 20, 1965.

Wu, C., 1976, "A Digital Approach to Produce Imagery from SAR Data," Paper No. 76-968, presented at the AIAA Syst. Design Driven by Sensors Conf., Pasadena, Ca, October 18–20, 1976.

Yoneyama, K., Koizumi, T., Suzuki, T., Kuramasu, R., Araki, T., Ishida, C., Kobayashi, M., and Kakuichi, O., 1989, "JERS-1 Development Status," 40th Congress of the International Astronautical Federation, 1989, Beijing, China, IAF-89-118.

2 Introduction of the SAR System

2.1 INTRODUCTION

Since its origin in the 1950s, the SAR system has continued to evolve. Performance has improved due to state-of-the-art hardware developed in tandem with other technological improvements. However, user requirements have become more precise and quantitative in order to enable geophysical parameter measurements. As a result, the SAR system and its processing algorithm have become very complicated and difficult to understand. In order to advance comprehension of these details, this chapter will introduce and explain the SAR hardware system, using the Japan Aerospace Exploration Agency's (JAXA) four generations of SARs as examples.

2.2 SAR SYSTEMS

As the SAR flies over a target, it repeatedly transmits microwave signals (pulses) to that target and receives corresponding backscattered signals from the target. Every target that stays on the ground is illuminated by the SAR signal as long as the SAR antenna beam (and its directional sensitivity) continues to cover it. The SAR records all signals returned as the phase history of radar backscatter from the targets. The two-dimensional correlation between the time-series received signals and the reference signals in range and azimuth produces a high-resolution SAR image, which is enhanced by the transmitted signal bandwidth and the Doppler bandwidth formed by the SAR-target relative motions within the SAR antenna pattern. In general, SAR imaging follows the sampling theorem on the target plane and limits the imaging swath and the resolution.

Because of advances in microwave hardware technology in the last several decades, the split-band allows wide-swath imaging by emitting the transmission energy diversely in a plane perpendicular to the direction of movement. Scanning synthetic aperture radar (ScanSAR) technology enables a swath of 350 km or 490 km, which is several times wider than the normal strip mode, as well as the use of dual receivers. The transmit-receive modules allow higher signal-to-noise ratio imaging, thereby suppressing electric discharging in space.

Most SARs typically are onboard flying carriers—either manned vehicles (aircraft) or unmanned vehicles (spacecraft and aircraft). An SAR captures the reflected signals from a target in response to its own transmission (and the temporal signal trajectory is extracted from the raw data sets and correlated with the reference signal in order to be transformed into the high-resolution image). However, some SARs sit on the ground to capture properties of moving targets: aircraft formation, rotation, and high-resolution imaging. Both types of SARs are equivalent because measurements are performed under relative motion. In order to coordinate relative motion, the SAR needs to measure precisely the range delay that corresponds to the distance between the target and the system. Figure 2-1 shows all the JAXA/SAR systems developed from the 1990s to the 2010s that have been used or currently are being used. Table 2-1 and Table 2-2 show the representative parameters of the international SARs onboard the spaceborne and airborne systems, respectively.

2.3 BLOCK DIAGRAM

In principle, the SAR consists of four major subsystems: transmitter, antenna, receiver, and signal processor. From the SARs that have been developed and operated (examples of which are listed in Tables 2-1 and 2-2), we have selected four Japanese SARs to introduce the sensor functionality

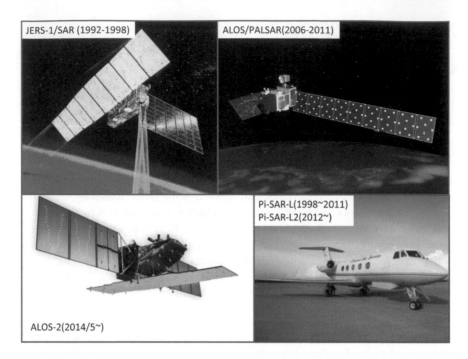

FIGURE 2-1 Illustrations of the three Japanese spaceborne SAR systems and one airborne SAR system. Upper left: JERS-1 SAR; upper right: ALOS/PALSAR; lower left: ALOS-2/PALSAR-2; and lower right: polarimetric interferometric airborne SAR in L band.

and its specialties: JERS-1 SAR, ALOS/PALSAR, ALOS-2/PALSAR-2, and the airborne SAR (Pi-SAR-L2). All of these examples are L-band systems developed beginning in the early 1990s on through the present day. Their measurement concepts obey the sampling theory on the radar returns from the target in frequency and time domains. An artist concept of JERS-1 and a simplified block diagram are shown in Figures 2-1 and 2-2, respectively. The key specifications for each of the four SAR systems are summarized in the following sections.

2.3.1 JERS-1 SAR (1992–1998)

In JERS-1, a transmitter created the cascaded transmission pulses, which were 325-W high-powered, frequency-modulated (chirp-modulated) transmissions with a 35-μs pulse width. These were produced by a pulse generator convolved with the surface acoustic wave (SAW) filter. Five pulse repetition frequencies (PRFs) of 1505.8 Hz, 1530.1 Hz, 1555.2 Hz, 1581.1 Hz, and 1606.0 Hz were prepared, and one of them assisted JERS-1 SAR in fully covering the latitude-dependent Doppler bandwidth. The relatively small transmission power of 325 W was selected in order to avoid the discharging at the edge of the antenna when activated with 1,300W (Nemoto et al. 1991).

The antenna consisted of eight 2.2 m by 2 m sub-antenna paddles, each of which contained 128 planar patches and generated the antenna gain directivity in space, with the HH polarization (horizontal transmit and horizontal receive) producing a 3-dB down beam width of 5.4 degrees in range and 0.98 degrees in azimuth. The signal from the transmitter was emitted into space through the antenna in the direction of interest. The antenna then received the scattered signals from the target and handed them over to the receiver. The antenna was 12.2 m long in the along-track direction (the direction in which the satellite was moving), 2.2 m wide in the range direction, and located on the right-hand side of the satellite body. This large-scale antenna was deployed in space through three-step procedures using mechanical springs and coils. Unfortunately, the

TABLE 2-1
SAR Systems (Spaceborne)

Name	Duration	Frequency	Polarization	Band Width (MHz)	Off-Nadir (deg.)	Pt (W)	Mode-Swath (km)	AD-(Gain)
Seasat (NASA)	1978/6/28-1978/10/10	L	S (HH)	19	20	1K	STRIP-100	FM
SIR-A (NASA)	1981	L	S (HH)	6 M	47(Mechanical)	1K	STRIP-50	
SIR-B (NASA)	1984	L	S (HH)	12 M	15–55 (Mechanical)	1.12 K	STRIP 30-60	3–6 bit
SIR-C (NASA)	1994	L-C	Quad		20–50(Electrical)	4.4/1.2 K	STRIP-Spotlight, ScanSAR-15-55	4 bit
X-SAR (DLR-ASI)	1994	X	S (VV)	10–20 MHz	20–49(Electrical)	5 K	STRIP-15-60	
ERS-1/2 (ESA)	1990/7/17	C	S (VV)	15.55 M	20	4.8 K	STRIP-100	51 + 5Q (MGC)
JERS-1 (JAXA)	1992/2/1-1998/10/12	L	S (HH)	15 MHz (17.076 MHz)	35.2	325	STRIP-75	31 + 3Q (AGC)
Radarsat-1 (CSA)	1995	C	S (HH)	11.6, 17.3, 30.0	20–49	5 K	Strip-ScanSAR 100-500	41 + 4Q (AGC)
Envisat-ASAR (ESA)	2000	C	D (HH + VV)	16 M	15–45		Strip-ScanSAR:100-400	41 + 4Q (BAQ)
ALOS-PALSAR (JAXA)	2006/1/24-2011/5/12	L	S, D, Q	28/14 MHz(32.0/16.0 MHz)	9.7–52	2 K	Strip-ScanSAR (70-350)	51 + 5Q (MGC)
Terra-SAR-X (DLR)	2007/06/15-	X	Q	300 M	-		Strip-ScanSAR-spotlight-10 ~ 100 km	
Radarsat-2	2007/12/14	C	Q	<100 M	10–60	1.65 K -2.28 K	Strip-ScanSAR-spotlight-10 ~ 100 km	
Sentinel 1A & B	2014/4/3	C	HH + HV, VV + VH, HH	<100 M	20–46	~ 4.4 K	Strip+Wideswath	10 bit
ALOS-2/PALSAR-2 (JAXA)	2014/5/24-	L	S-D-Q	84/42/28/14 MHz (103 MHz)	7–59	6 K	Strip-ScanSAR-Spotlight (50-350/490-25)	41 + 4Q (AGC + ABQ)

Source: https://directory.eoportal.org/web/eoportal/satellite-missions/s/seasat

Note: Q: Quad (HH + HV + VH + VV), D: Dual (HH + HV or VH + VV), S: Single (HH, HV, VH, VV)

TABLE 2-2
SAR Systems (Airborne), Samples

Name	Launched	Frequency	Polarization	Pt (W)	Bandwidth (MHz)	Mode-Swath	AD
UAVSAR (IPL)	2007~	L	Q	3.1K	80	Strip	12 bit-real
Pi-SAR-L2 (JAXA)	2012	L	Quad	3.5K	85	STRIP-20km	8I+8Q
Pi-SAR-2 (NICT)	2008	X	Quad + InSAR	5K	500/300/150	STRIP-20	8I+8Q
F-SAR (DLR)	2006	X-C-S-L-P	Quad	2.5/2.2/2.2/ 0.7/0.7K	800/400/300/ 150/100	STRIP-12.5	8 bit-real
RAMSES (ONERA)	2008	P-L-S-C-X- Ku-Ka-W	Quad		75/200/300/300/ 1200/1200/ 1200/500		

JERS-1 SAR observations were delayed in the very early phases due to an antenna deployment problem after the launch.

The receiver amplified the received signal using the low-noise amplifier (LNA) and demodulated (down converted) it using the coherent oscillator's output so that the signal could be handled at the signal processor adopted later. The signal level was always monitored at the short timing window and adjusted to about 0 dBm by changing the attenuator in one decibel step—the automatic gain control (AGC). The sensitivity time control (STC) was also applied in time to cancel the signal power variation within the observation window (360 µs), which was caused by the antenna elevation pattern and time variation. It should be noted that the AGC and STC made the radiometric calibration more difficult due to the lower SNR conditions.

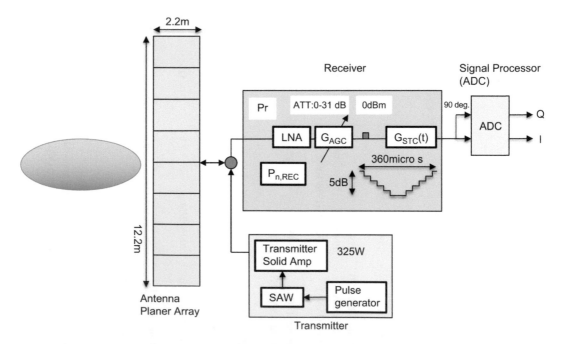

FIGURE 2-2   Simplified block diagram of JERS-1 SAR.

The signal processor divided the incoming signal into the in-phase (I-phase) component and the quad-phase (Q-phase) component. The latter phase was made 90 degrees ahead of the I-phase using a phase shifter. These signals were digitized in a 3-bit at 17.076 MHz sampling frequency. JERS-1 SAR was originally designed to find oil reservoirs in mountain folds by using the basic SAR system. The SAR could not be implemented on the yaw-steering satellite system; thus, the Doppler frequency varied greatly (within a range of +2,000Hz to approximately −2,000Hz) during the satellite's flight path around the Earth.

2.3.2 ALOS/PALSAR (2006–2011)

ALOS/PALSAR (Advanced Land-Observing Satellite/Phased-Array L-band Synthetic Aperture Radar) was the second-generation L-band SAR that appeared after JERS-1 SAR and incorporated high-performance technology (Itoh et al. 2001) such as:

1. A digital chirp-signal generator that produced a stable analogue chirp signal through the digital-to-analogue (DA) converter;
2. Eighty transmit-receive modules (TRMs) equipped with 5-bit phase shifters that produced the maximum 2,000 W of transmission power with a narrow beam width, which rapidly changed the bore-site elevation direction, enabling ScanSAR and strip imaging without discharging in space;
3. Polarimetric observation enabled by activating the H and V transmitting waves and cascading their reception;
4. A yaw steering mode always enabled to set the Doppler frequency nearly zero; and
5. Satellite timing fully synchronized with the GPS satellite.

In addition, PALSAR was operated in MGC mode and did not apply the STC mode so the received signal was not distorted by the phase variation that often occurred at a quick gain change in digital units (STC). Figure 2-3 shows the simplified block diagram of ALOS/PALSAR.

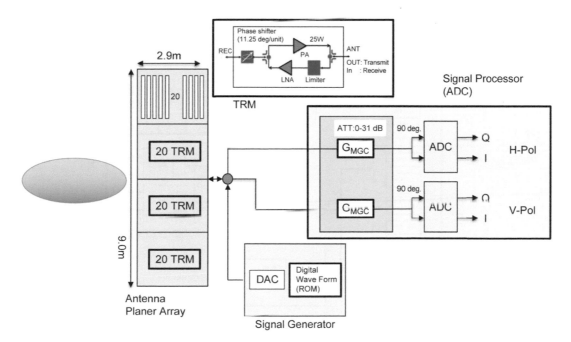

FIGURE 2-3 Simplified block diagram of PALSAR.

2.3.3   ALOS-2/PALSAR-2 (2014–)

ALOS-2/PALSAR-2 is the L-band SAR that followed ALOS/PALSAR. It incorporates contemporary state-of-the art technology (Kankaku et al. 2014), for example:

1. Dual receiver systems implemented with lower pulse repetition frequencies that are about half of what a single receiver requires, thereby allowing a relatively wider imaging swath while maintaining high resolutions in range and azimuth under the sampling theorem;
2. Azimuth phase coding that conducts the up-down chirp variation convolved with the pseudo-randomized 0 and π shifted in phase for suppressing the range ambiguities on the order of ten decibels;
3. An increase in the number of TRMs from 80 to 180 and in transmission power from 25W to 30W, thereby allowing beam steering in azimuth and range with a maximum transmission power of 6,000W for spotlight, strip, and ScanSAR modes;
4. Two data compression methods applied with block adaptive quantization (BAQ) and frequency-reduced BAQ;
5. Accurate attitude control using a star tracker system supported by a GPS system and autonomous orbit control for maintaining an orbital corridor with a radius of 500 m with 14-day repetitions.

In summary, the higher transmission power combined with the lower orbit height of 628 km improves the signal to noise ratio. A block diagram of the PALSAR-2 is shown in Figure 2-4.

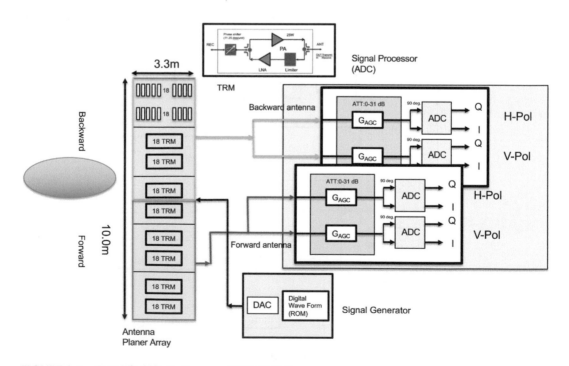

FIGURE 2-4   Simplified block diagram of PALSAR-2.

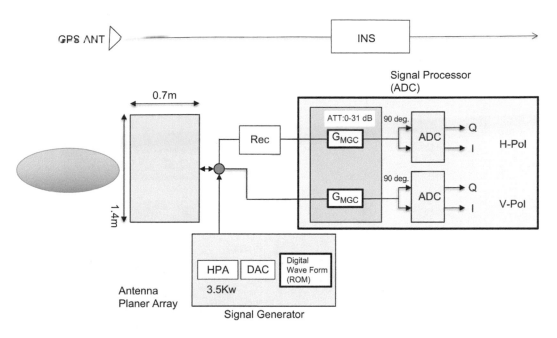

FIGURE 2-5   Simplified block diagram of Pi-SAR-L2.

2.3.4   Pi-SAR-L/L2 (1997–)

This airborne L-band SAR was developed as the pathfinder to the future spaceborne SAR after JERS-1 SAR. The first-generation SAR was operated between 1997 and 2012, and the second generation began operation in 2012. This SAR was designed as the most stable airborne SAR (Shimada et al. 2013) with features including:

1. Full polarimetry SAR equipped with two 8-bit AD converters,
2. Highly accurate integrated navigation system (INS),
3. High transmission power (3,500W), and
4. Installation on the Gulfstream II, jet stream airplane (see Figure 2-5).

2.4   HARDWARE COMPONENTS (TRANSMITTER, RECEIVER, ANTENNA, AND SIGNAL PROCESSOR)

Figure 2-6 has several photos of SAR system hardware components, using the ALOS/PALSAR as an example

2.5   SATELLITE SYSTEMS AND AIRBORNE SYSTEM

The three satellite systems, JERS-1, ALOS, and ALOS-2, were manufactured in different decades—the 1990s, 2000s, and 2010s, respectively, with the Pi-SAR systems beginning operation in 1997. Their performance differences are obvious, especially in datation, orbit determination accuracy, maneuvering, and attitude maintenance and determination. Each system is described in the following sections. Summaries of SAR satellite systems are listed in Table 2-1 (spaceborne) and Table 2-2 (airborne).

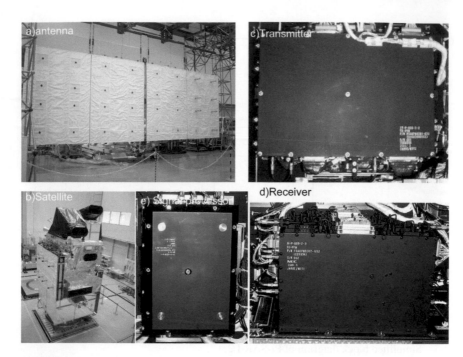

FIGURE 2-6   Hardware examples from ALOS/PALSAR: (a) antenna, (b) transmitter, (c) satellite, (d) signal processor, and (e) receiver. Images (a) and (c) are copyrighted by JAXA and (b), (d), and (e) by J-spacesystems/JAXA.

2.5.1   JERS-1

In the JERS-1 system, timing and datation relied on the internal crystal oscillator, the stability of which was not known. However, the timing accuracy was known to be 2^{-5} seconds and normally was shifted 2 seconds during a month. Calibration of the satellite time was mandatory to make the satellite operation consistent within the ground reference system. It was conducted as follows: First, the difference between the satellite time and the inter-range instrumentation group time codes-A (IRIG-A) of the ground station in Hatoyama were measured as the satellite passed over the ground station, and then the timing correction command was uploaded to the satellite. Orbital maintenance was performed on a weekly basis mainly by altitude maneuvering (thrustering, or lifting, the orbit height against the air drag caused by height descent), and the orbital determination was performed by range and range rate measurement and modeling. Orbit positioning was not accurate enough—the geolocation accuracy reported was larger than a hundred meters. The geometric application needed geometric calibration before using the DEM or ground control points.

Satellite attitude was maintained using the classical roll-pitch-yaw estimator. Yaw steering was not adopted. The antenna bore site direction in azimuth remained constant at 90 degrees, moving clockwise around the satellite (Figure 2-7).

2.5.2   ALOS/PALSAR

Datation was perfectly synchronized with the GPS time, which was provided from GPS satellites and received by ALOS. The timing accuracy was significantly improved over that of JERS-1, with that of ALOS being much less than 1.0 µs and that of JERS-1 being 10^{-5} to 10^{-6} seconds. Orbital determination was performed in three ways: (1) using the GPS positioning information with an accuracy of several tens of meters; (2) with the post-processed precise positioning information using the L1 and L2 carrier frequency synchronized with the IGS data corrected for the ionospheric delay

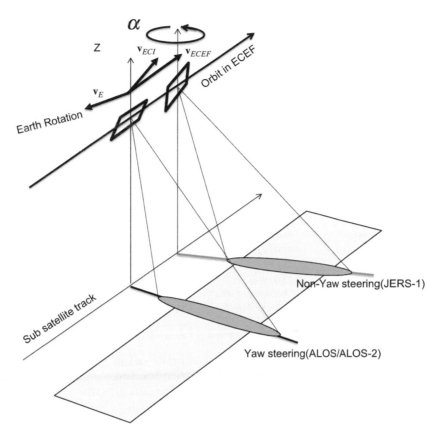

FIGURE 2-7 Yaw steering observation and non-yaw steering observation. Please refer to Chapter 3.4 for VE, VECI, and VECEF.

in achieving 30 cm of three sigmas, which was compared with the laser retro reflector; and (3) with orthodox range and range rate measurements using the S-band carrier associated with the modeling. The second method was recommended for using PALSAR data imaging and further analysis. Because of these two big improvements in timing and positioning, the ALOS/PALSAR showed significant improvement in SAR image accuracy with regard to focusing, radiometry, and location.

Orbit maintenance was performed weekly or biweekly based on the orbital drop measurement and estimations. Inclination maneuvering was not involved in the operation guidelines, and thus the inclination decreased slightly. Correction of the inclination took place every two-and-a-half years, and the resultant interferometric baseline showed a big jump at those times—especially for the high-latitude regions.

ALOS attitude measurement was performed using the accurate star tracker within the accuracy of 2.0×10^{-4} degrees determination for each of three axes. ALOS adopted the yaw steering method where the yaw angle slowly changed so as to cancel the local velocity at the beam illumination position, causing the zero-Doppler frequency. Zero Doppler operation fostered the imaging.

2.5.3 ALOS-2

Datation and positioning for ALOS-2 refer to GPS the same as for ALOS. Yaw steering mode is always adopted in the entire operation. Although the accuracy validation is limited due to a laser reflector not being installed on the satellite, the position accuracy seems improved (from the DinSAR results and the increase in the number of GPS receiver channels).

TABLE 2-3

JAXA Satellite and SAR Characteristics

Item	JERS-1/SAR	ALOS/PALSAR	ALOS-2/PALSAR-2	G-II/Pi-SAR-L2 (L1)
Altitude	568 km	691.5 km	628 Km	6 ~ 12 km
Ground speed	6.9 km/s	6.7 km/s	6.8 km/s	200–250 m/s
Orbit	44-day recurrent, 649 orbits in total	46-day recurrent, 671 orbits in total	14-days recurrent, 207 orbits	-
Inclination	97.67°	98.16°	97.92°	
Position accuracy	RARR	GNSS (L1 & L2 carrier), 10 cm	GNSS (L1 & L2 carrier), < 10 cm	2 cm, INS-POS610 (L1 & L2 carrier)
Timing	Internal crystal oscillator-calibrated with IRIG-A	GPS synchronized	GPS synchronized	GPS synchronized
Attitude	CO2 sensor	Star tracker	Star tracker	INS
Mass	1,350 kg	4,000 kg	2,000 kg	~ 500 kg (SAR)
Yaw steering	No	Yes	Yes	Yes yaw angle < 19°
Solar power	2,000 W at the beginning of life	7,000 W at the beginning of life	5,300 W at the end of life	-
Transmitter	Solid State Power Amp, 325 W	80 TR modules for 2.0 kW	180 TRM for 6.0 Kw	3.5 Kw
Antenna size	11.92 m in azimuth and 2.2 m in range	8.9 m in azimuth and 2.9 m in range	10 m × 3.5 m	1.4 m × 0.7 m
Frequency	1,275.0 MHz	1,270.0 MHz	1,215–1,300 MHz	1,215–1,300 MHz
Bandwidth	15.0 MHz	28.0 MHz, 14.0 MHz	84/42/28/14 MHz	85/50 (50/25) MHz
Sampling freq.	17.076 MHz	32.0 MHz, 16.0 MHz	104.8/52.4/34.9/17.5 MHz	100.0 (62.175) MHz
Pulse width	35.0 µs	27.0 µs	18.5 ~ 76.3 µs	28 µs
Mode and polarization	Strip	Strip (FBS, FBD, Pol), ScanSAR (Single)	Strip (UB, HB, FB, Quad), ScanSAR (dual)Spotlight (single)	Strip (Pol)
Number of beams	1	18 (strip) + Scan (3, 4, 5)	24 (strip:UB+HB), 22(strip:FB) + Scan (5×4+7×3), 14(Spot)	1
AD converters	I-Q, 3 bits	I-Q, 5 bits	8I + 8Q compressed to 4I + 4Q	8I + 8Q
Swath width	75 km at maximum	70 km – strip 350 km – Scan	50 km/70 km strip 350 km–490 km Scan25 km spotlight	20 km (15 km)
Gain control	AGC (Manual Gain Mode)	MGC	AGC	MGC
PRF	1505.8 Hz, 1530.1 Hz, 1555.2 Hz, 1581.1 Hz, and 1606.0 Hz	< 2,700 Hz	1,000 ~ 2,500 Hz	500–600 (400–500)
Polarization	HH	HH, HH + HV, Quad	HH, HH + HV, Quad	Quad
Off nadirs	35.1°	9.7 ~ 50.7°	7.3 ~ 58.8°	0–70°
Resolution (m) (azimuth/ Range)	6/10 (Strip)	5/5 (Strip) 25/10 (ScanSAR)	5/1.72 (Strip) 1/1.72 (Spotlight) 25/5 (ScanSAR)	0.8/1.72 (0.8/3)
Chirp Signal	D	D	UD, APC	D (UD)-(D)
Experimental mode			Compact-pol (Strip + Scan)	Compact-pol
NESZ (Spec)	−18 dB	−25 dB	−29 dB	−35 dB
InSAR orbit maintenance	Yes: Height Not for inclination	Yes: Height—biweekly Yes: Inclination— every 2.5 years	< 500 m Yes: Height and inclination—autonomously	< 5 m corridor-GPS guided

Note: D: Down chirp; UD: Up-down chirp; APC: Azimuth Phase Coding, FBS: Fine Beam single, FBD: Fine beam dual, UB: Ultra fine Beam, HB: High sensitivity Beam, FB: Fine Beam. Main mode for the receiver is written first, and second is in bracket. RARR: Range and Range Rate, GNSS: Global Navigation Satellite System, INS: Inertial Navigation System.

The ALOS-2 orbit is always maintained autonomously within the 500-m corridor criteria from the nominal orbit in monitoring the positions and in checking if they are within that criteria all over the globe. Otherwise, the altitude and/or the inclination is adjusted. As a result, the orbit is well maintained and, as such, the horizontal and vertical variations are less than 400 m and 100 m, respectively, from the nominal, and the inclination is maintained. Because of this, the ALOS-2 orbit is well controlled and preserved for Interferometric Synthetic Aperture Radar (InSAR) operations.

2.5.4 Pi-SAR-L1/L2 Systems

Grumman's Gulfstream II (GII) is the jet aircraft originally prepared for the Pi-SAR-L1 system and also used for the Pi-SAR-L2 system. Normally, it flies above 12,000 m, where the air turbulence is much less than at lower altitudes and therefore more appropriate for high-quality SAR imaging. Although smaller aircraft (including unmanned aerial vehicles, or UAVs) recently have become popular for airborne remote sensing, the ultra-stable motion statistics of the GII are quite appropriate for Pi-SAR-L2 standard observations as well as for the experimental challenges of aircraft movement. The GII carries the Inertial Measurement Unit (IMU), manufactured by the Applanix Co. and used to determine the 2-cm post-processing accuracy of the position determinations. The timing is also synchronized to the GPS information system. Aircraft information is summarized in Table 2-2.

2.6 SUMMARY

This chapter described the hardware configuration of the SAR systems developed by JAXA—three spaceborne SARs and one airborne SAR. Table 2-3 shows details of these four JAXA SARs: JERS-1 SAR, ALOS/PALSAR, ALOS-2/PALSAR-2, and Pi-SAR-L/L2.

REFERENCES

Ito, N., Hamazaki, T., and Tomioka, K., 2001, "ALOS/PALSAR Characteristics and Status," *Proc. of the CEOS SAR Workshop*, Tokyo, pp. 191–194.

Kankaku Y., Sagisaka, M., and Suzuki, S., 2014, "PALSAR-2 Launch and Early Orbit Status," *Proc. of the Geoscience and Remote Sensing Symposium (IGARSS), 2014 IEEE International*, pp. 3410–3412.

Nemoto Y., Nishino H., Ono M., Mizutamari H., Nishikawa K., and Tanaka K., 1991, "Japanese Earth Resources Satellite-1 Synthetic Aperture Radar," *Proc. of the IEEE*, Vol. 79, No. 6, pp. 800–809.

Shimada M., Kawano, N., Watanabe, M., Motohka, T., and Ohki, M., 2013, "Calibration and Validation of the Pi-SAR-L2", *Proc. of the APSAR2013*, Tsukuba, pp.194–197.

3 SAR Imaging and Analysis

3.1 INTRODUCTION

Similar to SAR system development, imaging and analysis algorithms have also evolved significantly because user requirements have become more detailed and quantitative in order to obtain more precise and reliable geophysical parameter measurements. Figure 3-1 depicts the conceptual process of SAR imaging, which transforms unfocused, raw data from a single small antenna to focused, high-resolution data from a synthesized larger-scale antenna. In Figure 3-1, the upper two images are examples of point target simulation and the lower two are real imaging. The location is the southern part of the island of Hawaii, containing Kona, Kilauea, and Mt. Mauna Loa, which the PALSAR FBS ("fine beam single" or single polarization) imaged by using continuous PALSAR illuminations. This image contains a lot of information that is helpful in understanding the land and its changes. In order to explain SAR imaging, this chapter will cover several basic and important issues, such as imaging principles, imaging algorithms, a definition of radar backscattering coefficients and their relationship with SAR images, the appropriate coordinate system, types of imaging, clarification of spaceborne and airborne imaging, and applications. There are many reference books that assist our understanding of SAR imaging (Curlander and McDonough 1991; Jin and Wu 1984; Smith 1990; Bamler 1992; Raney et al. 1994; and so on).

3.2 PROPAGATION OF ELECTROMAGNETIC WAVES IN A MEDIA

3.2.1 TRANSMISSION AND RECEPTION OF PULSES

When the electromagnetic wave satisfies the condition:

$$\mu\varepsilon\frac{\partial^2 E(r,t)}{\partial t^2} - \frac{\partial^2 E(r,t)}{\partial r^2} = 0, \tag{3.1}$$

where t is the time, r is the distance from the origin, E is the electric filed, ε and μ are the emissivity and permittivity of the free space, the microwave propagates in a media. The solution can be given by a linear summation of two waves propagating in the positive direction of r and the negative direction as

$$E(r,t) = a \cdot \exp\left\{(kr - \omega t)j\right\} + b \cdot \exp\left\{(kr + \omega t)j\right\}. \tag{3.2}$$

Here, k is the wave number $(2\pi / \lambda)$, ω is the angular frequency $(2\pi f)$, λ is the wave length, a and b are constants, and $kr - \omega t$ is the phase of the wave that propagates in the positive direction: meaning that one phase propagates in the positive direction with the speed of light, $c = 1/\sqrt{\varepsilon\mu}$ (Figure 3-2a).

Hereafter, we only consider the propagation of a pulse in the positive direction, which is transmitted by the radar, and a pulse is a signal with a relatively large amplitude during a very short time span. Figure 3-2b (upper right) shows the systematic flow of the radar system that communicates with the target in range "r" by using a pulse. In the image, the left triangle stands for the antenna, the white rectangle stands for the pulse propagating in the right direction that is emitted after the radar, and the dark rectangle in the left direction is that is returned from the target.

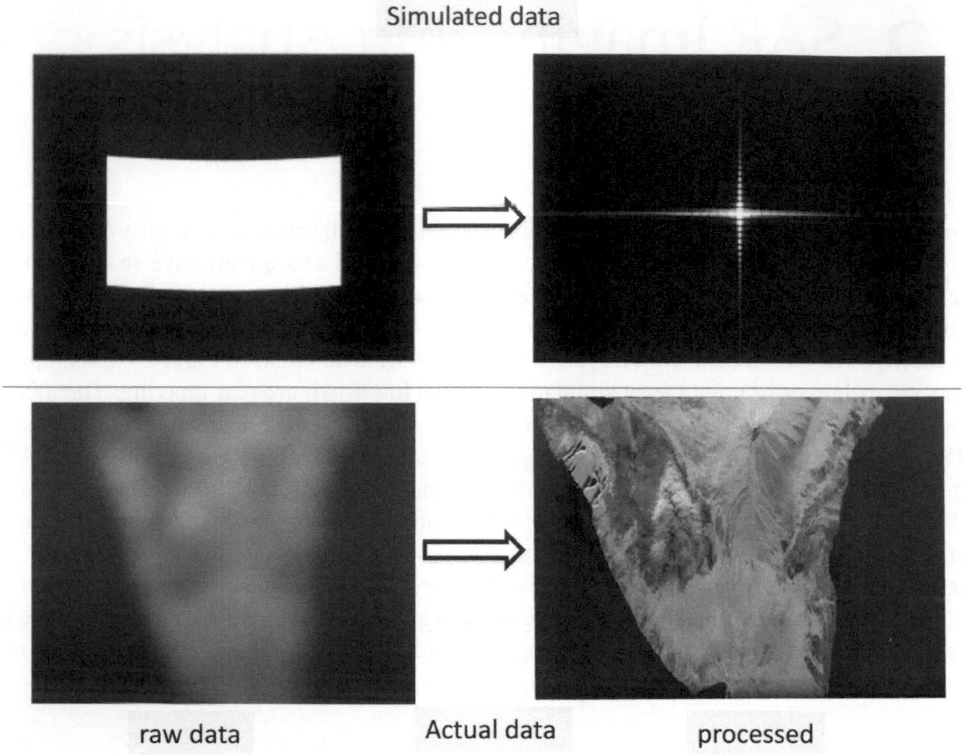

Simulated data

raw data   Actual data   processed

FIGURE 3-1   SAR imaging examples for (top) simulation data and (bottom) actual data for the island of Hawaii. The lower images were based on observations by ALOS/PALSAR.

A pulse is expressed by

$$f(t) = a \cdot \Pi\left(\frac{t}{\tau/2}\right) \exp\left(2\pi f_0 t j\right), \tag{3.3}$$

$$\Pi(x) = \begin{cases} 1 & |x| \leq 1, \\ 0 & |x| > 1, \end{cases} \tag{3.4}$$

Here, "a" is the amplitude, f_0 the carrier frequency, t the time, j the imaginary notation, and τ the pulse width. Function $\Pi(\cdot)$ is 1.0 during τ; otherwise, it is zero.

We consider a whole process that a signal leaves the radar, reaches the target that exists in range r, and returns to the radar; then we derive the resultant range resolution.

Figure 3-2c indicates that a pulse is emitted from $t = -\tau/2$ to $t = \tau/2$; other than that the radar listens to the returned signal from outer space including the target. The signal, emitted from the leading edge ($t = -\tau/2$), approaches the target as propagating on Line A: $r = c(t + \tau/2)$. Then, it returns to the radar propagating on Line B: $r = -c(t + \tau/2) + 2r_0$. All the later signals follow similar processes as does the last signal from the trailing edge ($t = \tau/2$). Some keep forwarding after hitting the target in r_0, but some return earlier from the nearer target in r_1 and could reach the radar simultaneously with the first signal returning from r_0. In summary, there is a range uncertainty as shown in Equation (3.5):

$$\delta r = r_0 - r_1 = \frac{c\tau}{2}. \tag{3.5}$$

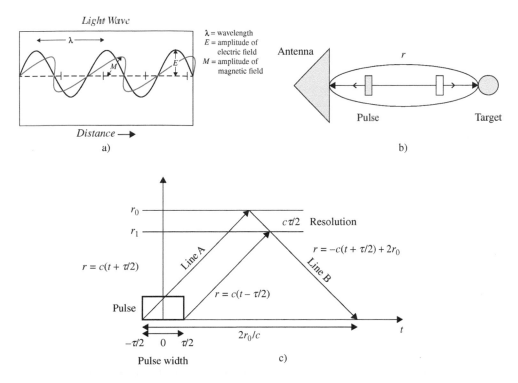

FIGURE 3-2 (a) Propagation of the electric and magnetic waves in the media, (b) signal transmit-receive concept, and (c) the range and time delay relationship at the pulsed radar.

Selecting $\tau = 15$ μs as a pulse width of the JERS-1 SAR, the range uncertainty is 2.25 km.

3.2.2 SAR Configuration in Space

Before moving to the imaging, we should know more about SAR geometry and the SAR illumination area. Figure 3-3 shows the simplified SAR imaging geometry: the satellite moves, the SAR continues to emit the radar pulses subsequentially, and the pulse illuminates the target, which is located at (r, x) on the Earth. While, in reality, many small scatterers and targets are distributed on the Earth's, we have simplify the problem here so that the target is a point: the point target.

To capture the signal scattered from the target, two parameters are important to discriminate between the scattered signal in azimuth and the range direction—the azimuth antenna pattern and the receiving window, respectively.

3.2.2.1 Antenna

The antenna pattern roughly discriminates the target in azimuth. The SAR antenna, represented by the planner array sized in L_A in azimuth and L_R in range, illuminates the Earth's surface with an intensive directionality due to the antenna pattern, which is modeled by

$$G(\phi,\varphi) = G_0 \left| \frac{\sin\{a(\theta-\theta_0)\}}{a(\theta-\theta_0)} \right| \left| \frac{\sin\{b(\phi-\phi_0)\}}{b(\phi-\phi_0)} \right|. \tag{3.6}$$

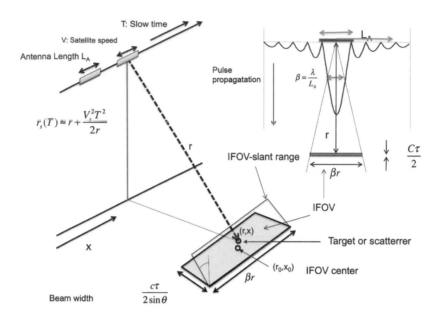

FIGURE 3-3   The SAR antenna and the emitted pulses illuminated on the Earth's surface.

Here, G_0 is the peak gain, θ the off-nadir angle, ϕ the azimuth angle, and the values with $_0$ are antenna bore site directions. Their main beam is approximated by the Gaussian function:

$$G(\theta,\phi) = G_0 \exp\left[-4\ln 2 \left\{ \left(\frac{\theta - \theta_0}{\theta_{BW}} \right)^2 + \left(\frac{\phi - \phi_0}{\phi_{BW}} \right)^2 \right\} \right] \tag{3.7}$$

$$\phi_{BW} = \lambda / L_R, \theta_{BW} = \lambda / L_A \tag{3.8}$$

where ϕ_{BW} and θ_{BW} are the beam widths in azimuth and range, respectively.

Based on the SAR observation scenario, the instantaneous field of view (IFOV) depicts a short pulse in a beam that dwells on the ground and is sized $r_0\lambda / L_A$ and $c\tau / 2$ in the azimuth and range directions, respectively. It moves in range direction with a light speed of $c / \sin\theta_I$, and subsequential pulses generate the series of IFOVs in azimuth direction at the satellite's ground speed.

3.2.2.2   Receiving Window

The received signals are detected and analog-to-digital (AD) converted within a sampling window at the sampling frequency (f_{sample}) in reference to the SAR leading edge so that the round-trip delay can be correctly measured by (Figure 3-4):

$$t_i = \frac{n}{f_{PRF}} + \frac{i}{f_{sample}} + \Delta t_{OFF}, \tag{3.9}$$

where: n is the number of delay pulses, f_{PRF} the pulse repetition frequency, i the sample address, and Δt_{OFF} the time offset between the pulse leading edge and the sampling window start time. Thus, r_0 is

$$r_0 = ct_i / 2. \tag{3.10}$$

Here, there are two notations, r and r_0. It should be noted that r is the exact distance between the point target on the Earth and the satellite, and r_0 is the distance between the satellite and the center position of the IFOV or the correlated pixel, which varies stepwise. It is important to distinguish these two ranges, which will be explained in the following sections.

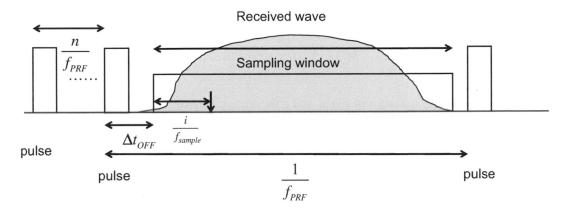

FIGURE 3-4 The sampling sequence of the SAR transmit-and-receive signal.

3.2.3 Pulse Compression (Range Compression)

For the returned signal, in L-band SAR, the carrier frequency of the received signal can be as high as 1.2 GHz. The signal cannot be processed unless the frequency is down-converted. Mixing the receiving signal with the transmission signal makes it time invariant and "r" dependent.

Let the transmission signal be

$$f_t = a \cdot \exp(2\pi f_0 tj), \tag{3.11}$$

then the received signal from the target at the distance r is given by

$$f_r = a \cdot \exp\left\{ 2\pi f_0 \left(t - \frac{2r}{c} \right) j \right\}. \tag{3.12}$$

Mixing these two modifies the received signal as a range-dependent function

$$f_r \cdot f_t^* = a \cdot \exp\left\{ 2\pi f_0 \left(t - \frac{2r}{c} \right) j \right\} \cdot \exp(-2\pi f_0 tj) = a \cdot \exp\left(-\frac{4\pi f_0 r}{c} j \right), \tag{3.13}$$

and maintaining the resolution of $\tau c / 2$ as before. Here, suffix t and suffix r repesent transmission and reception, f_0 the carrier frequency, and * the complex conjugate.

Next, we pick the FM chirp signal, which is quadratic and time dependent on phase within a pulse because the such pulse normally is used as the SAR transmission signal. Similar to Equation (3.11), the processes at transmission, reception, and their mixing are manipulated and converted to Equations (3.14–3.16)

$$f_t(t) = a \cdot \Pi\left(\frac{t}{\tau / 2} \right) \exp\left\{ 2\pi \left(f_0 t + \frac{k}{2} t^2 \right) j \right\}, \tag{3.14}$$

$$f_r(t) = a \cdot \Pi\left(\frac{t - 2r / c}{\tau / 2} \right) \exp\left\{ 2\pi \left(f_0 \cdot (t - 2r / c) + \frac{k}{2} (t - 2r / c)^2 \right) j \right\}, \tag{3.15}$$

and,

$$f_r \cdot f_t^* = a \cdot \Pi\left(\frac{t - 2r / c}{\tau / 2} \right) \exp\left\{ 2\pi \left(-f_0 \frac{2r}{c} + \frac{k}{2} \left(t - \frac{2r}{c} \right)^2 \right) j \right\}. \tag{3.16}$$

The exponent of Equation (3.16) contains r and the chirp property.

The following correlation focuses the received signal:

$$h(t) = \int_{-\tau/2}^{\tau/2} f(t') \cdot g*(t+t')dt',$$

(3.17)

where h is the output, f the received signal, and g the reference signal.

In principle, the received signal is a time-shifted transmit signal multiplied by the free space decay and the target backscatter. However, some more complex phase modulation might occur and interact with the target or the propagation media of the atmosphere and ionosphere.

By selecting the transmit signal as the reference, we can calculate the correlation (as the range compression),

$$P(t') = \int_{-\infty}^{\infty} \mathbf{S}_r(t) \cdot \mathbf{S}_{r,ref}^*(t+t')dt$$

$$= \int_{2r/c-\tau/2}^{2r/c+\tau/2} \Pi\left(\frac{t-2r/c}{\tau/2}\right)\Pi\left(\frac{t+t'-2r/c}{\tau/2}\right)\exp\left\{-j2\pi\left(\frac{k}{2}\left(t-\frac{2r}{c}\right)t'+\frac{k}{2}t'^2\right)\right\}dt$$

(3.18)

where

$$S_r(t) = a \cdot \Pi\left(\frac{t-2r/c}{\tau/2}\right)\exp\left\{2\pi\left(-f_0\frac{2r}{c}+\frac{k}{2}\left(t-\frac{2r}{c}\right)^2\right)j\right\},$$

(3.19)

$$S_{r,ref}(t+t') = a \cdot \Pi\left(\frac{t+t'-2r/c}{\tau/2}\right)\exp\left\{2\pi\left(-f_0\frac{2r}{c}+\frac{k}{2}\left(t+t'-\frac{2r}{c}\right)^2\right)j\right\},$$

(3.20)

And k is the chirp rate as $k = B_c / \tau, \tau$ is the pulse width, and B_c is the transmission bandwidth. Equation (3.18) can be calculated to

$$P(t') \cong \exp\left(-\frac{4\pi r}{\lambda}j\right)\int_{2r/c-\tau/2}^{2r/c+\tau/2}\exp\left\{-j2\pi kt'\left(t-\frac{2r}{c}\right)\right\}dt$$

$$= \exp\left(-\frac{4\pi r}{\lambda}j\right)\left[\exp\left\{-j2\pi kt'\left(t-\frac{2r}{c}\right)\right\}/(-j2\pi kt')\right]_{2r/c-\tau/2}^{2r/c+\tau/2}$$

(3.21a)

$$= \frac{\sin\{\pi B_c t'\}}{\pi B_c t'}\tau\exp\left(-\frac{4\pi r}{\lambda}j\right).$$

Here, t' varies stepwise as Equation (3.9), and only the scatterer locating the vicinity of $ct'/2$ is visible. Letting r be the distance to the scatterer, r_0 the distance to the pixel center, and $t' = 2(r-r_0)/c$, we have the following response as the range correlation:

$$P(t') \cong \frac{\sin\{2\pi B_c(r-r_0)/c\}}{2\pi B_c(r-r_0)/c}\tau\exp\left(-\frac{4\pi r}{\lambda}j\right)$$

$$= \sin c\left(\pi\frac{r-r_0}{c/2B_c}\right)\tau\exp\left(-\frac{4\pi r}{\lambda}j\right).$$

(3.21b)

Taking the absolute value, we have

$$|P(r)| = \mathrm{sinc}\{a(r - r_0)\}\tau, \tag{3.22}$$

$$a = \frac{2\pi B_C}{c}. \tag{3.23}$$

The second equality of Equation (3.21b) used the sinc function as the expression. Again, it should be noted that r is the satellite-point target distance and r_0 is the satellite-pixel center distance.

Thus, the range correlation amplitude has been highly sharpened (high resolution), and the phase is preserved as the same as the raw data.

Equation (3.22) expresses the range response of the correlated (compressed) signal and shows what the target brightness looks like as a function of range.

Defining a resolution (ρ) as the spatial or timing width that lessens the peak intensity 3 dB, we have

$$\frac{\sin(ar_1)}{ar_1} = \frac{\sin(ar_2)}{ar_2} = \frac{\sin(a\rho/2)}{a\rho/2} = \frac{1}{2}, \tag{3.24}$$

and, finally, we have the following resolution:

$$\rho = r_2 - r_1 = \frac{c}{2B_c}. \tag{3.25}$$

Here, r_1 and r_2 are the nearer and farther ranges, respectively, meeting the 3-dB beam width. This indicates that the raw data resolution of $c\tau/2$ is significantly improved to $c/2B_C$ by a correlation process, and the pulse compression ratio (PCR) is defined as the resolution improvement factor:

$$PCR = \frac{c\tau}{2} / \frac{c}{2B_c} = \tau B_c. \tag{3.26}$$

Figure 3-5 shows the representative responses to the typical L-band bandwidths of 14, 28, 42, and 85 MHz.

The raw data compression of Pi-SAR-L2 and the improved time resolution of τ to $1/B_C$ is shown in Figure 3-6.

3.2.4 AZIMUTH COMPRESSION

The illumination (dwelling) time in azimuth depends on the satellite (ground) speed, the azimuth antenna beam width, and the beam steering mechanism. If we select the strip SAR mode (see Section 3.3) as an example, which does not change the antenna bore site direction during the observation, the azimuth dwelling time or synthetic aperture time, T_A, is given by

$$T_A = \frac{r\lambda}{LV_s}, \tag{3.27}$$

which is a duration when the azimuth antenna gain is higher than 3 dB below the peak. Here, V_s is the satellite ground speed. Figure 3-7 shows an observation scenario where a target is being azimuthally observed by the SAR seven times during the satellite passage. Each observation is from a different distance, and the point scatterer locates at (r, x).

Since the range-compressed data preserve the distance between the satellite and the target over the T_A, the correlation on this phase could be utilized.

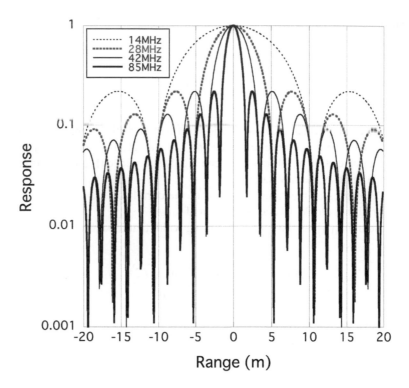

FIGURE 3-5 The impulse response function on four bandwidths (85, 42, 28, and 14 MHz).

From here, we use "r_s" as a distance between the scatterer and the satellite, and its time dependency can be calculated from the orbital data and the target position as shown in Equation (3.28a):

$$r_s = \sqrt{r^2 + (V_s T)^2}. \tag{3.28a}$$

Here, we use r as the closest distance between the satellite and the target. When the beam width is small enough, the second term within the root is much smaller than the first term, and Taylor expansion approximates it as a quadratic function of T. It can also be approximated by expansion around the distance to the center of the pixel, r_0:

$$r_s \cong r + \frac{V_s^2}{2r} T^2$$
$$\cong r + \frac{V_s^2}{2r_0} T^2. \tag{3.28b}$$

Here, we simplified the case where the antenna beam exactly directs the 90 degrees rotated around the z-axis from the satellite's direction as it moved around the Earth. The resultant Doppler frequency is zero, such as in the yaw steering case shown in Figure 2-7. (The coordinate system and definition of the satellite direction is introduced in Section 3.4.2.) The second equality of Equation (3.28b) means that the time dependency of distance can be represented by the pixel center and precisely modeled by the Doppler frequency or Doppler chirp as discussed in Section 3.2.8.

Figure 3-8 shows movement of a distance, r_s, over the range (vertical)–azimuth time (horizontal) plain, where the solid vertical line (L_1) corresponds to the range at time T_1 and solid line (L_2) shows the corresponding real distance. The solid line (L_3) is the locus of the distance in the plain on which the data exist. This line is called the migration, and its data reallocation is inevitable for SAR imaging.

Range compression

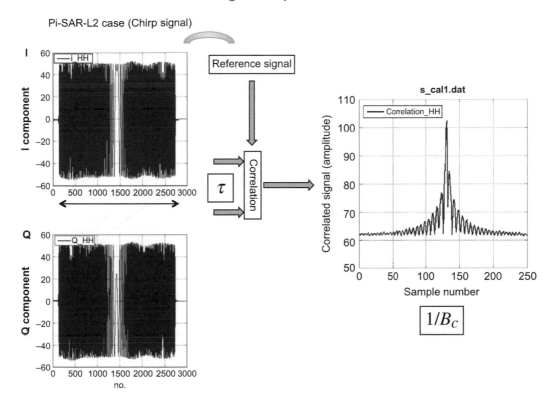

FIGURE 3-6 Pulse compression.

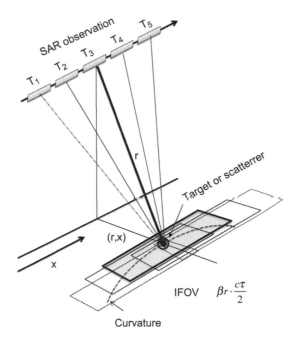

FIGURE 3-7 Target and pulse illuminations during the satellite's passage.

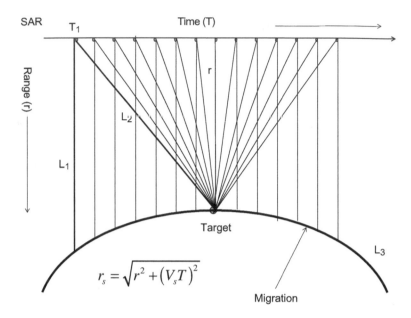

FIGURE 3-8   The migration curve where the horizontal axis is the azimuth time and the vertical axis is the slant range coordinate.

The non-zero Doppler case can be expanded to include a term proportional to T in Equation (3.28b). Using this, we reconstruct the raw data (3.16) as follows, which are modified by using λ instead of c/f_0,

$$f(t,T) = a\Pi\left(\frac{t - 2r_s/c}{\tau/2}\right)\Pi\left(\frac{T}{T_A/2}\right)\exp\left\{2\pi\left(-\frac{2r_s}{\lambda} + \frac{k}{2}\left(t - \frac{2r_s}{c}\right)^2\right)j\right\}. \qquad (3.29)$$

The first term of the exponent in Equation (3.29) can be given by

$$f(T) = \exp\left\{-\frac{4\pi}{\lambda}\left(r + \frac{V_s^2}{2r_0}T^2\right)j\right\}, \qquad (3.30)$$

and Equation (3.29) is rewritten as

$$f(t,T) = a\Pi\left(\frac{t - 2r_0/c}{\tau/2}\right)\Pi\left(\frac{T}{T_A/2}\right)\exp\left\{2\pi\left(-\frac{2r}{\lambda} + \frac{k}{2}\left(t - \frac{2r}{c}\right)^2 - \frac{V_s^2}{\lambda r_0}T^2\right)j\right\}, \qquad (3.31)$$

where the second exponent, as the quadratic term of T, is not expanded using Equation (3.29) because the term only lasts a pulse duration, and it does not create the T effect. Equation (3.31) has quadratic functions on two time variables, t and T, which are $k/2$ and $V_s^2/\lambda r0$, respectively. When we use $Vs \sim 7.0$ km/s, $\lambda \sim 23$ cm, and $r \sim 700$ km, they have a totally different order of factors $\sim 10^{12}$ and $\sim 10^2$, and they can be handled as different processes even in the same time coordinate. In this case, t is only effective within a pulse width, and T is effective over several seconds. Thus, "t" and "T" are called the "fast time" and "slow time," respectively.

As a simulation of the range variation over the synthetic aperture time of the ALOS-2/PALSAR: 4 s is shown in Figure 3-9a and its cosine component in Figure 3-9b is shown as the enlarged part for 0.8 s; the range varies 140 m during this time, and the cosine of the range–phase varies the same as the range compression of Figure 3-7, with the only difference being the time scale.

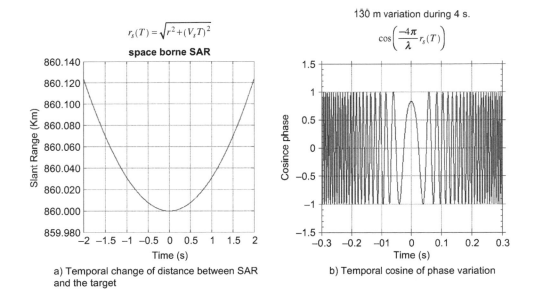

a) Temporal change of distance between SAR and the target

b) Temporal cosine of phase variation

FIGURE 3-9   Range (a) and phase (b) variations during a synthetic aperture time (4 s), assuming the ALOS/PALSAR orbit.

After the range compression in Equation (3.31), we have

$$f_R(r_0, T) \cong a \operatorname{sinc}\left(\pi \frac{r - r_0}{c/2B_c}\right)\Pi\left(\frac{T}{T_A/2}\right)\exp\left\{2\pi\left(-\frac{2r}{\lambda} + \frac{V_s^2}{r_0\lambda}T^2\right)j\right\}. \qquad (3.32)$$

For T around T_0, a correlation process of T over the synthetic aperture time (T_A) can be obtained the same way as the earlier range compression process:

$$h(r_0, T) = \int_{-T_A/2}^{T_A/2} f(T') \cdot g^*(T + T')dT'$$

$$\cong a \cdot \operatorname{sinc}\left(\pi \frac{r - r_0}{c/2B_c}\right)\exp\left(-\frac{4\pi r}{\lambda}j\right)\int_{-T_A/2}^{T_A/2}\exp\left\{2\pi\left(\frac{V_s^2}{r_0\lambda}T^2\right)j\right\}\exp\left\{-2\pi\left(\frac{V_s^2}{r_0\lambda}(T+T')^2\right)j\right\}dT'$$

$$= a \cdot \operatorname{sinc}\left(\pi \frac{r - r_0}{c/2B_c}\right)\exp\left(-\frac{4\pi r}{\lambda}j\right)\int_{-T_A/2}^{T_A/2}\exp\left\{-2\pi\left(\frac{V_s^2}{r_0\lambda}(2TT' + T'^2)\right)j\right\}dT'$$

$$\cong a \cdot \operatorname{sinc}\left(\pi \frac{r - r_0}{c/2B_c}\right)\exp\left(-\frac{4\pi r}{\lambda}j\right)\left[\frac{\exp\left\{-2\pi\left(\frac{V_s^2}{r_0\lambda}(2TT')\right)j\right\}}{-2\pi\left(\frac{V_s^2}{r_0\lambda}(2T)\right)j}\right]_{-T_A/2}^{T_A/2} \qquad (3.33)$$

$$= a \cdot \operatorname{sinc}\left(\pi \frac{r - r_0}{c/2B_c}\right)T_A\frac{\sin 2\pi\left(\frac{V_s^2}{r_0\lambda}T_A T\right)}{2\pi\left(\frac{V_s^2}{r_0\lambda}T_A T\right)}\exp\left(-\frac{4\pi r}{\lambda}j\right)$$

$$= a \cdot \operatorname{sinc}\left(\pi \frac{r - r_0}{c/2B_c}\right)T_A \cdot \operatorname{sinc}\left(2\pi \frac{V_s}{L_A}T\right)\exp\left(-\frac{4\pi r}{\lambda}j\right)$$

TABLE 3-1

Improvement of the Resolutions

	Raw Data	Correlated	SPECAN
Range resolution	$\dfrac{c\tau}{?}$	$\dfrac{c}{2B}$	$\dfrac{c}{2B}$
Azimuth resolution	$\dfrac{r\lambda}{L_A}$	$\dfrac{L_A}{2}$	$\dfrac{f_{PRF}v_g}{N_{az}f_{DD}}$
Range spacing	$\dfrac{c}{2f_{sample}}$	$\dfrac{c}{2f_{sample}}$	$\dfrac{c}{2f_{sample}}$
Azimuth spacing	$\dfrac{v_g}{f_{PRF}}$	$\dfrac{v_g}{f_{PRF}}$	$\dfrac{f_{PRF}v_g}{N_{az}f_{DD}}$

Here, we assumed that all the range compressed data were rearranged on a migration curve. We used the second equality of Equation (3.28b) in Equation (3.33). This is the time domain processing, which clarifies the meaning of azimuth correlation.

If we use $T = (x - x_0)/V_s$, we have

$$h(r,x) = a \cdot \mathrm{sinc}\left(\pi \frac{r - r_0}{c/2B_c} \right) T_A \cdot \mathrm{sinc}\left(\pi \frac{x - x_0}{L_A/2} \right) \exp\left(-\frac{4\pi r}{\lambda} j \right), \qquad (3.34)$$

as a two-dimensional response of a point target. Thus, the similar phase variation in the quadratic function leads to the compression in azimuth.

From this equation, the resolutions for the range and azimuth (ρ_r and ρ_a respectively) are derived as

$$\rho_r = \frac{c}{2B_c}$$
$$\rho_a = \frac{L_A}{2} \qquad (3.35)$$

Thus, the SAR image resolution is improved by two one-dimensional correlation processes as summarized in Table 3-1. (The migration curve will be discussed in Section 3.2.8.4.) The pixel spacing of the data can be given as

$$s_r = \frac{c}{2f_{sample}}$$
$$s_a = \frac{v_g}{f_{PRF}}. \qquad (3.36)$$

where, s_r and s_a are pixel spacings in range and azimuth respectively. From this analysis, we can interpret that a pixel contains one scatterer, and the intensity relies only on the shift from the pixel center. The phase is EXACTLY the distance from the radar NOT the pixel center. Figure 3-10a shows the conceptual structure of a point target in a pixel.

3.2.5 DISTRIBUTED TARGET CASE

In reality, there are many scatterers located at the different positions on the ground, and they also constitute an SAR pixel whose size is now assumed to be several meters in range and azimuth. After

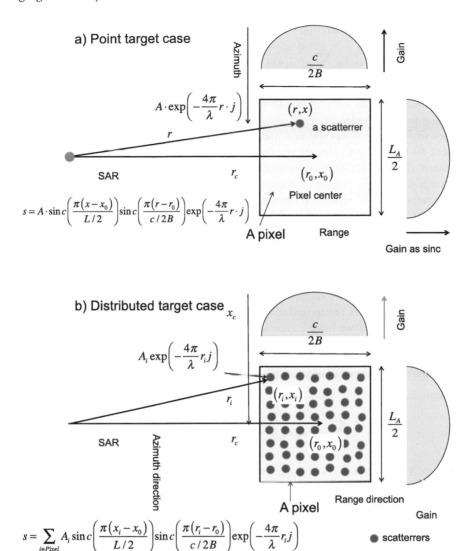

FIGURE 3-10 (a) Details of a single look image containing a single scatter in a resolution cell. (b) Distributed target expression for the SAR image pixel.

being illuminated by the SAR beam, they all retransmit the signals following the targets' directional sensitivity.

The backscattering electromagnetic waves received by radar are weighted primarily by distance with the antenna pattern in azimuth and range secondary. The received signals (after down converting) are summed up linearly for all the signals over an illumination area:

$$f(t) = \sum_{i=0}^{N} a_i \Pi\left(\frac{t - 2r_{si}/c}{\tau/2}\right) \exp\left\{2\pi\left(-f_0 \frac{2r_{si}}{c} + \frac{k}{2}\left(t - \frac{2r_{si}}{c}\right)^2\right)j\right\}, \tag{3.37}$$

where i is the integer number assigned to each scatterer, and a_i is the amplitude lasting in τ. It should be noted that the scatterers' have wide-ranging locations on the ground and are not limited by resolution.

The range correlation of Equation (3.37) gives the following response,

$$f_R(r_0) = \sum_{i=0}^{N} a_i \cdot \mathrm{sin}\,c\left(\pi\frac{r_{si}-r_0}{c/2B_c}\right)\exp\left\{2\pi\left(-f_0\frac{2r_{si}}{c}\right)j\right\},\tag{3.38}$$

where the ith scatterer has a distance r_i from the radar, and r_0 is the distance corresponding to the data sampling from the radar.

Equation (3.38) implies that the raw data are received during a time window in such a way so as to collect all the scattered signals weighted by the antenna elevation pattern and pulse width τ at the different timings, and they are screened at the sampled range with a weighted resolution ρ at the slant range difference. This is the range compression process that simply and linearly sums up all the associated scatters' signals at the amplitudes and the phase. This point is very important in interpreting the SAR image and the interferometric SAR.

Resolution ρ depends on the bandwidth of the radar system—especially for the L-band SARs—for example, 10.7 m (14 MHz), 5.4 m (28 MHz), 3.6 m (42 MHz), and 1.72 m (85 MHz), and it is not dependent on the radar wavelength. The signals discriminated for one resolution cell are exactly scattered from all the scatterers within the cell. Normally, the size of the resolution cell is larger than the radar wavelength. One resolution cell contains a larger number of scatterers, and the scattered signal components are composed of the linear summation of all the individual components.

From Equation (3.28b), we have

$$r_{si} \cong r_i + \frac{V_s^2}{2r_0}T^2.\tag{3.39}$$

Inserting Equation (3.39) into Equation (3.38), we have

$$f_R(r_0) \cong \sum_{i=0}^{N} a_i \cdot \mathrm{sin}\,c\left(\pi\frac{r_i-r_0}{c/2B_c}\right)\exp\left\{2\pi\left(-\frac{2r_i}{\lambda}+\frac{V_s^2}{r_0\lambda}T^2\right)j\right\}.\tag{3.40}$$

Applying the correlation process in time, we have

$$h(r_0,x_0) = \int_{-T_A/2}^{T_A/2} f(T')\cdot g^*(T+T')dT'$$

$$= \int_{-T_A/2}^{T_A/2}\sum_{i=0}^{N} a_i \cdot \mathrm{sin}\,c\left(\pi\frac{r_i-r_0}{c/2B_c}\right)\exp\left\{2\pi\left(-\frac{2r_i}{\lambda}+\frac{V_s^2}{r_0\lambda}T^2\right)j\right\}\exp\left\{-2\pi\left(\frac{V_s^2}{r_0\lambda}(T+T')^2\right)j\right\}dT'$$

$$= \sum_{i=0,i\in Cell}^{N} a_i \cdot \mathrm{sin}\,c\left(\pi\frac{r_i-r_0}{c/2B_c}\right)\cdot \mathrm{sin}\,c\left(\pi\frac{x_i-x_0}{L_A/2}\right)\exp\left(-\frac{4\pi r_i}{\lambda}j\right)\tag{3.41}$$

Figure 3-10b also depicts the conceptual structure of a pixel where all the scatterers independently contribute to its scattering coefficient in amplitude and to the wavelength fraction of round-trip distance to EACH scatterer in the phase. This is just the linear summation of the point targets involved in a pixel. The behavior of the SAR image very much depends on the number and distribution function of the scatterers.

In reality, the SAR receiver is one of the noise sources: the thermal noise generator. This thermal noise degrades the SAR image performance at the noise equivalent signal sensitivity. Thus, Equation (3.42) is the realistic SAR image model:

$$S(r_0,x_0) = \sum_{i=0}^{N} a_i \cdot \mathrm{sin}\,c\left(\pi\frac{r_i-r_0}{c/2B_c}\right)\cdot \mathrm{sin}\,c\left(\pi\frac{x_i-x_0}{L_A/2}\right)\exp\left(-\frac{4\pi r_i}{\lambda}j\right)+N\tag{3.42}$$

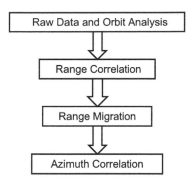

FIGURE 3-11   Simplified flow chart of SAR imaging.

This is the time domain processing, which fosters the meaning of azimuth correlation even when losing speed. The processing routine is summarized in Figure 3-11.

3.2.6   FREQUENCY DOMAIN PROCESSING

Correlation (and convolution) are convertible between time and frequency domains. We consider the correlation in the frequency domain to gain the processing efficiency (fast processing). The following equation shows the equality of the time and frequency domain processes:

$$\int_{-\infty}^{\infty}\left[f(T')\right]_M \cdot g^*(T+T')dT' = \int_{-\infty}^{\infty}\left[F(\omega)\right]_M \cdot G^*(\omega)e^{j\omega T}, \tag{3.43}$$

where $F(\omega), G(\omega)$, and $[\ ]_M$ are the frequency spectrum of $f(t)$, $g(t)$, and migration curve, respectively.

To perform the correlation in the frequency domain, we need the time and frequency domain conversion for the distance between the satellite and the target: the migration.

For the observation target, the distance shows the trajectory as

$$r_s = r + \frac{V_s^2}{2r_0}T^2 + O\!\left(T^3\right), \tag{3.44}$$

and this is the range curvature or migration (Figure 3-8) on which the data need to be resampled from the nearby range-aligned data by using the sinc interpolation or Fourier shift and finally over which the data is azimuthally compressed. Interpolation using the Fast Fourier Transform (FFT) is more accurate than interpolation at the expense of the processing cost.

In general, r can be Taylor expanded around the origin of azimuth time $T = T_0$ as

$$r_s = r + \dot{r}_s\left(T - T_0\right) + \frac{1}{2}\ddot{r}_s\left(T - T_0\right)^2 + \frac{1}{2\cdot3}\dddot{r}_s\left(T - T_0\right)^3 + \tag{3.45}$$

When inserted into the exponent term of Equation (3.37), we have

$$\exp\!\left(-\frac{4\pi r_s}{\lambda}j\right) = \exp\!\left(-j\frac{4\pi}{\lambda}\!\left(r + \dot{r}_s\left(T - T_0\right) + \frac{1}{2}\ddot{r}_s\left(T - T_0\right)^2 + \frac{1}{6}\dddot{r}_s\left(T - T_0\right)^3\right)\right)$$

$$= \exp\!\left(j2\pi\!\left(-\frac{2}{\lambda}r + f_D\left(T - T_0\right) + \frac{1}{2}f_{DD}\left(T - T_0\right)^2 + \frac{1}{6}f_{DDD}\left(T - T_0\right)^3\right)\right). \tag{3.46}$$

Here, f_D is the Doppler, f_{DD} is the Doppler chirp rate, f_{DDD} is the third derivative, and so on. Here, the important factors for SAR imaging are the range and azimuth dependency of the Doppler frequency, Doppler chirp rate, third-order time derivatives, and so on. We will derive the Doppler frequency and the Doppler chirp rate in the next section.

3.2.7 SPECTRUM ANALYSIS (SPECAN) METHOD

Other than time domain, and frequency domain processings, spectrum analysis (SPECAN) is often used, especially for scanning synthetic aperture radar (ScanSAR) imaging.

If we express $f(T)$ as the range-compressed and range-migrated data,

$$f(T) = \sum_{i \in Cell} \exp\left(j2\pi \left(-\frac{2}{\lambda} r_i + f_D (T - T_i) + \frac{1}{2} f_{DD} (T - T_i)^2 \right) \right). \tag{3.47}$$

The reference signal is

$$f_A(T) = \exp\left(j2\pi \left(-\frac{2}{\lambda} r_0 + f_D T + \frac{1}{2} f_{DD} T^2 \right) \right). \tag{3.48}$$

Multiplying both signals gives

$$f(T) f_R^*(T) = \sum_{i \in Cell} \exp\left(j2\pi \left(-\frac{2}{\lambda} r_i + f_D (T - T_i) + \frac{1}{2} f_{DD} (T - T_i)^2 \right) \right) \exp\left(-j2\pi \left(-\frac{2}{\lambda} r_0 + f_D T + \frac{1}{2} f_{DD} T^2 \right) \right)$$

$$= \sum_{i \in Cell} \exp\left(j2\pi \left(-\frac{2}{\lambda} \Delta r_i + f_D (-T_i) + \frac{1}{2} f_{DD} \left(-2TT + T_i^2 \right) \right) \right) \tag{3.49}$$

Here, $\delta r_i = r_i - r_0$. The FFT of this equation will be

$$\int_{-T_A/2}^{T_A/2} f(T) e^{-2\pi fTj} dT$$

$$= \sum_{i \in Cell} \int_{-T_A/2}^{T_A/2} \exp\left(j2\pi \left(-\frac{2}{\lambda} \Delta r_i + f_D (-T_i) + \frac{1}{2} f_{DD} \left(-2TT + T_i^2 \right) \right) \right) \exp(-j2\pi fT) dT$$

$$= \sum_{i \in Cell} A_i \int_{-T_A/2}^{T_A/2} \exp(-j2\pi (f_{DD} TT_i)) \exp(-j2\pi fT) dT \tag{3.50}$$

$$= \sum_{i \in Cell} A_i \int_{-T_A/2}^{T_A/2} \exp(-j2\pi (f_{DD} T_i + f) T) dT$$

$$= -\sum_{i \in Cell} A_i T_A \mathrm{sin}\, c\left(\frac{\pi (f_{DD} T_i + f)}{1/T_A} \right)$$

This means that the frequency component f has the peak for the scatterer at $-T_i f_{DD}$ with the resolution of $1/T_A$, and its geometric resolution is obtained by multiplied by $-v_g / f_{DD}$

$$\rho_{AZ} = \frac{v_g}{T_A f_{DD}} = -\frac{f_{PRF} v_g}{N_{az} f_{DD}} \tag{3.51}$$

The spacing of the data is the same for the resolution

$$s_{AZ} = \frac{v_g}{T_A f_{DD}} = -\frac{f_{PRF} v_g}{N_{az} f_{DD}}, \tag{3.52}$$

where v_g is the ground speed. In the same way, the azimuth length of the imaged burst is

$$L = s_{AZ} N_{AZ} = -\frac{f_{PRF} v_g}{f_{DD}} \tag{3.53a}$$

where L contains the synthetic aperture length, and the effective azimuth length should be

$$L' = -\left(\frac{f_{PRF}}{f_{DD}} + \frac{N_{az}}{f_{PRF}} \right) v_g. \tag{3.53b}$$

Resolution improvement by SPECAN is also summarized in Table 3-1.

3.2.8 DOPPLER PARAMETERS

3.2.8.1 Doppler Frequency

The keys to SAR imaging and target pixel determination are range determination and Doppler frequency measurements. We will describe the Doppler frequency measurement and its model in this section.

Doppler frequency is expressed theoretically by

$$f_D = \frac{2}{\lambda} \left(\mathbf{u}_s - \omega \times \mathbf{r}_p \right) \cdot \frac{\left(\mathbf{r}_p - \mathbf{r}_s \right)}{\left| \mathbf{r}_p - \mathbf{r}_s \right|}, \tag{3.54}$$

where $\mathbf{u}_s$ is the satellite velocity vector, ω the earth rotation angular speed, $\mathbf{r}_p$ the target vector on the Earth, $\mathbf{r}_s$ the satellite position vector, and $\times$ the vector product, for which the velocities and position vectors are likely to be expressed in the Earth-centered inertial (ECI) coordinate, and the cross-product term is raised from the conversion of the ECI to the Earth-centered-Earth-fixed (ECEF) coordinate (formerly known as the Earth-centered rotational, or ECR, coordinate) where the coordinate rotates around the Earth's pole with the average rotation speed of the Earth (ω).

Position vector $\mathbf{r}_p$ could be obtained iteratively by solving Equations (3.55), (3.56), and (3.57):

$$\mathbf{r}_p = r \cdot \mathbf{a} \mathbf{E}_y \mathbf{E}_p \mathbf{E}_y + \mathbf{r}_s \tag{3.55}$$

$$r = \left(\frac{n}{f_{PRF}} + \frac{i}{f_{sample}} + \Delta_{off} \right) \frac{c}{2} \tag{3.56}$$

$$\left(\frac{x_p}{R_a} \right)^2 + \left(\frac{y_p}{R_a} \right)^2 + \left(\frac{z_p}{R_b} \right)^2 = 1 \tag{3.57}$$

where $\mathbf{E}_R$, $\mathbf{E}_P$, and $\mathbf{E}_Y$ are the Euler matrix on the roll, pitch, and yaw (including the steering angle), respectively; $\mathbf{a}$ is the unit vector directing the line of sight in the antenna beam plane; n is the integer number of the delay pulses; i is the sampling address of the data; f_{PRF} is the pulse repetition frequency; f_{sample} is the sampling frequency of the AD converter; Δ_{OFF} is the time offset between the pulse leading edge and the sampling window start time; x_p, y_p, and z_p are the components of $\mathbf{r}_p$, and R_a and R_b the equatorial and polar radius, respectively (Figure 3-12).

Here, roll, pitch, and yaw (angles) are the measured data from the satellite attitude sensors. Once these values are known, the aforementioned equation can determine the target point in principle.

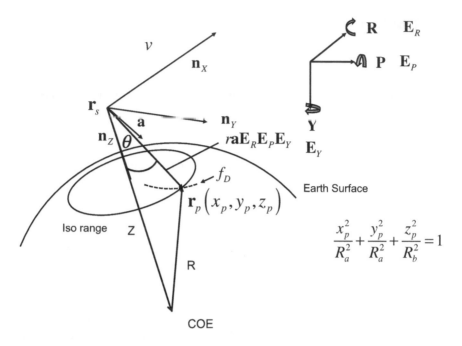

FIGURE 3-12   Doppler frequency and the target point.

However, their accuracies are not good enough to determine the radar line of sight vector with the required accuracy. In that case, the target area can be determined by using the range-Doppler method, which requests the predetermination of the Doppler frequency, for which the following section introduces the operational methods: iteratively solving Equations (3.54), (3.56), and (3.57) where the Doppler center frequency, f_D, is predetermined in the SAR imaging.

Doppler frequency is range and azimuth dependent. When the targets distribute heterogeneously in a scene (i.e., the many small islands in a sea image), Doppler frequency often differs from Equation (3.54), as seen in JERS-1 or even in ALOS in the 1990s or 2000s.

3.2.8.2   Doppler Frequency Measurement

a. Clutter Lock

If the SAR observed the Amazon Rainforest and the uniform ocean area, the azimuthal frequency spectrum of the raw or range-correlated data could show the azimuth antenna pattern as a function of the Doppler frequency, and the peak might give the Doppler center frequency or the azimuth beam direction, which is the important parameter in creating a well-focused SAR image. We consider this parameter estimation. First, the azimuth Fourier transform of the signal can be written as

$$\int_{-\infty}^{\infty} A(T)\exp\left(\left(-\frac{4\pi r_0}{\lambda} + 2\pi f_D T + \pi f_{DD} T^2\right)j\right)\exp(-j\omega T)\,dT. \tag{3.58}$$

The first term in the exponent is a constant and can be written as

$$\propto \int_{-\infty}^{\infty} A(T)\exp\left((2\pi f_D T + \pi f_{DD} T^2 + \omega T)j\right)dT. \tag{3.59}$$

This term is also modified to

$$= \int_{-\infty}^{\infty} A(T)\exp\left(\pi f_{DD}T^2 j\right)\exp\left(-(\omega - 2\pi f_D)Tj\right)dT \tag{3.60}$$

$$\propto \overline{A}(\omega - 2\pi f_D).$$

A () is the power spectrum of the first two terms $A(T)\exp\left(\pi f_{DD}T^2 j\right)$ and its peak position provides the best estimate of the Doppler: f_D. Two examples in Figure 3-13a and b show the Doppler frequencies in near and far ranges, separated 50 km in range. Thin busy lines are the measured data, and the solid line represents their moving average. Both show similarities in the azimuth direction while the intensity decreases with range. Doppler frequency with the peak value is the estimated Doppler and is empirically modeled as a range function:

$$f_D = a + b \cdot r + c \cdot r^2 + \text{higher order terms} \tag{3.61}$$

Here, a, b, and c are the constants. Generally, the range is linear with two constants. Figure 3-13c and d compare the range dependence of the Doppler frequency for two SARs: JERS-1 SAR and ALOS/PALSAR.

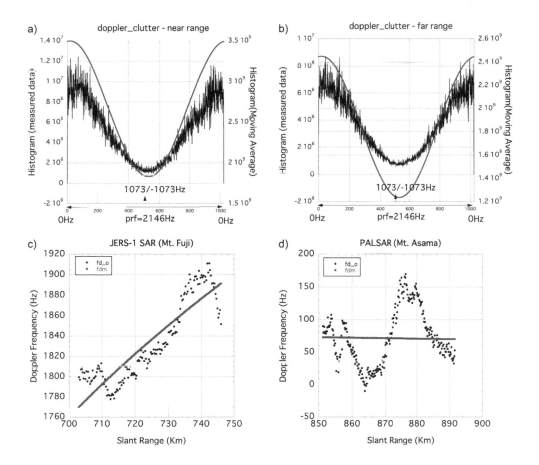

FIGURE 3-13 Doppler frequency dependence of the power spectrum: (a) near range pattern and (b) far range pattern. Where busy lines are measured data and solid line is the weighted average. Slant range dependence of Doppler center frequency: (c) JERS-1 SAR and (d) ALOS/PALSAR. Where fd_O is the observed data and fdm is the doppler model.

b. **Adaptive Method**

The previous discussion can be extended more generally considering the relationship between the power spectrum and the auto-correlation of the stochastic signal processes (Madsen 1989). In general, the azimuth spectrum can be written as

$$W(f) = W(f - f_D),\tag{3.62}$$

where W is the azimuth power spectrum of the data, f is the frequency, and f_D is the Doppler frequency. For the stocastic processes, with which the SAR data are often associated, there is an expression between the power spectrum and auto correlation:

$$W(f) = \int R(t)\exp(-2\pi ftj)dt,\tag{3.63}$$

$$R(t) = \int h(t')h^*(t+t')dt',\tag{3.64}$$

where R is the auto-correlation function and h is the complex raw data.

Equation (3.63) tells us

$$R(t) = \exp(2\pi f_D j) \cdot R_0(t).\tag{3.65}$$

Here, the auto-correlation suffixed 0 is the reference. Thus, we have

$$R_0(k) = E\{h_0(k+m)h_0^*(m)\}\tag{3.66}$$

$$\overline{f}_D = \frac{1}{2\pi kT}\arg(\overline{R}_h(k)).\tag{3.67}$$

For these theoretical basics, there are two data selections and two representations, for a total of four existing combinations. As for data selection, there are raw data and range-compressed data. As for the representations, one option is to use the data as they are, and the other is to use the sign of the data, which is 1.0 if the real or imaginary part is greater than 0.0. Otherwise, it is −1.0. The latter is the so-called sign function. Using the sign function is more appropriate for estimating f_D, even for heterogeneous data, which are often observed in inland sea areas. In addition, the empirical modeling of the unwrapped f_D in the range with the least square sense estimates the Doppler more accurately, especially for non-yaw steering satellites such as JERS-1/SAR and RADARSAT-1.

3.2.8.3 Doppler Chirp Rate

Next, we will derive the Doppler chirp rate and higher order derivatives. Using the same coordinates as in the Doppler calculation, the line of sight vector to the target can be given by

$$\mathbf{r} = \mathbf{r}_p - \mathbf{r}_s.\tag{3.68}$$

Here, $\mathbf{r}_p$ is the target vector to the point p, and $\mathbf{r}_s$ is the satellite position vector. Denoting the distance r as

$$r_s \equiv |\mathbf{r}|,\tag{3.69}$$

the n-th order Doppler time derivatives can be given by

$$f_{nD} = -\frac{2}{\lambda}\frac{d^n}{dt^n}|\mathbf{r}|.\tag{3.70}$$

We show the third order derivatives in Equation (3.71) by using the "dot" notations:

$$\dot{r}_s = \frac{d}{dt}(\mathbf{r} \cdot \mathbf{r})^{1/2} = (\mathbf{r} \cdot \mathbf{r})^{-1/2}(\mathbf{r} \cdot \dot{\mathbf{r}}) = a_1^{-1/2} \cdot a_2. \tag{3.71}$$

Here, we defined

$$\begin{aligned} a_1 &\equiv \mathbf{r} \cdot \mathbf{r} \\ a_2 &\equiv \mathbf{r} \cdot \dot{\mathbf{r}} \end{aligned} \tag{3.72}$$

In same way, we have the second and third order derivatives as follows:

$$\begin{aligned} \ddot{r}_s &= -a_1^{-3/2}a_2^2 + a_1^{-1/2}(a_3 + a_4) \\ a_3 &= \dot{\mathbf{r}} \cdot \dot{\mathbf{r}} \\ a_4 &= \mathbf{r} \cdot \ddot{\mathbf{r}} \\ \dddot{r}_s &= 3a_1^{-5/2}a_2^3 - 3a_1^{-3/2}a_2(a_3 + a_4) + a_1^{-1/2}(3a_5 + a_6) \\ a_5 &= \ddot{\mathbf{r}} \cdot \dot{\mathbf{r}} \\ a_6 &= \mathbf{r} \cdot \dddot{\mathbf{r}} \end{aligned} \tag{3.73}$$

It should be noted that the time derivative, denoted by "dot", is operated only on the ECI (the inertial coordinate). Next, we express the same point on the Earth in ECEF as $\mathbf{r}_R$ and ECI as $\mathbf{r}_I$, and these connected as follows:

$$\mathbf{r}_R = \mathbf{A} \cdot \mathbf{r}_I$$

$$A = \begin{pmatrix} \cos\omega t & \sin\omega t & 0 \\ -\sin\omega t & \cos\omega t & 0 \\ 0 & 0 & 1 \end{pmatrix}. \tag{3.74}$$

Here, ω is the Earth's rotation's angular speed, and t is the time difference in the two-coordinate system. Then, we have

$$\mathbf{r}_I = \mathbf{A}^{-1} \cdot \mathbf{r}_R. \tag{3.75}$$

The derivative of the target fixed on the Earth's surface is

$$\dot{\mathbf{r}}_I = \dot{A}^{-1} \cdot \mathbf{r}_R$$

$$\dot{A}^{-1} = \omega \begin{pmatrix} 0 & -1 & 0 \\ 1 & 0 & 0 \\ 0 & 0 & 0 \end{pmatrix} = \omega E_1, \tag{3.76}$$

with $t = 0$. In the same way, the second derivative can be

$$\ddot{\mathbf{r}}_I = \ddot{A}^{-1} \cdot \mathbf{r}_R$$

$$\ddot{A}^{-1} = \omega^2 \begin{pmatrix} -1 & 0 & 0 \\ 0 & -1 & 0 \\ 0 & 0 & 0 \end{pmatrix} = \omega^2 E_2. \tag{3.77}$$

Thus, the first and second order derivatives of Equation (3.68) can be given as follows:

$$\dot{\mathbf{r}} = \dot{\mathbf{A}}^{-1} \cdot \mathbf{r}_{pR} - \dot{\mathbf{r}}_{sl}, \tag{3.78}$$

$$\ddot{\mathbf{r}} = \ddot{\mathbf{r}}_{pI} - \ddot{\mathbf{r}}_{sI} = \ddot{\mathbf{A}}^{-1} \cdot \mathbf{r}_{pR} - \ddot{\mathbf{r}}_{sI}, \tag{3.79}$$

Recent technology provides accurate state vectors, x-y-z, at a given time, and their velocities, or JERS-1 measured them using the range and range-rate technologies: r_{pR} and r_{sl}; thus Equation (3.78) can be determined. The remained component is the accelerator of r_{sl} in Equation (3.79). This component could be obtained by the numerical differentiation of the state vectors through interpolation or high-order polynomial approximation. But none give accurate accelerations other than the method explained in the next paragraph.

This method uses the Newtonian motion equation under the geopotential function. Today, the geopotential is modeled as accurately as possible by using the 180-th order spherical functions, but the largest components are J_0 and J_2 terms, which implies the total mass of the Earth and the ellipsoidal component of the Earth. Such a geopotential function (V) can be represented by

$$V = \frac{GM}{R} \left\{ 1 + C_2 \frac{R_a^2}{R^2} P_2 (\sin \phi) \right\}. \tag{3.80}$$

Here, $GM = 398,600.5 \text{ Km}^3\text{s}^{-2}$; R is the distance between the satellite and the geo center, $C_2 = -1.082637032e-3$; ϕ is the right ascension (latitude); and P_2 is the second order associated Legendre's function. Differentiating the aforementioned potential on x-y-z, we have the following acceleration terms (Hagiwara 1982):

$$\ddot{\mathbf{r}}_{s,I} = -\nabla V = -\frac{GM}{R^3} \begin{pmatrix} x \cdot \left(1 - \frac{3C_2 R_a^2}{2R^2} \left(1 - \frac{5z^2}{R^2} \right) \right) \\ y \cdot \left(1 - \frac{3C_2 R_a^2}{2R^2} \left(1 - \frac{5z^2}{R^2} \right) \right) \\ z \cdot \left(1 - \frac{3C_2 R_a^2}{2R^2} \left(3 - \frac{5z^2}{R^2} \right) \right) \end{pmatrix} \tag{3.81}$$

$$\nabla \equiv \begin{pmatrix} \dfrac{\partial}{\partial x} & \dfrac{\partial}{\partial y} & \dfrac{\partial}{\partial z} \end{pmatrix}$$

And, Equation (3.79) then can be determined.

The important point of this section is that the Doppler chirp rate, f_{DD}, and higher derivatives (the most important parameters for SAR focusing) can be precisely calculated from the higher derivatives of the geopotential function and the measured state vectors: the position and velocity vectors. The high accuracy of the state vectors was achieved by improved GPS technology, and knowledge about the geopotential functions was acquired by the end of 1990s.

3.2.8.4 Migration in Time and Frequency Domains

The migration or range curvature is a key point in both time and frequency domain processings and it can be formulated as a behavior of the material point in the range Doppler or azimuth time plane under the given Doppler and Doppler chirp rate.

The differential equation for the migration is given by

$$\ddot{r}_s = -\frac{\lambda}{2} f_{DD}. \tag{3.82}$$

We integrate this twice and obtain the following:

$$\dot{r}_s = -\frac{\lambda}{2} f_{DD}T + C,$$

$$r = -\lambda f_{DD}T^2 + CT + D.$$

(3.83)

As the initial condition $T = 0$,

$$\dot{r}_s = -\frac{\lambda}{2} f_D,$$

$$r = r_0;$$

(3.84)

next, we can determine the unknown parameters:

$$C = -\frac{\lambda}{2} f_D,$$

$$D = r_0.$$

(3.85)

Then, we finally have the migration curve in time domain:

$$r = -\lambda f_{DD}T^2 + \frac{\lambda}{2} f_D T + r_0.$$

(3.86)

The migration curve in the frequency domain will start from the Doppler frequency and the Doppler chirp relationship:

$$f = f_{DD}T + f_D.$$

(3.87)

Inserting Equation (3.87) into Equation (3.86), we have

$$r = -\lambda f_{DD}\left(\frac{f - f_D}{f_{DD}}\right)^2 + \frac{\lambda}{2} f_D \frac{f - f_D}{f_{DD}} + r_0.$$

(3.88)

This is the migration curve in the frequency domain. These two expressions can be selectively implemented for SAR imaging.

Thus, the SAR processing in the frequency domain can be summarized as follows:

$$h(r,T) = \int_{-\infty}^{\infty} \left[F(r,\omega)\right]_{M(r,\omega)} \cdot G^*(r,\omega)e^{j\omega T} d\omega$$

$$G(\omega) = \int_{-\infty}^{\infty} g(T)e^{-j\omega T} dT.$$

(3.89)

Here, $g(T)$ is the phase history during the synthetic aperture length, and $[]M(r,\omega)$ is the migration process by the interpolation at the r, ω coordinates.

3.2.9 SAR IMAGING SUMMARY

We can summarize the SAR processing as follows:

1. Raw data:

$$S_r(t) = S_t\left(-\frac{4\pi r}{\lambda}\right)A_{att} \exp(-j\delta)$$

(3.90)

2. Range compression:

$$S_{rc}(r_0) = \frac{1}{\tau} \int_{-\tau/2}^{\tau/2} S_r(t+t')S_t^*(t')dt' \tag{3.91a}$$

$$= \operatorname{sinc}\left(\frac{\pi(r'-r_0)}{c/2B_C}\right)\exp\left\{-j\left(\frac{4r'}{c}f_0\pi+\delta\right)\right\}A_{att}$$

3. Azimuth compression:

$$S_{rca}(r_0,x_0) = \operatorname{sinc}\left(\pi\cdot\frac{r-r_0}{c/2B_C}\right)\operatorname{sinc}\left(\pi\frac{x-x_0}{L/2}\right)\exp\left(-\frac{4\pi r}{\lambda}j\right)A_{att}e^{-j\delta} \tag{3.91b}$$

This means that the point target locates at (r, x), and the reference signal—assuming that the target locates at (r_0, x_0)—can be compressed in this equation. To summarize, the correlated SAR image can be expressed in two ways: as a single scatter or as distributed targets.

3.2.9.1 Single Scatter
With a single scatter (a point target),

$$S_{ra}(r_0,x_0) = \tau T_A A(r_i,x_i)\operatorname{sinc}\left\{\frac{\pi(r_i-r_0)}{c/2B_C}\right\}\operatorname{sinc}\left\{\frac{\pi(x_i-x_0)}{L_A/2}\right\}\exp\left(-\frac{4\pi r'_i}{\lambda}j\right)+\mathbf{N} \tag{3.92}$$

where r is the distance to a scatterer from the SAR, x is the coordinate in azimuth, r_0 is the range at the pixel center, and x_0 is the azimuth coordinate at the pixel center.

3.2.9.2 Distributed Targets
For distributed targets,

$$S_{ra}(r_0,x_0) = \tau T_A \sum_{i=0}^{N} A(r_i,x_i)\operatorname{sinc}\left\{\frac{\pi(r_i-r_0)}{c/2B_C}\right\}\operatorname{sinc}\left\{\frac{\pi(x_i-x_0)}{L_A/2}\right\}\exp\left(-\frac{4\pi r_i}{\lambda}j\right)+\mathbf{N} \tag{3.93}$$

and the suffix i means the ith scatterer.

Thus, an SAR image pixel consists of a linear summation of all the scatterers in a pixel. And, the process flow can be summarized as the five individual processes shown in Figure 3-14.

This flowchart follows the range-Doppler method and often is used because it is easy to understand. However, the wider bandwidth, or Doppler, has offset (beam offset), and the SAR may lose resolution and focus. Recently, other methods such as chirp scaling or migration methods have been implemented. The behavior of scatterers in the resolution cell is similar, and the SAR pixel is composed of all the range information associated with the all the scatterers in the resolution cell. Phase is only related to the distance between the SAR and all the scatterers, while their azimuth locations are only considered in the weighting function. This is a feature of SAR imaging and SAR images. The phase component is applied for SAR interferometry, which will be described in chapter 12.

Again, Figure 3-1 shows the two schematic process results for the raw data and processed SAR imagery for the simulated point target and the actual SAR data from ALOS/PALSAR over the island of Hawaii; in each, the left image is before the compression and the right is after the processing.

A summary of the resolution improvement by the correlation and SPECAN is given in Table 3-1.

Equations (3.92) and (3.93) both imply that the SAR correlation amplitude is gained by the pulse width (τ) multiplied by the synthetic aperture time (T_A), while the power may be double the

Process Flow

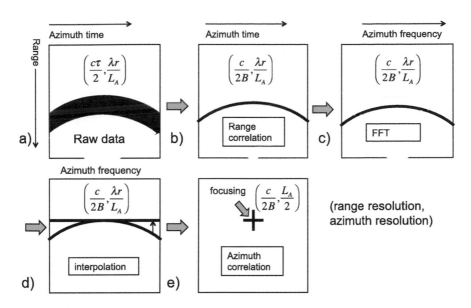

FIGURE 3-14 Processing flowchart for SAR image focusing: (a) Raw data arraigned in the range–azimuth time coordinate, (b) range correlation, (c) azimuth time-Doppler conversion (azimuth FFT), (d) migration process, and (e) azimuth correlation. Range and azimuth resolution at each processing step is shown in each bracket.

gain $(\tau T_A)^2$. On the other hand, noise components cannot be correlated with the reference function, but the power is gained by τT_A, which will be introduced in Chapter 4.

3.2.10 GENERATION OF THE CHIRP SIGNAL

In this section, we consider the generation of the chirp signal. Recent chirp signal generators is so free of error that a digital signal form recorded on a Read-Only-Memory (ROM) is reproduced and converted to the analogue signal repeatedly without any degradation. Originally, the chirp signal is produced by a surface acoustic wave filter triggered by a short pulse income using the depth-dependent propagation speed. The low-frequency signal propagates with slower speed at deeper positions, and the higher frequency signal propagates with faster speed at shallower positions. Thus, a pulse is converted to linear FM signals with lower amplitude based on the energy preservation. After the amplified FM signal is emitted from the SAR, the scattered signal is received by the SAR and reproduced to the original pulse through pulse compression. Figure 3-15 shows a block diagram of how a chirp signal is generated.

3.3 SAR IMAGING METHODOLOGY

Table 3-2 shows SAR imaging categories. There are two sensor carriers: satellites and aircraft including uninhabited aerial vehicles (or UAVs). Three imaging modes are implemented in most SAR instruments: the strip mode, ScanSAR mode, and the spotlight mode. In this book, the first two are described in detail, and final one is described in connection with the observation principle.

3.3.1 IMAGING THE STRIP MODE

This imaging method often is applied in focusing the strip data (Figure 3-16a) and rewriting it once more. One antenna beam is selected and used to continuously observe the 50- to 75-km swath area

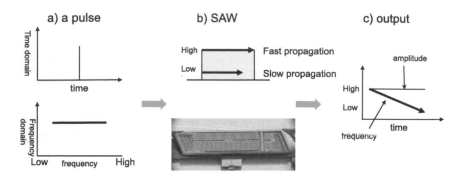

FIGURE 3-15 Chirp signal generator. a) a short pulse is given in time at the above, and given in frequency domain, b) the signal is ingested to the SAW which allows higher frequency component propagates on the surface with faster speed and lower component does deeper part with a slower speed, and c) as a result higher frequency component outputs earlier and lower one later. Thus, the chirp signal is generated.

with one or several PRFs, depending on Doppler frequency bandwidth in the target area, and the raw data are compressed two-dimensionally or twice one-dimensionally. Imaging of multiple PRFs over a long strip (path) as a gapless imaging method will be discussed in Chapter 11.

3.3.2 IMAGING THE SCANSAR MODE

The ScanSAR mode was experimentally verified in the 1990s and implemented in most SARs after the late 1990s. This mode became available because of antenna arrays with phase shifters (such as in RADARSAT-1) or active arrays by all SARs after the 1990s. This mode performs the periodic activation of the antenna beams, each of which occurs shortly and sequentially, in hundreds of milliseconds, and then quickly is switched to the neighboring beam, finally widening the swath and sacrificing the resolution.

ScanSAR prepares several beams in the range direction. One beam with N contiguous pulses is selected, for which short-term observation is called a burst. The same procedure continues with the neighboring beams having different pulses and then returns to the original beam, for which one cycle is called a period. Since the ScanSAR data have fewer pulses within a synthetic aperture length than the strip mode, imaging and radiometry differ (Figure 3-16b).

ScanSAR imaging consists of range correlation, migration, and azimuth correlation (Cumming and Wong 2005). There are two methods of final azimuth correlation: SPECAN that images every burst data, and full aperture that images a complied several burst data using a method introduced at the early part of this chapter. Differing from correlation processing, SPECAN is obtained as a mixture of the SAR data and the reference function and the FFT because the first step provides the location-frequency conversion, and FFT converts the imaging. The processing flow is shown in Figure 3-17 and its formation is detailed in the next section.

TABLE 3-2
SAR Imaging Mode Identification on Spacecraft and Aircraft

Carrier	Strip	ScanSAR	Spotlight
Satellite	X	X	X
Aircraft	X	-	-

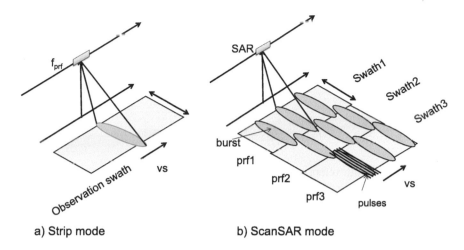

a) Strip mode   b) ScanSAR mode

FIGURE 3-16   Artistic view of the (a) strip SAR and ScanSAR imaging modes.

3.3.2.1   Range Correlation

In range correlation,

$$g = f \oplus f_{r,ref} = F_r^{-1}\left(F_r(f) \cdot F_r(f_{r,ref}) \cdot W_r\right) \tag{3.94}$$

$$f_{r,ref} = \exp\left\{-2\pi j\left(\frac{k}{2}t^2\right)\right\} \tag{3.95}$$

where f is the raw data, $f_{r,ref}$ is the reference signal, $F_r(f)$ is the Fourier transform in range, W_r is the window function in range, $\oplus$ is the correlation, k is the chirp rate, t is the short time, and g is the range-correlated signal.

3.3.2.2   Range Migration

In range migration,

$$g'(T) = F^{-1}\left(F_a(g) \cdot F_a(C) \cdot W_a\right) \tag{3.96}$$

where $F_a(g)$ is the Fourier transform in azimuth, C is the curvature in the range Doppler plane, W_a is the window function in azimuth, and g' is the range curvature signal in T.

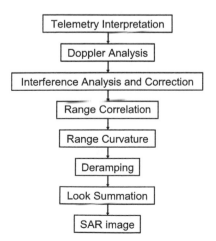

FIGURE 3-17   Flowchart of SPECAN SAR imaging.

3.3.2.3 SPECAN

In SPECAN,

$$g'' = F_a^{-1} \left(\int_{-T_a/2}^{T_a/2} g'(T) \cdot f_{a,ref}^*(T) dT \right) \tag{3.97}$$

$$f_{a,ref} = \exp\left\{ -2\pi j \left(\frac{f_{DD}}{2} T^2 + f_D T \right) \right\} \tag{3.98}$$

where $f_{a,ref}$ is the azimuth reference function, f_{DD} is the Doppler chirp rate (Hz/s), f_D is the Doppler frequency (Hz), T is the time in azimuth direction, and g'' is the final output. Here, in the deramping process, the chirp-z transformation is implemented instead of a Fourier transformation due to the higher resolution and the fact that any number can be the base number. The fan beam property of ScanSAR processing should be corrected in multi-look processings.

Figure 3-18 shows the schematic process flow as to how the SPECAN images the ScanSAR. Burst 1 images the larger area, A, bordered by solid line, and Burst 2 covers Area B and is shifted in a moving direction with $-f_{prf} v_g / f_{dd}$; then Burst 3 does the same for Area C. Thus, the surface can be covered repeatedly by the shifted burst images. The top-right image shows the ten burst images with the ten rectangular images, each of which is shifted in the moving direction. Collecting the co-registered pixels from all the associated burst images forms the final ScanSAR image, as shown in the bottom-right image in Figure 3-18.

This type of imaging has three big problems: scalloping, truncation noise, and stripe noise across the beam. A detailed explanation of ScanSAR imaging is provided in Chapter 5.

3.3.3 AIRBORNE SAR IMAGING

JAXA's airborne SAR (represented by Pi-SAR-L2 after it replaced Pi-SAR-L in 2012), specifications for which are shown in Table 2-3, fly at a height of 6,000 m to 12,000 m looking at the left-hand side of a moving track using the L-band frequency for a duration of around 10 minutes

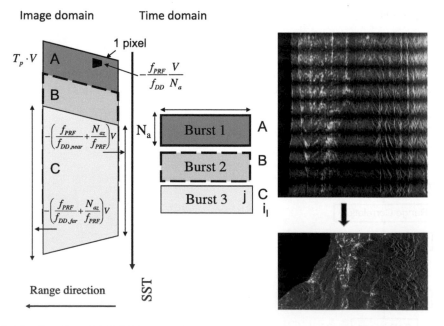

FIGURE 3-18   Conceptual ScanSAR process image.

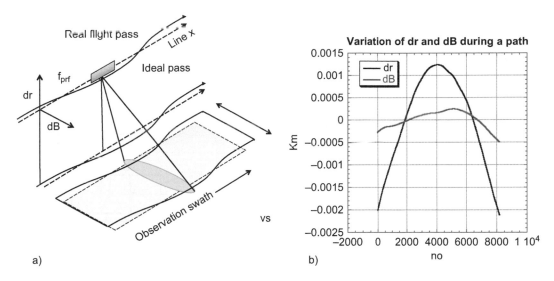

FIGURE 3-19 (a) Observation concept of the airborne SAR (the solid lines and the solid-line covered area are the actual flight trajectory, sub-satellite point, and the observed area; dotted lines and areas are the ideal trajectory and the area). (b) Vertical and horizontal variation of the flight trajectory experienced by Pi-SAR-L, where the horizontal variation is the component parallel to the Earth's surface and the vertical variation is the zenith direction.

or so (a longer period is possible within the recorder capacity). The typical observation mode is a full polarimetric strip mode supplemented by ScanSAR and a compact polarization strip mode. Essentially, the Pi-SAR-L2, as well as most of the airborne SARs, is difficult to fly on a straight path due to unexpected wind speed variations. Figure 3-19a shows the airborne SAR observation concept where the sold lines and solid covered area represent the actual flight pass, the corresponding sub-airplane track, and the observation area, respectively; the dotted lines show the ideal (straight flight case) flight track, sub-airplane track, and the observation area, respectively. This means that the airplane's flight path varies its positions in time, but the spacecraft is stable. This is the biggest difference between them. Figure 3-19b shows a real example of a Pi-SAR-L flight pass and its vertical and horizontal variations (r and b) from the average straight flight pass. Observing 300 cm vertically and 75 cm horizontally, it is much larger than the radar wavelength of 23.6 cm (L band). Because the ground trajectory is curved and the ground speed varies in time, the target is not illuminated with an equal space on the ground in time, although the pulses are emitted at an equal time period. The data collection takes place because the PRF, set much higher than the Doppler bandwidth of IFOV and meeting the sampling theorem, succeeds in the reconstruction of the received signal as if the airplane flew a straight pass using the interpolation or Fourier shift method.

There are two imaging methods for this "nonuniformly sampled SAR data": time domain processing and frequency domain processing. The former pertains to the processing load, and the latter performs the data reconstruction in range and azimuth.

3.3.3.1 Time Domain Processing
Time domain processing simply collects all the range-correlated data along the trajectory as shown in Figure 3-20. If the target point is expressed by $\mathbf{r}_p$ and the SAR position when the pulse was emitted on time, T, by $\mathbf{r}_s(T)$, the received signal will be obtained $-2r/c$ sec later. The SAR flies on the curved orbit shown in Figure 3-20a and receives the signal from the target contiguously as long as the SAR antenna beam dwells the target. The trajectory in the data plane (range-time coordinate) is the curved line, C_2, as shown in Figure 3-20b.

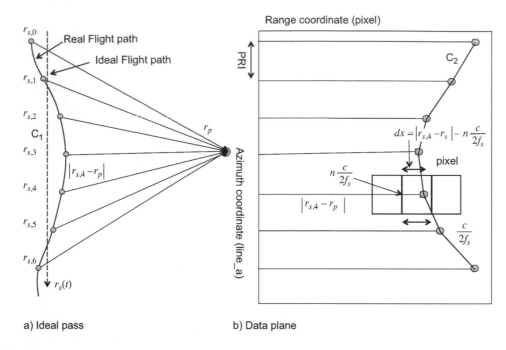

a) Ideal pass   b) Data plane

FIGURE 3-20   Concept of the time domain correlation. (a) Airborne SAR flies from the top to the bottom in observing left side by transmitting the cascaded pulses on the ground with constant PRF. The corresponding azimuth range plane is shown in (b).

Those signals are detected at a sampling frequency of f_{sample}. The corresponding data are received at time, T_4, as a distance of

$$r = \left| \mathbf{r}_{s,4} - \mathbf{r}_p \right|,$$ (3.99)

and are involved at the nth address,

$$n = \left[\frac{2r}{c} f_{sample} \right],$$ (3.100)

where $[\ ]$ stands for the integer address. More precisely, the data exist at Δi in this pixel:

$$\Delta i = \frac{2r f_{sample}}{c} - n.$$ (3.101)

Thus, the data corresponding to this time fraction can be extracted by the interpolations at Δi: CC, sinc, or FFT or by simply phase sifting in Δi. Now, we introduce FFT interpolation. Let $f(t)$ represent the range correlated data, $F(\cdot)$ the FFT, and $F^{-1}(\cdot)$ the inverse FFT. Its shift theorem gives the appropriate values for this real point at Δi:

$$f(i - \Delta i) = F^{-1} \left\{ e^{j\omega\Delta i} \cdot F(f(i)) \right\}.$$ (3.102)

Then, the azimuth correlation can be expressed by

$$p_{i,j} = \sum_{k=0}^{N-1} f(r_k) e^{\frac{4\pi r_k}{\lambda}}.$$ (3.103)

Here, N represents the synthetic aperture samples. In this way, the correlations are conducted for all the pixels because it takes a large computation load. When the orbital information is accurate, this produces a high-quality SAR image.

3.3.3.2 Frequency Domain Processing

Cost-effective processing is available in the frequency domain. As long as the range correlation is performed first, all data are reconstructed in two ways as if the SAR flies on a straight path with a constant speed so that the range-Doppler-type of SAR algorithm can be applied.

The process for procedure is as follows (see Figure 3-21):

1. Ideal orbit preparation: Determine an ideal flight pass that is curvilinear with height, h, parallel to the ellipsoidal Earth or a linear line connecting the entrance and the exit of the real flight corridor, both determined in the least square sense (with Line X as the ideal path in Figure 3-19a).
2. Range arrangement: Calculate the distance between the target and the ideal orbit at each pulse transmit timing between the target and the real orbit and differentiate them as expressed by Δr:

$$\Delta r_{i,j} = \left| r_{s,j} - r_{i,j} \right| - \left| \overline{r}_{s,j} - r_{i,j} \right| \tag{3.104}$$

Here, $\overline{r}_{s,j}$ is the ideal path. The data will be resampled or shifted in time using the interpolation or FFT. FFT shift will be performed using the following:

$$f(i - \Delta i) = F^{-1}\left\{ e^{j\omega\Delta i} \cdot F\big(f(i)\big) \right\} \tag{3.105}$$

$$\Delta i = 2\Delta r_{i,j} f_s$$

where i is the address number, Δi the fraction of shift (real number), and $F(\cdot)$ the FFT. Equation (3.104) needs to be calculated for all the pixels on a line of sight—all the data on the received data on this pulse transmit time.

3. Azimuth arrangement: The aforementioned data are rearranged in azimuth direction keeping an equal spacing on the ground. The relationship between the equally spaced azimuth address and the real address can be given by a solution of the nonlinear equation;

$$\frac{L}{N}i = \sum_{j=0}^{i-1} \hat{\mathbf{v}}_I \cdot \mathbf{v}(j); \tag{3.106}$$

$$L = \sum_{j=0}^{N-1} \hat{\mathbf{v}}_I \cdot \mathbf{v}(j). \tag{3.107}$$

Here, L is the ground length along the ideal orbit, N the number of the samples (pulses) used for the observation, $\mathbf{v}(j)$ the velocity vectors measured at the pulse transmit time, and $\hat{\mathbf{v}}_I$ is the ideal unit velocity vector. The azimuth rearrangement will be done using the interpolation or FFT shift as well. A more detailed procedure is shown in Figure 3-22. This way of data rearrangement is called motion compensation.

Azimuth compression follows the method introduced in the strip mode using the Doppler chirp rate

$$f_{DD} = -\frac{2}{\lambda} \frac{\overline{v}^2}{r}. \tag{3.108}$$

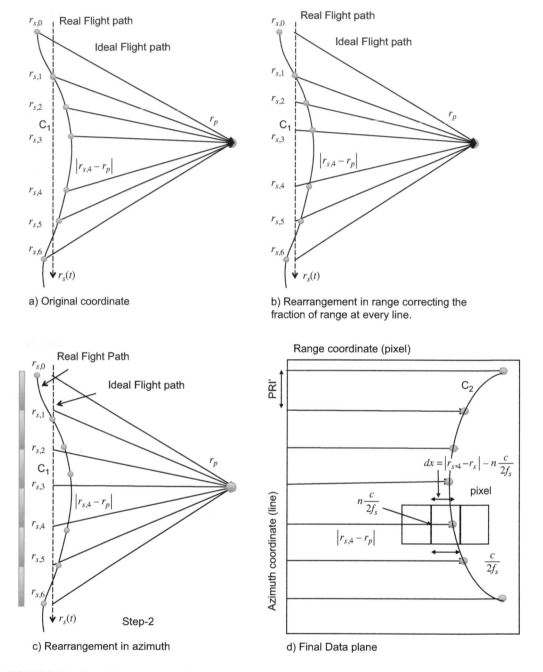

a) Original coordinate

b) Rearrangement in range correcting the fraction of range at every line.

c) Rearrangement in azimuth

d) Final Data plane

FIGURE 3-21 Raw data acquired in the range-azimuth coordinate: (a) Original status of the data acquisition with regard to the target on the ground is depicted. (b) All the data are rearranged (shifted and phase rotated) in range direction based on the range difference between the averaged and real orbit. (c) The data are rearranged in azimuth direction as if the ground speed is constant. (d) Data are reconstructed as if the spaceborne SAR had acquired it.

Figure 3-23 depicts three examples from the inland area of Japan's Miyagi prefecture observed by an X-band SAR on August 26, 2016, such that the time domain processing is shown in (a), the frequency domain processing with motion compensation in (b), and the frequency domain processing without motion compensation in (c). Figure 3-24 shows the associated dr / dB variations of the flight trajectory in (a) and along-track variation of the sub-satellite track in (b), where the total

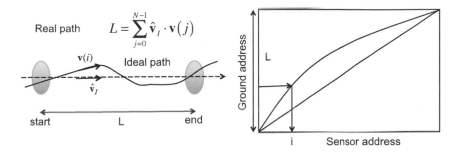

FIGURE 3-22 Correction factor for the azimuth motion compensation. The left image shows the total length of the trajectory projected on the Earth's surface, and the right image shows a graph used to obtain the azimuth address or pulse number corresponding to the address of the given (equally spaced) point on the Earth.

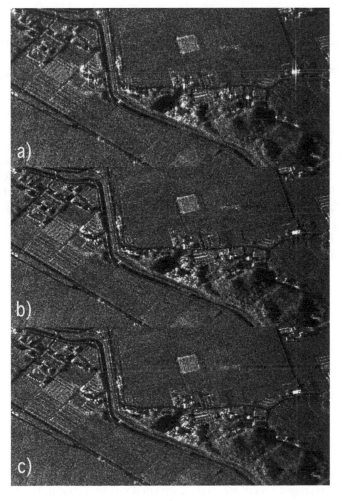

FIGURE 3-23 (a) Time domain processing, (b) frequency domain processing, and (c) without motion compensation.

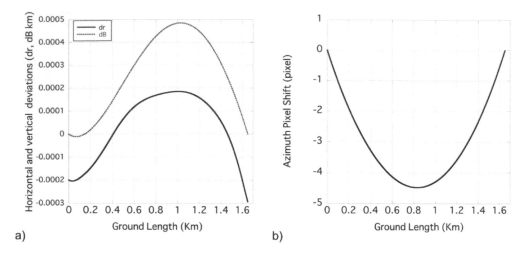

FIGURE 3-24   (a) The vertical and horizontal deviations of the orbit within a scene, and (b) the azimuth reconstruction factor used to linearly align the SAR data.

flight distance was 1.6 km. This image example was acquired in relatively stable conditions, where the maximum pixel dislocation in azimuth was 4 pixels and the dr/dB dislocation was ±30-cm to 50-cm deviations.

3.3.3.3   Airborne SAR and Integrated Navigation System

In order to fully reconstruct the airborne SAR data, highly accurate state vectors are mandatory. Recent GPS data provide accurate state vectors but the updating frequency (or sampling frequency) is too low to assist the data interpolation in time. To this end, an integrated navigation system (INS), which is combined with the GPS for position and speed measurement, three axes accelerators, and rate gyros are required to precisely generate the state vector at the frequency of the PRF. Figure 3-25 shows a block diagram of the INS system that was implemented onboard the Pi-SAR-L1, where the INS was made by Litton Aero Products (Woodland Hills, CA) in 1989.

Referring to the accelerators' sensitivity to the motions around the three axes shown in Figure 3-25, we have motion equations for governing the INS output:

$$\mathbf{A}_2(t) = \mathbf{M}_2^{-1}\left(A_m - \mathbf{N} \cdot \mathbf{M}_0 \cdot \mathbf{g}_0 + \mathbf{a}\right) + \boldsymbol{\omega} \times \mathbf{v}_{ECR} \tag{3.109}$$

$$\mathbf{M}_2^{-1} = \mathbf{M}_\lambda^{-1} \cdot \mathbf{M}_\varphi^{-1} \cdot \mathbf{M}_\chi^{-1} \cdot \mathbf{M}_r^{-1} \cdot \mathbf{M}_p^{-1} \cdot \mathbf{M}_y^{-1} \tag{3.110}$$

$$\overline{v}(t) = \int_0^t A_2(t')dt' + \mathbf{v}_0 + \mathbf{b} \tag{3.111}$$

$$\overline{r}(t) = \int_0^t \overline{v}(t')dt' + r_0 + \mathbf{c} \tag{3.112}$$

where $\mathbf{A}_m$ is the acceleration; $\mathbf{N}$ is the body normal vector in Z; $\mathbf{M}_0$ is the matrix for the incidence angle; $\mathbf{g}_0$ is the gravity vector; $\mathbf{M}_\lambda$, $\mathbf{M}_\varphi$, $\mathbf{M}_\chi$, $\mathbf{M}_r$, $\mathbf{M}_p$, and $\mathbf{M}_y$ are the rotation matrices for longitude, latitude, azimuth angle measured in the clockwise direction, roll, pitch, and yaw rotations, respectively; $\overline{v}(t')$ is the averaged speed; $r(t)$ is the estimated position; and $\mathbf{a}$, $\mathbf{b}$, and $\mathbf{c}$ are the unknown offset vectors for accelerator, velocity, and position, respectively, that the accelerometer could contain. Three unknowns can be estimated by minimizing the scalar residual Equation (3.113) of the

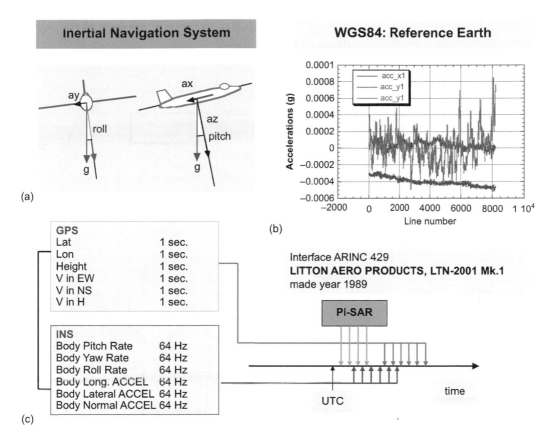

(a)

(b)

(c)

FIGURE 3-25 Gulfstream II (GII) accelerators mounted on three axes with the sample accelerations. Timing block diagram of inertial measurement unit (IMU) combining the GPS system and the INS system. a) Three accelerators installed on the aircraft can detect the gravities associated with pitch and roll in addition to its original accelarations as shown in b). c) GPS, INS and SAR data are connected using the time (UTC).

squared speed difference between the GPS velocity and model velocity and the differential positions between the GPS and model as a least squared sense:

$$E(\mathbf{a}, \mathbf{b}, \mathbf{c}) \equiv \sum \left\{ \mathbf{r}_i - \mathbf{r}(t \mid \mathbf{a}, \mathbf{b}, \mathbf{c}) \right\}^2 + \left\{ \mathbf{V}_i - \mathbf{V}(t \mid \mathbf{a}, \mathbf{b}) \right\}^2 \mapsto \min. \tag{3.113}$$

This can be Taylor expanded as

$$\mathbf{M}_{i,j} \cdot \Delta a_j = \mathbf{N}_i \tag{3.114}$$

$$\mathbf{M}_{i,j} = \frac{\partial^2 \mathbf{E}}{\partial a_i \, \partial a_j} = 2 \sum_{l=1,3, i=1,9, j=1,9} \left(\frac{\partial \overline{r_l}}{\partial a_i} \cdot \frac{\partial \overline{r_l}}{\partial a_j} + \frac{\partial v_l}{\partial a_i} \cdot \frac{\partial v_l}{\partial a_j} \right)$$

$$\mathbf{N}_i = \frac{\partial \mathbf{E}}{\partial a_i} = -2 \sum_{l=1,3, i=1,9} \left\{ (r_l - \overline{r_l}) \frac{\partial \overline{r_l}}{\partial a_i} + (v_l - v_l) \frac{\partial \overline{v_l}}{\partial a_i} \right\}, \tag{3.115}$$

and the solutions obtained iteratively. Some results are shown in Figure 3-26.

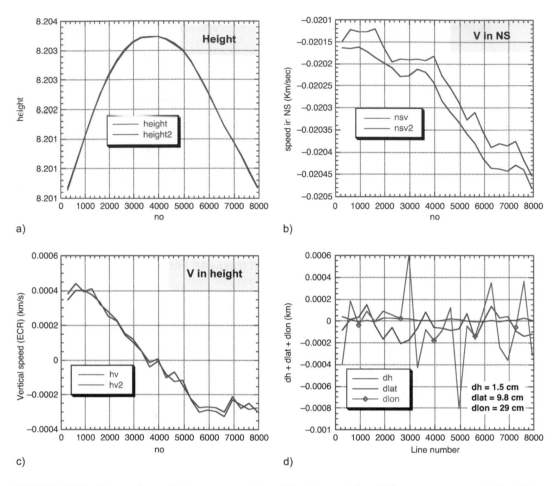

FIGURE 3-26 Comparison of the regenerated model value and the GPS measurements of (a) height, (b) navigation system speed, (c) speed in Z, and (d) total error of the positions.

3.4 SAR GEOMETRY AND THE COORDINATE SYSTEM

3.4.1 ORBITAL EXPRESSION

State vectors (positions and speeds) of all the objects on Earth (and in space) are described in the International Terrestrial Reference Frame (ITRF), the unified time-space coordinate system, originally established in 1992 and updated in 2014 (http://itrf.ensg.ign.fr). It provides the information detailed in the following paragraph.

The time coordinate is the GPS time coordinate, that is, International Atomic Time (TAI), with the Universal Time Coordinate (UTC) coordinating the leap second. The time code in the SAR data is expressed by GPS week and GPS second, and it is converted to the UTC coordinate. GPS time is recorded in the SAR telemetry data.

With regard to the space coordinate, the ITRF consists of the solar coordinate, ECI coordinate, and the ECEF coordinate, which considers the Earth's rotation, precession, and nutation movement. In particular, satellite movement is expressed by the ECEF coordinate. The Japan Geodetic System adopted the world geodetic system instead of the Tokyo datum in 2002. Before that time, the satellite state vector was composed of an ECEF and ECI coordinate. Satellite motion was described in

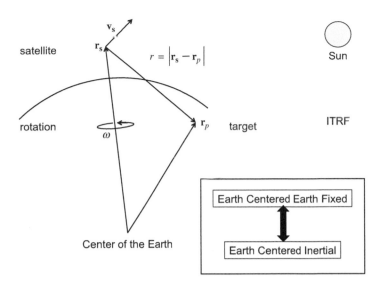

FIGURE 3-27 Calculation of the Doppler-related values.

the ECI and converted by the ECR coordinate. Satellite state vector is represented by six components and time. Following is a summary:

- In the 1990s: $(x, y, z, v_x, v_y, v_z, t)$: x, y, z were in ECEF; v_x, v_y, v_z were in ECI; and t was in UTC. They were produced every one minute and one image contained 28 elements, which enabled an errorless expression using the Hermite interpolation,

$$\mathbf{v}_{ECR} = \mathbf{v}_{ECI} + \omega \times \mathbf{r}_s, \tag{3.116}$$

which is the conversion formula from ECEF and ECI for velocity. Different coordinate systems for velocity and position were not known.

- In the 2000s, satellite positions fully depend on the ITRF, and all the components are expressed in ECR. The time coordinate is synchronized to the GPS satellite and converted to the UTC time. They are in operation for ALOS and ALOS-2. UAVSAR and Pi-SAR-L1/L2 use the real-time kinematic (RTK) and precise state vector (Figure 3-27).

3.4.2 SAR COORDINATE SYSTEM

The SAR coordinate system adopts the following conditions: (1) the SAR flies an almost circular orbit around the Earth's equator; (2) the Earth's surface is expressed by the Earth ellipsoid (i.e., Geodetic Reference System 1980, or GRS80), on which the SAR image is projected; (3) the Earth's surface is modeled by the summation of topography, the geoid, and the Earth ellipsoid; and (4) to enable the SAR images on the right and left sides of the direction in which the satellite is moving to be processed (Figure 3-28), for this book, the range direction is defined in the x positive direction, and azimuth direction is taken in the $-y$ direction. As such, the following provides the SAR coordinate system.

3.4.2.1 Orbit Coordinate System

In an orbit coordinate system,

- z axis: $-z$ direction of the orbit data (z) directed toward the Earth's equator
- x' axis: along the vx in ECI

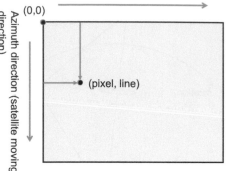

FIGURE 3-28 SAR coordinate system.

- y axis: right-handed coordinate made by z-x'
- x axis: z-y in the right-hand coordinate

For this x-y-z coordinate, the final coordinate is Euler rotated in R-P-Y because R-P are very small and Taylor expanded. The yaw angle will be within ± 4 degrees over the equator and zero degrees in the polar region. In summary,

$$
\mathbf{n}_z = -\frac{\mathbf{z}}{|\mathbf{z}|}
$$

$$
\mathbf{n}'_x = -\frac{\mathbf{v}}{|\mathbf{v}|}
$$

$$
\mathbf{n}_y = \mathbf{n}_z \times \mathbf{n}'_x \tag{3.117}
$$

$$
\mathbf{n}_x = \mathbf{n}_y \times \mathbf{n}_z
$$

where n_x, n_y, n_z, and n_x' are the unit vectors. Next, the Euler rotation around each axis can be considered:

$$
\mathbf{n}'_b = \mathbf{M}_Y \mathbf{M}_P \mathbf{M}_R \cdot \mathbf{n}_b
$$

$$
\mathbf{M}_R = \begin{pmatrix} 1 & 0 & 0 \\ 0 & \cos R & \sin R \\ 0 & -\sin R & \cos R \end{pmatrix}
$$

$$
\mathbf{M}_P = \begin{pmatrix} \cos P & 0 & \sin P \\ 0 & 1 & 0 \\ -\sin P & 0 & \cos P \end{pmatrix} \tag{3.118}
$$

$$
\mathbf{M}_Y = \begin{pmatrix} \cos Y & -\sin Y & 0 \\ \sin Y & \cos Y & 0 \\ 0 & 0 & 1 \end{pmatrix}
$$

Here, the n_b, n'_b are the line-of-sight vectors (before the Euler rotation) and after the Euler rotation; R, P, and Y are the roll angle, pitch angle, and yaw angle, respectively.

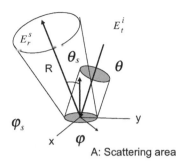

FIGURE 3-29   Normalized radar cross section as a definition. Signal is input from the right top to the surface then is scattered to the right up direction.

3.4.2.2   Earth's Surface

GRS80 as the earth ellipsoid can be expressed by the following model:

$$\frac{x^2}{R_a^2} + \frac{y^2}{R_a^2} + \frac{z^2}{R_b^2} = 1 \tag{3.119}$$

where Ra = 6,378.137 km, Ra = 6,356.7523141 km, flattening $(Ra-Rb) / Ra$ = 1 / 298.257222101.

3.5   NORMALIZED RADAR CROSS SECTION AND OTHER EXPRESSIONS

SAR measures the radar backscattering coefficient of the target, which is the normalized radar cross section, or NRCS (3.120) (Ruck et al. 1970, Ulaby et al. 1982). This can be interpreted as the ratio of unit solid-angler energy outflow to the direction of ϕ_s and θ_s to the unit solid-angler uniformly scattered energy flow, which all the energy input from the direction of ϕ_i and θ_i over the area, A (Figure 3-29):

$$NRCS(\phi_s, \theta_s; \phi_i, \theta_i) = \frac{R^2 \left\langle \left| E_s(\phi_s, \theta_s) \right|^2 \right\rangle}{\dfrac{\left\langle \left| E_i(\phi_i, \theta_i) \right|^2 \right\rangle}{4\pi} A}, \tag{3.120}$$

where E_s is the scattering electric field at the distance R from the target; E_i is the incoming electric field over the target; θ_i (θ_s) is the incidence angle for the incoming wave (scattering wave); and ϕ_i (ϕ_i) is the azimuth angle for the scattering. This definition is similar to the antenna pattern: the directional sensitivity of the antenna radiation normalized by the uniformly emitted energy. While depending on the target, the forward scattering $(\phi_i = \phi_s + \pi, \theta_i = \theta_s)$ shows the largest intensity; the backscattering at $(\phi_i = \phi_s, \theta_i = \theta_s)$ is not the largest (Figure 3-30), in general.

The Amazon Rainforest can be assumed to be a fully distributed target. Thus, the scattering directionality is incidence angle independent, and the theoretical backscattering coefficient can be calculated in such a way that all the energy of the incoming signal is scattered uniformly

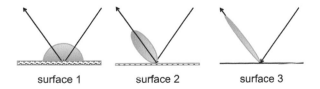

surface 1   surface 2   surface 3

FIGURE 3-30   Scattering properties of the target and its roughness.

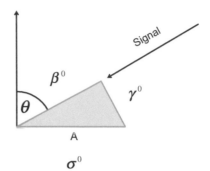

FIGURE 3-31 Coordinates of the three backscattering coefficients.

within a half sphere unless absorption occurred, and its normalized radar cross section can be expressed by

$$NRCS_{forest} = 10\log_{10}\frac{1}{2\pi} = -7.9dB. \tag{3.121}$$

Here, 2π is the solid angle for a half sphere. At first the measured NRCS for the Amazon shows similar numbers. However, more detailed analysis shows that the measured value is larger, and this means that uniform scattering could occur over a smaller solid angle than a half sphere.

NRCS is denoted in multiple ways: by the backscattering coefficient, sigma-naught, sigma-zero, and σ^0. Other than these, there are two more expressions referring to the area $A\cos\theta$ or $A\sin\theta$ rather than A. As shown in Figure 3-31, this depends on whether the area is parallel to the surface, perpendicular to the incoming wave direction, or parallel to the incoming wave direction. Two expressions are

$$\gamma^0 = \frac{\sigma^0}{\cos\theta} \tag{3.122}$$

$$\beta^0 = \frac{\sigma^0}{\sin\theta} \tag{3.123}$$

where γ^0 is gamma-zero, and β^0 is the beta-zero. The use of gamma-zero is preferred for forest monitoring because the terrain modulation is corrected.

3.6   DISTRIBUTION FUNCTION OF THE SIGMA-NAUGHT AND SPECKLE NOISE

In a normal situation, the SAR has the bandwidth of less than 85 MHz and an antenna length of 10 m or less, the one-pixel size is several meters, and the number of scatterers involved reaches more than several hundred. This condition allows the distributed target approximation for each pixel to be expressed by

$$\mathbf{x} = \sum_{i=0}^{N-1} \sin c(r_i - r_0)\cdot A_i \exp(\phi_i \cdot j), \tag{3.124}$$

$$\phi_i = -4\pi r_i/\lambda, \tag{3.125}$$

where r_i is the distance between the satellite and the ith scatterer; r_0 is the distance to the center of the pixel; A_i is the amplitude of the ith backscatter; and φ_i is its phase. Because each scatterer could locate at r_i randomly, ϕ_i then distributes uniformly. Although there are several distribution functions, such as the binomial, Poisson, normal, Weibull, and Rayleigh functions, and so on, the central limit theorem for a large N number and ignorance of the sinc function in Equation (3.124) gives **x** as the complex Gaussian distribution function:

$$p(x,y) = \frac{1}{2\pi\sigma^2}\exp\left(-\frac{x^2+y^2}{2\sigma^2}\right), \tag{3.126}$$

where x and y are the cosine and sin component of the **x**, and σ is the standard deviation. This distribution function is decomposed to amplitude and the phase as follows:

$$p(A,\phi) = \frac{A}{2\pi\sigma^2}\exp\left(-\frac{A^2}{2\sigma^2}\right)$$

$$p(A) = \begin{cases} \dfrac{A}{2\pi\sigma^2}\exp\left(-\dfrac{A^2}{2\sigma^2}\right) & A >= 0 \\ 0 & A < 0 \end{cases} \tag{3.127}$$

$$p(\phi) = \frac{1}{2\pi}$$

In addition, if we calculate the distribution of the power, we have the following exponential distribution function:

$$p(p) = \frac{1}{\bar{P}}\exp\left(-\frac{p}{\bar{P}}\right) \tag{3.128}$$

where we use $\bar{P}$ as the averaged power. As a result, the single look complex (SLC) data are distributed in Rayleigh for amplitude and uniformly for phase. A SAR image, especially the SLC or lower number of the image viewed, often looks noisy because of the amplitude distribution. This is called speckle noise. Speckle noise sometimes needs to be reduced or eliminated in order to improve the interpretation accuracy.

To reduce the speckle noise, N-look summation normally is performed. Many adaptive despeckle filters have been developed. However, we will consider the simple averaging filter (the box car filter). The cross-distribution function of the two-summation variable is generated as follows (Papoulis 1977, Ulaby et al. 1982):

$$z = x_1 + x_2$$

$$P(z) = \int_{-\infty}^{\infty} P(x_1)P(z-x_1)dx_1 \tag{3.129}$$

$$P(x) = \frac{1}{\mu}\exp\left(-\frac{x}{\mu}\right)$$

Here, x_1 and x_2 are the independent variables, which distribute under the same distribution function; z is their sum, and μ is the average. Then, the two-look distribution function will be

$$P(z) = \frac{2^2 \cdot z}{\bar{z}^2}e^{-2\frac{z}{\bar{z}}}. \tag{3.130}$$

We can conduct the calculation as the total number of looks that reached N. As a result, the N-look intensity has the following distribution function:

$$P(z_N) = \frac{N^N \cdot z_N^{N-1}}{(N-1)! \cdot \overline{z_N}^N} e^{-N \frac{z_N}{\overline{z_N}}} \tag{3.131}$$

where $\overline{z_N}$ is the average of the N-look summed intensity, that is, N times the SLC's pixel power $(\overline{z_N} = N \cdot \overline{z_1})$. Ratio of the variance to the square of the average can be given by

$$RA = \frac{\overline{x^2}}{\overline{x}^2} = \frac{N+1}{N}. \tag{3.132}$$

Equation (3.131) is the chi-square distribution function with a freedom of $2N$. The greater N approaches the normal distribution by the central limit theorem. One example of actual data is shown in Figure 3-32, where the SIR-B L-band data are shown. Here, the three-look and four-look images have good agreement with the real SAR data.

As a result, the pixel intensity follows a probability distribution function depending on the number of scatterers involved. Even a uniform target, where a large number of the same amplitude scatterers randomly exist in a pixel, the pixel intensity follows a distribution function just like the speckle. The number of scatterers depends on the target property and the SAR resolution. If a stronger scatterer is added to many weak scatterers in a pixel, the pixel intensity may not distribute, and the scattering property will only depend on the one of the strongest targets. A high-resolution pixel reduces the number of scatterers, and the distribution function is different from the normal and becomes Weibull or binomial.

Thermal noise is generated at a receiver. This noise behaves independently from the SAR signal but is mixed with the SAR image as a power. Thermal noise is interpreted as the brown motion of the small particles as follows:

$$\mathbf{n} = A_N \exp(\phi_N \cdot j); \tag{3.133}$$

ϕ_N is the uniform distribution, and A_N is the Rayleigh distribution. N-look data have a function form that is similar to Equation (3.131). When the thermal noise is less than the SAR image from the target, it is negligible. The thermal noise has a function form similar to the speckle.

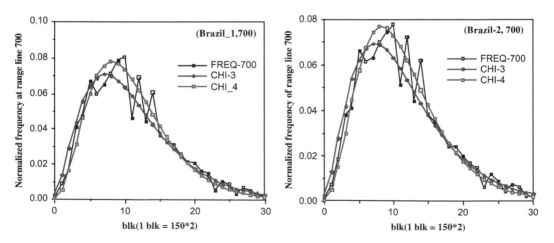

FIGURE 3-32 Histogram of SAR-B data over the Amazon. FREQ-700 is the measured histogram at line 700, CHI-3 is chi-square distribution with 3 looks summation, and CHI-4 with four look summation.

3.7   SOME TECHNIQUES TO IMPROVE SAR IMAGE QUALITIES

ALOS-2/PALSAR-2 is a third-generation JAXA SAR that enables high-resolution imaging with wide swath and lower ambiguities. Two major improvements were involved in its development: a dual receiver system and azimuth phase coding.

3.7.1   DUAL RECEIVER SYSTEM (DRS)

High-resolution imaging can be done by a large bandwidth and corresponding AD sampling. The sampling theorem governs to preserve the nature of the signal and to reconstruct the SAR data without loss, where the signal with bandwidth B should be sampled at least at a 2-B frequency. The azimuth resolution is theoretically half of the azimuth antenna length, and PRF should be greater than the corresponding Doppler bandwidth. By the ALOS/PALSAR era, one antenna was used both to transmit and receive in timesharing, and the simple observation concept was advantageous. High resolution has been the most important goal in recent years. Increasing the radar bandwidth is a simpler solution for range high resolution, but the high azimuth resolution is difficult and likely suffered from the narrower swath that was due to the higher PRF for the larger Doppler bandwidth. To solve this dilemma, the dual receiver system (DRS) was proposed (see Chapter 2).

DRS is an application of generalized multiple sampling introduced by Papoulis, although the implementing algorithm was not given. Brown (1981) was the first to prepare the operational algorithm, and Krieger et al. (2004) modified it for satellites. DRS was adopted by RADARSAT-2, TerraSar-X, and COSMO-SkyMed as well.

In this system, each channel receives the returned signal by a half of a bandwidth; the signal is "under sampled," and the processed image suffers severely from the ambiguity. However, the algorithm reconstructs the "fully sampled" data by combining two channels, each of which is slightly separated in azimuth.

Figure 3-33 shows the signal reconstruction where the original signal $x(T)$ is observed by two receivers with H_1 and H_2 transfer functions at a sampling interval of T, processed at two reconstruction

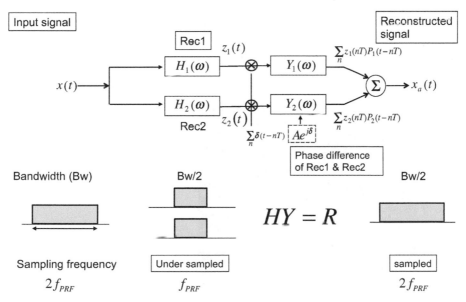

H=transfer function of DR, Y the unknown transfer function of the reconstruction filter, and R is the condition. Z_1 and Z_2 are both under sampled, x_a is reconstructed and sampled.

FIGURE 3-33   Configuration of signal reconstruction.

filters (P_1 and P_2), and summed up to determine the fully sampled signal $x_a(T)$. Signal reconstruction is expressed in the frequency domain:

$$H(\omega) \cdot Y(\omega) = R(\omega), \tag{3.134}$$

where H is the frequency spectrum of the two receivers, Y is the response function of the dual receivers, and R is the condition; all are complex. The core part of the filter is given as follows:

$$Y_1(f) = \begin{cases} \dfrac{e^{\frac{j\pi\Delta x_1^2}{2\lambda r_0} + \frac{j\pi\Delta x_1 f}{v}}}{1 - e^{\frac{j\pi PRF(\Delta x_2 - \Delta x_1)}{v}}} & -PRF < f < 0 \\[3em] \dfrac{e^{\frac{j\pi\Delta x_1^2}{2\lambda r_0} + \frac{j\pi\Delta x_1 f}{v}}}{1 - e^{\frac{j\pi PRF(\Delta x_1 - \Delta x_2)}{v}}} & 0 < f < PRF \end{cases} \quad Y_2(f) = \begin{cases} \dfrac{e^{\frac{j\pi\Delta x_2^2}{2\lambda r_0} + \frac{j\pi\Delta x_2 f}{v}}}{1 - e^{\frac{j\pi PRF(\Delta x_1 - \Delta x_2)}{v}}} & -PRF < f < 0 \\[3em] \dfrac{e^{\frac{j\pi\Delta x_2^2}{2\lambda r_0} + \frac{j\pi\Delta x_2 f}{v}}}{1 - e^{\frac{j\pi PRF(\Delta x_2 - \Delta x_1)}{v}}} & 0 < f < PRF \end{cases}$$

$$\tag{3.135}$$

Here, Δx_1 and Δx_2 are the relative locations of antenna$_1$ and antenna$_2$ in azimuth direction, v is the satellite speed, and PRF is the pulse repetition frequency of the single receiver.

Let us consider the PALSAR-2, which adopts DRS at most modes. A reconstruction process is performed by (1) raw data compression using each chirp signal, (2) FFT in azimuth direction with full band width Bw that is zero padded every other pulse, (3) multiplying the reconstruction filters, and 4) inverting FFT. Then, the "full sampled" azimuth data in time domain will be created. The R-D algorithm embeds the migration process after Step (3), and the additional processing cost is not an issue.

The two receivers are different from one another and need phase and amplitude calibration; the amplitude is equalized for both receivers' output, and the optimal phase is obtained in a way that minimizes the azimuth ambiguity of the brighter target selected from near the coastal area, such as Figure 3-34 shows the PALSAR-2 42-MHz dual polarization at Hokkaido (HH for solid and HV for dotted). Minimal ambiguities are obtained at −135.0 degrees for HH and 95.0 degrees for HV. We confirmed that the right image is uncalibrated, showing the ambiguities of the oil reservoir in the

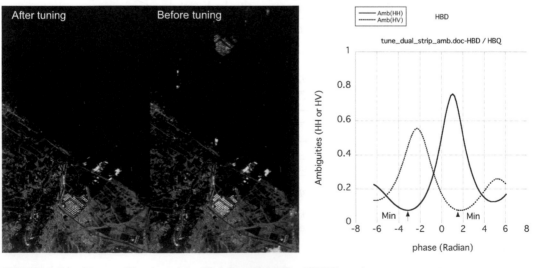

FIGURE 3-34   Phase calibration of the High Sensitivity Dual(HBD) mode.

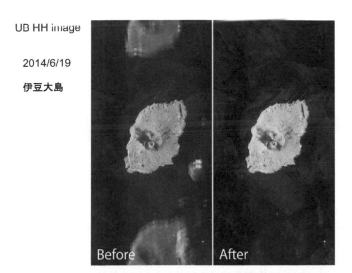

UB HH image

2014/6/19

伊豆大島

Before   After

FIGURE 3-35 Improvement in the dual receiver systems: (left) uncalibrated and (right) calibrated.

coastal area, while the left image suppresses these ambiguities. Figure 3-35 shows the Izu-Ohshima volcano image from PALSAR-2 UB, where the left image is uncalibrated and largely suffers from ambiguities, and the right image is the calibrated image. We again confirmed the effectiveness of the calibration and the DRS.

3.7.2 AZIMUTH PHASE CODING REDUCING THE RANGE AMBIGUITY

The range ambiguity (RA) of a spaceborne SAR is the signal contamination from the radar backscatter at the nearby regions illuminated by the antenna beams of the sequential pulses. RA essentially depends on the SAR range antenna pattern. Most SARs consist of a planar array antenna, and PALSAR has a sinc antenna elevation pattern that covers a 70-km swath. Range compression by a chirp signal is not sharp enough to fully suppress the RA. Figure 3-36 depicts the schematic flow as to how the nth beam is contaminated from $n-1$th and $n+1$th beam, and their range correlation power can be given by

$$
\begin{aligned}
F_R(t) &= \sum_{k=-1}^{1} G_R\left(t - \frac{n}{f_{PRF}} - \frac{i}{f_{sample}}\right) f_R\left(t - \frac{n}{f_{PRF}} - \frac{i}{f_{sample}}\right) \oplus f_R^{\star}(t') \\
&\approx \sum_{k=-1}^{1} G_R\left(t - \frac{n}{f_{PRF}} - \frac{i}{f_{sample}}\right).
\end{aligned}
\tag{3.136}
$$

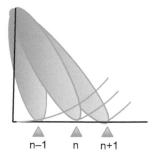

n−1   n   n+1

FIGURE 3-36 Location of the three contiguous pulses.

The first term is the beam of interest, and the second and the third are the unwilling's. If we let h be the satellite altitude and n the number of delayed pulses, the off-nadir angles for three succeeding pulses are

$$\theta_{off+k} = \cos^{-1}\left(\frac{(R+h)^2 + \left(r_s + k\dfrac{c}{2f_{PRF}}\right)^2 - R^2}{2\left(r_s + k\dfrac{c}{2f_{PRF}}\right) - (R+h)}\right) \tag{3.137}$$

$$r_s = \frac{c}{2}\left(i + \frac{n}{f_{PRF}} + \Delta t_{off}\right).$$

Here, $k = -1, 0, 1$. Inserting $h = 628$ km, and $f_{PRF} = 2,000$ Hz, the angular difference between the off-nadir of the pixel in the main beam and the off-nadir of one pulse after or before is almost 5 degrees. Although most of the antenna beam width is 5 degrees, the bigger signal component may leak to the main beam, depending on the location of the pixel.

By changing the chirp rate alternately up or down associated with the pseudo randomized 0-pai phase shifts, the range compression intensively differentiates the process gain depending on the pulse shift within the number of combinations and reduces the range ambiguity. This example has four combinations. Following is the range compression using the proper chirp signals as

$$F_R(t,m) = \sum_{k=-1}^{1} G_R\left(t - \frac{n+k}{f_{PRF}} - \frac{i}{f_{sample}}\right) f_{R,m+k}\left(t - \frac{n+k}{f_{PRF}} - \frac{i}{f_{sample}}\right) \oplus f^*_{R,m}(t')$$

$$\approx G_R\left(t - \frac{n}{f_{PRF}} - \frac{i}{f_{sample}}\right). \tag{3.138}$$

Here, $f_{R1}, f_{R2},$ and f_{R3} are the scattered signals and refer to the transmit sequence of the right pulse, one pulse before, and one pulse after.

Let us look at the image examples. Figure 3-37a shows that the PLASAR image suffered from range ambiguity, especially at the far range. The image depicts the coastal region of the Chosi peninsula in Japan's Chiba prefecture. PALSAR uses only a down-chirp transmission and elevation antenna width of 5 degrees. Several vertical short lines are the range ambiguities. Russia's Kamchatka Peninsula in Figure 3-37b is very contaminated near the peninsula, which is one pulse closer.

PALSAR-2 implements the azimuth phase coding (APC) method, which restrains the up-down/ chirp alternately and pseudo-randomizes the 0-pai shift in transmission signal; proper extraction of the intense chirp information perfectly suppresses the ambiguity, as shown in Figure 3-37c and d. We confirm the effectiveness of the APC.

3.7.3 SATURATION CORRECTION

SAR data are a bit saturated, but the saturation rate has decreased in SARs over time, owing to the technological revolution in preserving the phase and amplitude as much as possible. The correction can be simply implemented in the processing routine. This method will be described in Chapter 4.

3.8 SAR IMAGE QUALITY

There are several issues with SAR image quality, which are represented by the ambiguities (range and azimuth components), geometric distortion, radiometric distortion, and the interactions from the internal and outer signal sources (radio frequency interference, ionospheric disturbances, and so on) or speckle noise associated with the coherent processing.

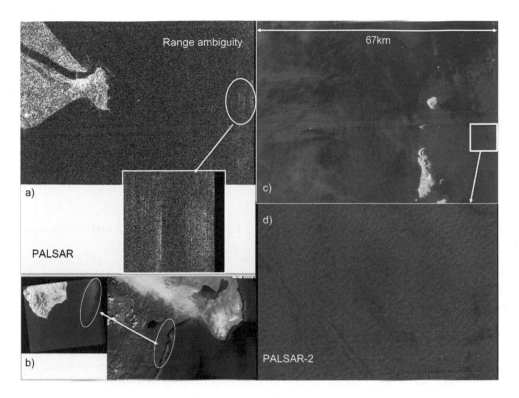

FIGURE 3-37   PALSAR and PASAR-2 coastal region images (a) PALSAR far range image with range ambiguity; (b) PALSAR with large ambiguity at Russia's Kamchatka Peninsula; (c) PALSAR-2 full swath image over the Niijima of Tokyo Japan; and (d) zoomed image of the far range confirmed no ambiguity.

3.8.1   Ambiguities

SAR fully relics on the sampling theorem that observes the IFOV, with f_{PRF} fully exceeding its Doppler bandwidth and range bandwidth by f_{sample}. However, excess Doppler bandwidth caused by the remaining azimuth antenna pattern is truncated and processed as under-sampled data. This component appears as the blurred image (the noise). We consider such a point target that is located far in the azimuth direction and the signal as range correlated. The range compressed signal for such a target is expressed by

$$f(r_s) \cong a \sin c\left(\frac{\pi(r_s - r_{s0})}{c/2B} \right) \exp\left\{ -2\pi\left(\frac{2r_s}{\lambda} + f'_{DD} T^2 \right) j \right\} \qquad (3.139)$$

$$f'_{DD} = -\frac{4\pi}{\lambda r_s} V^2. \qquad (3.140)$$

Expanding

$$r_s = \sqrt{r^2 + (V_s T + L_x)^2} \qquad (3.141)$$
$$\cong r + \frac{V_s T + L_x}{2r_0}.$$

The reference function is prepared for the closest target at $T = 0$ the same as before:

$$f'_{DD} = -\frac{4\pi}{\lambda r_0} V_s^2. \qquad (3.142)$$

These two f_{DD} differ greatly and their azimuth correlation should be blurred. Moreover, the correlation process is applied to the data aligned on the migration. If the signal is truncated by PRF and aligned on the migration, the data still will be correlated. All the undesired signals whose Doppler frequencies exceed PRF/2 are truncated within the azimuth frequency band by plus or minus shifting of f_{PRF} for f_D and are correlated, although perfect compression cannot be obtained.

Thus, the f_{PRF} is the ambiguous frequency and $-f_{PRF}/f_{DD}$ is the time shift. The location shift in azimuth is given by

$$-\frac{f_{PRF}}{f_{DD}} v_g. \tag{3.143}$$

3.8.2 Geometric Distortions

There are three well-known geometric distortions—foreshortening, layover, and shadowing, which are discussed in the following sections.

3.8.2.1 Foreshortening

All the non-zero height targets are shifted to the nadir because the SAR imaging is the geometric conversion preserving the distance and projecting it onto the Earth's ellipsoid.

3.8.2.2 Layover

The foreshortening moves the target as close as the satellite nadir. This means that the higher the position is, the closer the target will be on the SAR image. The higher the shift, the closer it is to the nadir than to the neighboring lower height target. This change of the projection order is called the layover. So many scatterers concentrate on the layover points, the radar backscatter becomes extremely bright.

3.8.2.3 Shadowing

In this case, the target area is not illuminated due to occultation. High mountains and distant areas are often shadowed.

3.8.3 Radiometric Distortions

Radiometric distortion occurs from the distortion of the pixel area. The radar cross section is defined as the unit aerial radar backscatter. This unit area is severely modulated by the surface terrain. This normalization will be discussed in Chapter 8.

3.8.4 Interaction from Internal or External Signal Sources

There are several types of interactions, such as radio frequency interference, ionospheric disturbance, atmospheric disturbance, saturation noise, thermal noise, and so on. These also degrade the SAR image quality. Chapter 13 will describe these examples and the theoretical background of these noises.

3.9 SAR PROCESS FLOW

3.9.1 SAR Thematic Process Flow

The SAR process flow is summarized in Figure 3-38, which depicts the ALOS-2/PALSAR-2 systems. This process covers the reconstruction of the raw data for SAR imaging followed by the calibration and correction of all the necessary processes.

3.9.2 General Process Flow of SAR Imaging to the Final Products

Figure 3-39 shows the general flow of SAR imaging and calibration of the applications.

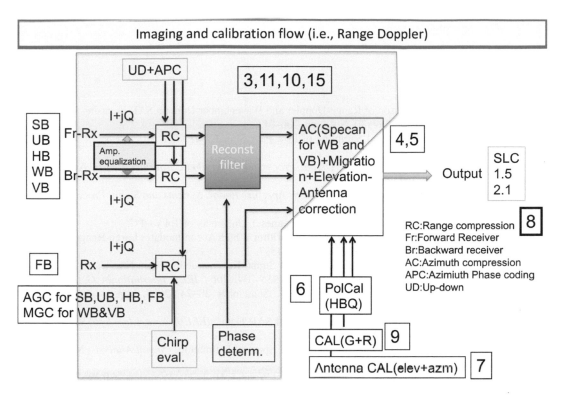

FIGURE 3-38 General process flow of the PALSAR-2 data. The boxed numbers indicate the chapter that deals with each topic.

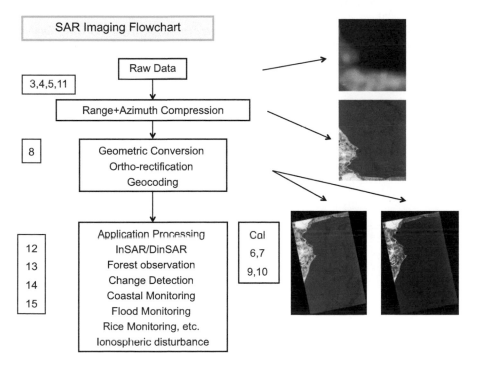

FIGURE 3-39 Thematic process flow of SAR processing. Boxed numbers indicate the corresponding chapter numbers where these subjects are discussed.

3.10   SUMMARY

This chapter introduced and briefly discussed the basic concepts of SAR imaging and analysis.

REFERENCES

Bamler, R., 1992, "A Comparison of Range-Doppler and Wavenumber Domain SAR Focusing Algorithms," *IEEE T. Geosci. Remote*, Vol. 30, No. 4, pp. 706–713.

Brown, J. L., 1981, "Multi-Channel Sampling of Low-Pass Signals," *IEEE T. Circuits Syst.*, Vol. CAS-28, No. 2, pp. 101–106.

Cumming, I. G. and Wong, F. G., 2005, *Digital Processing of Synthetic Aperture Radar*, Artech House, Norwood, MA, pp. 369–423.

Curlander, J. C. and McDonough, R., 1991, *Synthetic Aperture Radar: Systems and Signal Processing*, Wiley, Hoboken, NJ.

Hagiwara, Y., 1982, *Introduction to Geodetics* (in Japanese), University of Tokyo, PC.

Jin, M. Y. and Wu, C., 1984, "A SAR Correlation Algorithm Which Accommodates Large-Range Migration," *IEEE T. Geosci. Remote*, Vol. GE-22-6, pp. 592–597.

Krieger G., Gebert, N., and Moreira, A., 2004, "SAR Signal Reconstruction From Non-Uniform Displaced Phase Centre Sampling," *Proceedings of IGARSS '04, 2004 IEEE International Geoscience and Remote Sensing Symposium*, Anchorage, AK, September 20–24, 2004, http://dx.doi.org/10.1109/IGARSS.2004.1370674

Madsen, S., 1989, "Estimating the Doppler Centroid of SAR Data," *IEEE T. Aero. Elec. Sys.*, Vol. 25, No. 2, pp. 134–140, http://dx.doi.org/10.1109/7.18675

Papoulis, A., 1977, "Part Three: Data Smoothing and Spectral Estimation," *Signal Analysis*, McGraw-Hill, New York.

Raney, R. K., Runge, H., Bamler, R., Cunning, I. G., and Wong, F. H., 1994, "Precision SAR Processing Using Chirp Scaling," *IEEE T. Geosci. Remote*, Vol. 32, No. 4, pp. 786–799.

Ruck, G. T., Barrick, D. E., Stuart, W. D., and Kric, C. K., 1970, *Radar Cross Section Handbook, Volume 2*, Plenum, New York, pp. 588.

Smith, A. M., 1990, "A New Approach to Range-Doppler SAR Processing," *Int. J. Remote Sens.*, Vol. 12, No. 2, pp. 235–251.

Ulaby, F., Moore, R., and Fung, A., 1982, *Microwave Remote Sensing, Active and Passive, Volume II: Radar Remote Sensing and Surface Scattering and Emission Theory*, Addison-Wesley, Boston, pp. 767–779.

4 Radar Equation for SAR Correlation Power—Radiometry

4.1 INTRODUCTION

SAR imaging is a linear process as long as the signal is within the dynamic range of the SAR instrument; it is not saturated. When the signal exceeds the dynamic range of the analog-to-digital converter (ADC) or the signal is saturated, SAR imaging amplifies the correlative signal and the noise differently, thereby reducing the correlation power. This chapter describes how SAR imaging accumulates scatterers distributed in a resolution cell and how it reduces the correlation power. A radar equation is derived for an SAR image that considers saturation noise, A/D conversion noise, and thermal noise. Three radar equations are also derived for real aperture radar (RAR), strip SAR mode, and scanning SAR (ScanSAR) mode. (The strip SAR and ScanSAR modes will be compared later in this chapter, see Figure 4-1.) A radiometric correction method is proposed to recover the reduced correlation in such a way that it amplifies the range correlation or raw signal by 1/(1-saturation rate) (Shimada 1999a).

4.2 THEORETICAL DESCRIPTION OF SAR IMAGING

4.2.1 ASSUMPTIONS

From the data reception and digitization points of view, an SAR consists of three units: the low-noise amplifier (LNA), the intermediate amplifier, and the ADCs (Figure 4-2). The LNA amplifies the signals from the antenna to an intermediate power level and also generates thermal noise. The intermediate amplifier adjusts the LNA output to a final level, in which the gain is selected manually by manual gain control (MGC) or automatically by automatic gain control (AGC). The noise generated in the intermediate amplifier is negligibly smaller than the thermal noise. Redundancy and saturation noises probably generate at the ADC as a difference between input and digitized signal. Although redundancy noise occurs only when the input is within the ADC range, saturation noise occurs when the input exceeds the ADC range. Thus, redundancy and saturation noises exist complementally.

Within an SAR observation scheme, a series of pulses are transmitted to the target as the SAR moves. These signals are then scattered from the target and received by the SAR. In each process, from pulse transmission to reception, there are three modulations: a gradual phase change that is due to the relative movement of the SAR and the scatterer, an amplitude change that depends on the scatterer's reflectivity, and a quick phase change that depends on the scatterer's complex reflectivity. Because each modulation is a linear process, summation of the processes over all the scatterers in the illuminated area expresses the signal that will be received by the SAR. The scatterers are distributed randomly in space, not in time, so the scattered signals can be approximated by a random process. Note that the received signal is a stationary and independent Gaussian process in space, and the thermal noise is a stationary Gaussian process in time.

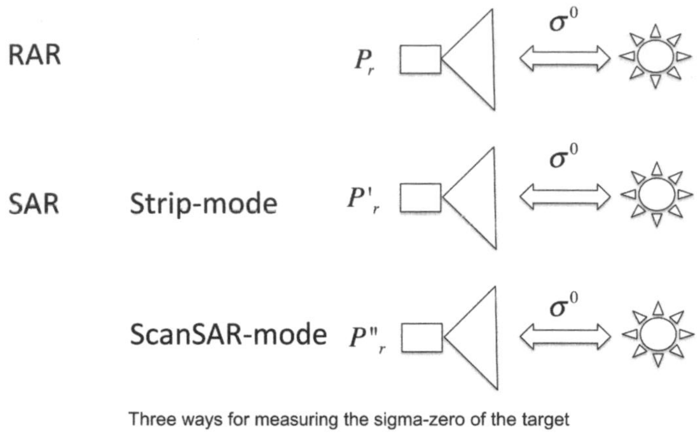

RAR P_r σ^0

SAR   Strip-mode P'_r σ^0

           ScanSAR-mode P''_r σ^0

Three ways for measuring the sigma-zero of the target

FIGURE 4-1   Three ways to measure the sigma-zero of the target.

We assume the following for the ADC and the receiver:

i. The ADC has a quantization interval, h, and a saturation level, C, of $h*2^{(L-1)}$, where L is the number of bits.
ii. The ADC causes redundancy noise and saturation noise; the former is Gaussian with a zero mean, and the latter is not Gaussian (as will be discussed later in this chapter).
iii. A receiver is a linear instrument and does not cause any nonlinear noise.
iv. The LNA generates Gaussian thermal noise.

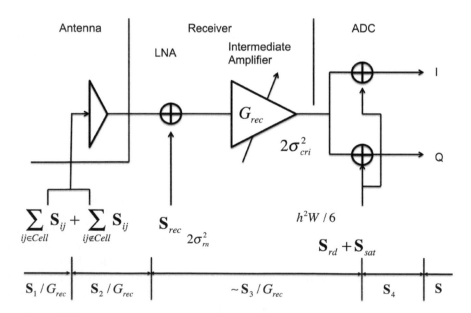

FIGURE 4-2   Simplified block diagram of SAR. Gaussian noise is generated in the LNA and ADC; saturation noise (S_4) is generated in the ADC.

4.2.2 SAR Raw Data Expression

Instantaneous SAR raw data at the ADC output can be expressed by (Figure 4-3):

$$\mathbf{S} = \sqrt{G_{rec}(t_r, t_a)} \cdot \sum_{ij \in Cell} \mathbf{S}_{ij} + \sqrt{G_{rec}(t_r, t_a)} \cdot \sum_{ij \notin Cell} \mathbf{S}_{ij} + \sqrt{G_{rec}(t_r, t_a)} \cdot \mathbf{S}_{rec} + \mathbf{S}_{rd} + \mathbf{S}_{sat}, \quad (4.1)$$

where

$$\mathbf{S}_{ij} = a_{ij} e^{j\delta_{ij}} e^{2\pi j \left(\frac{k_r}{2} t_r^2 + \frac{k_a}{2} t_a^2 \right)} \quad (4.2)$$

is the received signal from the ijth scatterer in an instantaneous field of view (IFOV); $G_{rec}(t_r, t_a)$ is the receiver gain at an azimuth time, t_a, when pulse is transmitted, and at a range time, t_r, when the pulse is received; $\mathbf{S}_{rec}$ is the thermal noise of the receiver; $\mathbf{S}_{rd}$ is the redundancy noise of the ADC; $\mathbf{S}_{sat}$ is the saturation noise; k_r and k_a are the range and azimuth chirp rate, respectively; a_{ij} is the amplitude of the received signal from the ijth scatterer; δ_{ij} is the phase for the ijth scatterer; and δ_{ij} contains information on the distance from the satellite and the scatterer's physical characteristics (Figure 4-4). The first term on the right side of Equation (4.1) sums up all the correlative scatterers in a resolution cell (Cell). If we focus on the scatterers in a Cell, the remaining four terms should be non-correlative components or noise. The second, third, and fourth terms are two-dimensional Gaussian signals, whose variances are the amplified signal, amplified thermal noise, and ADC redundancy noise. The fifth term represents nonlinear noise with an unknown distribution function.

Equation (4.1) is rewritten under the aforementioned assumptions as

$$\mathbf{S} = \sum_{i=1}^{4} \mathbf{S}_i, \quad (4.3)$$

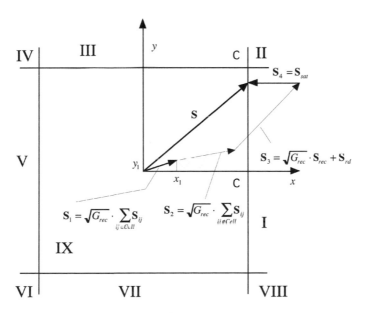

FIGURE 4-3 Coordinate system of A/D conversion. $\mathbf{S}_1$, $\mathbf{S}_2$, $\mathbf{S}_3$, and $\mathbf{S}_4$ are the amplified signal from the resolution cell, the amplified signal outside the resolution cell, the amplified thermal and ADC redundancy noise, and the ADC saturation noise, respectively. C is the maximum input voltage. ADC saturation noise differs in each region (I to IX).

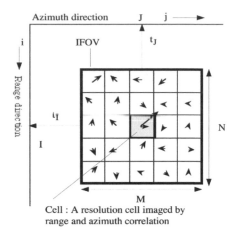

FIGURE 4-4 Coordinate system of SAR imaging and the composition of the illuminated area. IFOV is composed of M and N pixels in the azimuth and range direction respectively. Each one is represented by a different scatterer, which is shown as an arrow.

where $\mathbf{S}_1$ is the amplified correlative signal from the scatterers that locate within a Cell; $\mathbf{S}_1$'s x-y component is (x_1, y_1). Its power, P_1, is expressed by using the signal-to-clutter ratio (SCR) and the averaged background power, a_d^2, from each resolution cell as:

$$\mathbf{S}_1 \equiv \sqrt{G_{rec}\left(t_r, t_a\right)} \sum_{ij \in Cell} \mathbf{S}_{ij} \tag{4.4}$$

$$P_1 = a_d^2 \cdot SCR \cdot G_{rec} \tag{4.5}$$

$$SCR \equiv \frac{\sum_{ij \in Cell} a_{ij}^2}{a_d^2} \tag{4.6}$$

where $\mathbf{S}_2$ is the amplified non-correlative signal from an IFOV, excluding $\mathbf{S}_1$. Its power, $2\sigma_n^2$, is given by using M and N, which are the sizes of the IFOV divided by the pixel spacing in the azimuth and range directions:

$$\mathbf{S}_2 \equiv \sqrt{G_{rec}\left(t_r, t_a\right)} \sum_{ij \notin Cell} \mathbf{S}_{ij} \tag{4.7}$$

$$2\sigma_n^2 = a_d^2 \cdot \left(MN - 1\right) \cdot G_{rec} \tag{4.8}$$

where $\mathbf{S}_3$ is the summation of the amplified thermal noise and ADC redundancy noise. Its Gaussian component is (x_3, y_3) and its power is $2\sigma_{rn}^2$

$$\mathbf{S}_3 = \sqrt{G_{rec}\left(t_r, t_a\right)} \cdot \mathbf{S}_{rec} + \mathbf{S}_{rd} \tag{4.9}$$

$$2\sigma_{rn}^2 \cong 2\sigma_{rn0}^2 \cdot G_{rec}\left(t_r, t_a\right) + \frac{h^2}{6} W \tag{4.10}$$

where $2\sigma_{rn0}^2$ is the thermal noise power. The second term on the right side of Equation (4.10) is the ADC redundancy noise, which depends on the power input to the ADC (see Appendix 4A-1);

W is a weighting factor. Combining Equations (4.5), (4.8), and (4.10), the SNR of the raw data can be expressed by

$$SNR = \frac{(MN - 1 + SCR) \cdot a_d^2 \cdot G_{rec}}{2\sigma_{rn}^2 \cdot G_{rec} + \frac{h^2}{6} W} \cong \frac{(MN) \cdot a_d^2 \cdot G_{rec}}{2\sigma_{rn}^2}. \tag{4.11}$$

Usually, SCR is less than 10,000—even for the brightest target. M and N are typically several hundred to a thousand or more. Thus, the approximation in Equation (4.11) is valid; S_4 is the ADC saturation noise (x_4, y_4) generated when real, x, and/or imaginary, y, components of $S_1 + S_2 + S_3$ exceed C. The saturation noise, S_4, introduced here is a nonstationary noise because it is driven by the stationary noise, S_3, and the time-dependent signal, $S_1 + S_2$.

4.2.3   RANGE AND AZIMUTH CORRELATION POWER

4.2.3.1   General Approach for Model Expression

SAR imaging, which means range and azimuth correlation processing, is performed in the range and azimuth directions with different gains. Saturation in the ADC affects both correlations, causing loss of amplitude and phase information. Correlation of Equation (4.1) can be expressed as follows:

$$S_C(t', t'') = \sum_{m=1}^{4} S_{cra,m}(t', t'') \tag{4.12}$$

$$S_{cra,m}(t', t'') \equiv S_m \oplus \mathbf{g}_r^* \oplus \mathbf{g}_a^*$$

$$= \frac{1}{\Delta\tau\Delta T} \int_{-T/2}^{T/2} \int_{\tau/2}^{\tau/2} \sqrt{G_{rec}(t_1 + t', t_2 + t'')} \frac{S_m(t_1 + t', t_2 + t'')}{\sqrt{G_{rec}(t_1 + t', t_2 + t'')}} \mathbf{g}_r^*(t_1) \mathbf{g}_a^*(t_2) dt_1 dt_2 \tag{4.13}$$

$$= \frac{\overline{\sqrt{G_{rec}(t', t'')}}}{\Delta\tau\Delta T} \int_{-T/2}^{T/2} \int_{-\tau/2}^{\tau/2} \frac{S_m(t_1 + t', t_2 + t'')}{\sqrt{G_{rec}(t_1 + t', t_2 + t'')}} \mathbf{g}_r^*(t_1) \mathbf{g}_a^*(t_2) dt_1 dt_2$$

$$\overline{\sqrt{G_{rec}(t', t'')}} = \frac{1}{\Delta\tau\Delta T} \int_{-T/2}^{T/2} \int_{-\tau/2}^{\tau/2} \sqrt{G_{rec}(t_1 + t', t_2 + t'')} dt_1 dt_2 \tag{4.14}$$

$$g_r(t) = \left(\begin{array}{cc} e^{-2\pi j \frac{k_r}{2} t^2} & |t| \leq \tau/2 \\ 0 & else \end{array} \right) \tag{4.15a}$$

$$g_a(t) = \left(\begin{array}{cc} e^{-2\pi j \frac{k_a}{2} t^2} & |t| \leq T/2 \\ 0 & else \end{array} \right) \tag{4.15b}$$

where S_C is the SAR correlation output; $\mathbf{g}_r$ and $\mathbf{g}_a$ are the range and azimuth reference functions; "*" is the complex conjugate; $\oplus$ is the correlation operation; τ and T are the range and azimuth correlation durations; $\Delta\tau$ and ΔT are the range and azimuth sampling intervals; and $S_{cra,m}$ is the SAR correlation output for the mth signal normalized by the receiver gain; $G_{rec}()$ is not correlated with the reference functions, so the moving average of receiver gain centered at t' and t'' is taken out of the integration. Then, the remaining term is almost separated from the receiver gain; (t_1, t') and (t_2, t'') are the range and azimuth times. Note that the range curvature in azimuth processing is ignored for simplicity.

Here, $\mathbf{S}_3$ is stationary Gaussian, and $\mathbf{S}_2$ is Gaussian in space but correlative in time; $\mathbf{S}_4$ and $\mathbf{S}_C$ are random processes; $\mathbf{S}_C$'s power, denoted by P_C, is given with two step expectations: (1) an expectation over $\mathbf{S}_3$ leaving $\mathbf{S}_1 + \mathbf{S}_2$ fixed, and (2) an expectation over $\mathbf{S}_2$ as:

$$P_C\left(t',t''\right) = \left\langle \mathbf{S}_C\left(t',t''\right) \cdot \mathbf{S}_C^*\left(t',t''\right)\right\rangle = \sum_{m=1}^{4}\sum_{n=1}^{4}\left\langle \mathbf{S}_{cra,m}\left(t',t''\right) \cdot \mathbf{S}_{cra,n}^*\left(t',t''\right)\right\rangle \tag{4.16}$$

where $\langle\cdot\rangle$ stands for the two step expectations mentioned earlier. Hereafter, "an expectation" is used for this meaning. Note that $\mathbf{S}_2$ and $\mathbf{S}_3$ are essentially different processes: $\mathbf{S}_2$ correlates with the SAR reference signal; $\mathbf{S}_3$ is white noise. Introducing $\mathbf{R}_{mn}$, an expectation of the mutual correlation of $\mathbf{S}_m$ and $\mathbf{S}_n^*$, which is a mutual correlation of $\mathbf{S}_{cra,m}$ and $\mathbf{S}_{cra,n}^*$, denoted by $\mathbf{P}_{mn}$, can be expressed more effectively (Papoulis 1977) as:

$$P_{mn}\left(t',t''\right) \equiv \left\langle \mathbf{S}_{cra,m}\left(t',t''\right) \cdot \mathbf{S}_{cra,n}^*\left(t',t''\right)\right\rangle$$

$$= \frac{\overline{\sqrt{G_{rec}\left(t',t''\right)}}^2}{\left(\Delta\tau\Delta T\right)^2}\int_{-T/2}^{T/2}\int_{-T/2}^{T/2}\int_{-\tau/2}^{\tau/2}\int_{-\tau/2}^{\tau/2}\mathbf{R}_{mn}(t_1+t',t_2+t',t_3+t'',t_4+t'')\mathbf{g}_r^*(t_1)\mathbf{g}_r(t_2)\mathbf{g}_a^*(t_3)\mathbf{g}_a(t_4)dt_1dt_2dt_3dt_4$$

$$\tag{4.17}$$

$$\mathbf{R}_{mn}(t_1,t_2,t_3,t_4) = \frac{\left\langle \mathbf{S}_m(t_1,t_2) \cdot \mathbf{S}_n^*(t_3,t_4)\right\rangle}{G_2 \cdot G_4}, (m,n \geq 2) \tag{4.18}$$

where $G_2 \equiv \sqrt{G_{rec}\left(t_2\right)}$ and $G_4 \equiv \sqrt{G_{rec}\left(t_4\right)}$. The expectations are calculated for m, $n \geq 2$. In the following, the variables t_1, t_2, t_3, and t_4 are not written except when important. SAR correlation power can be calculated term-by-term.

4.2.3.2 Calculation of P_{11}

In this calculation,

$$\mathbf{R}_{11} = \frac{\mathbf{S}_1 \cdot \mathbf{S}_1^*}{G_2 \cdot G_4}$$

$$= \sum_i\sum_j a_{ij}e^{2\pi j\frac{k_r}{2}(t_1-t_i)^2 + 2\pi j\frac{k_a}{2}(t_2-t_j)^2 + j\delta_{ij}} \cdot \sum_k\sum_l a_{kl}e^{-2\pi j\frac{k_r}{2}(t_3-t_k)^2 - 2\pi j\frac{k_a}{2}(t_4-t_l)^2 - j\delta_{kl}}, \tag{4.19}$$

where t_i is the time delay in range and t_j is the azimuth time. An expectation is not applied because $\mathbf{S}_1$ does not depend on $\mathbf{S}_2$ or $\mathbf{S}_3$. Substituting Equation (4.19) into Equation (4.17), we have

$$P_{11} = \frac{\overline{\sqrt{G_{rec}\left(t',t''\right)}}^2}{\left(\Delta\tau\Delta T\right)^2}\sum_i\sum_j\sum_k\sum_l a_{ij}a_{kl}\tau^2 T^2 \frac{\sin\left\{\pi k_r\tau(t_i+t')\right\}}{\pi k_r\tau(t_i+t')} \cdot \frac{\sin\left\{\pi k_r\tau(t_k+t')\right\}}{\pi k_r\tau(t_k+t')} \cdot$$

$$\frac{\sin\left\{\pi k_a T(t_j+t'')\right\}}{\pi k_a T(t_j+t'')} \cdot \frac{\sin\left\{\pi k_a T(t_l+t'')\right\}}{\pi k_a T(t_l+t'')} \cdot e^{2\pi j\frac{k_r}{2}(t_i^2-t_k^2)+2\pi j\frac{k_a}{2}(t_j^2-t_l^2)+j(\delta_{ij}-\delta_{kl})}. \tag{4.20}$$

This is valid only when $k = i$ and $l = j$ as:

$$P_{11} = \overline{\sqrt{G_{rec}}}^2 \sum_i \sum_j a_{ij}^2 (MN)^2 \left[\frac{\sin\{\pi k_r \tau(t_i + t')\}}{\pi k_r \tau(t_i + t')} \right]^2 \left[\frac{\sin\{\pi k_a T(t_j + t'')\}}{\pi k_a T(t_j + t'')} \right]^2 \quad (4.21)$$

t_i and t_j distribute around t' and t'', respectively. Using SCR, Equation (4.21) is modified to

$$P_{11} = \overline{\sqrt{G_{rec}}}^2 SCR \cdot a_d^2 \cdot (MN)^2 \cdot D \quad (4.22)$$

$$D = \sum_i \sum_j \left[\frac{\sin\{\pi k_r \tau(t_i + t')\}}{\pi k_r \tau(t_i + t')} \right]^2 \left[\frac{\sin\{\pi k_a T(t_j + t'')\}}{\pi k_a T(t_j + t'')} \right]^2 \Bigg/ \sum_i \sum_j 1 \quad (4.23)$$

Here, D is the weighting function for forming a resolution cell.

4.2.3.3 Calculation of $P_{14} + P_{41}$

S_1 is not a random process, so $\mathbf{R}_{14} + \mathbf{R}_{41}$ is given by:

$$\mathbf{R}_{14} + \mathbf{R}_{41} = \frac{S_1 \cdot \langle S_4^* \rangle + \langle S_4 \rangle \cdot S_1^*}{G_2 \cdot G_4} \quad (4.24)$$

$$\langle S_4^* \rangle = \langle x_4 \rangle - j \langle y_4 \rangle \quad (4.25)$$

$$\langle x_4 \rangle = \frac{C - x_1}{2} Erfc\left(\frac{C - x_1}{\sqrt{2}\sigma} \right) - \frac{C + x_1}{2} Erfc\left(\frac{C + x_1}{\sqrt{2}\sigma} \right) - \frac{\sigma}{\sqrt{2\pi}} e^{-\frac{(C - x_1)^2}{2\sigma^2}} - \frac{\sigma}{\sqrt{2\pi}} e^{-\frac{(C + x_1)^2}{2\sigma^2}} \quad (4.26)$$

$$Erfc(x) = \frac{2}{\sqrt{\pi}} \int_x^\infty e^{-t^2} \, dt \quad (4.27)$$

The derivation of $\langle x_4 \rangle$ is given in Appendix 4A-2, together with the other expectations; $\langle y_4 \rangle$ is obtained by replacing x_1 of $\langle x_4 \rangle$ with y_1. $Erfc()$ is the Gaussian complementary error function. Because x_1 in $\langle x_4 \rangle$ is much smaller than C, $\langle x_4 \rangle$ can be Taylor expanded in x_1, and the first order terms are mixed up as,

$$\langle x_4 \rangle \cong E \cdot x_1 \quad (4.28)$$

$$E \equiv -Erfc\left(\frac{1}{\sqrt{2}\eta} \right) \quad (4.29)$$

where $\eta = \sigma / C$. Then, Equation (4.25) is rewritten as the correlative form.

$$\langle S_4^* \rangle = E \cdot S_1^* \quad (4.30)$$

Furthermore, Equation (4.17) is rewritten as

$$\mathbf{R}_{14} + \mathbf{R}_{41} = 2 \cdot E \cdot \mathbf{R}_{11}. \quad (4.31)$$

Finally, we have

$$P_{14} + P_{41} = 2a_d^2 \cdot SCR \cdot (MN)^2 \cdot D \cdot E \cdot \overline{\sqrt{G_{rec}}}^2. \quad (4.32)$$

The reduction of the correlation power due to the saturation is clearly expressed in Equation (4.32). The negative sign, contained in E, in this term arises from $\langle \mathbf{S}_4 \rangle$ and is reasonable because this instantaneous $\mathbf{S}_4$ generated in the saturation process is generally in the negative direction of $\mathbf{S}_1 + \mathbf{S}_2 + \mathbf{S}_3$; $\mathbf{S}_3$ is the Gaussian process with a zero mean. An expectation of $\mathbf{S}_4$ over $\mathbf{S}_3$ becomes proportional to $-\mathbf{S}_1$, where the proportional factor is related to the saturation.

4.2.3.4   Calculation of P_{44}

$\mathbf{S}_4$ consists of a deterministic component and a random component. The random component may make $\mathbf{R}_{44}$ behave as a delta function only at around $t_1 = t_3$ and $t_2 = t_4$. The deterministic component behaves the same as the square of $\langle \mathbf{S}_4 \rangle$ because $\mathbf{S}_4(t_1,t_2)$ and $\mathbf{S}_4^*(t_3,t_4)$ are independent of each other except when $t_1 = t_3$ and $t_2 = t_4$. $\mathbf{R}_{44}$ is given by

$$\mathbf{R}_{44}(t_1,t_2,t_3,t_4) = \frac{\langle \mathbf{S}_4 \cdot \mathbf{S}_4^* \rangle (t_1,t_2,t_3,t_4)}{G_2 G_4} + \frac{\langle \mathbf{S}_4 \cdot \mathbf{S}_4^* \rangle \big|_{t_1=t_3, t_2=t_4}}{G_2 G_4} \delta(t_1 - t_3)\delta(t_2 - t_4)$$

(4.33)

$$= E^2 \cdot \mathbf{R}_{11} + \frac{\langle \mathbf{S}_4 \cdot \mathbf{S}_4^* \rangle \big|_{t_1=t_3, t_2=t_4}}{G_2 G_4} \delta(t_1 - t_3)\delta(t_2 - t_4).$$

The expectation power is then given by

$$P_{44} = \overline{\sqrt{G_{rec}}}^2 \left\{ E^2 \cdot SCR \cdot a_d^2 (MN)^2 \cdot D + \frac{\langle x_4^2 \rangle + \langle y_4^2 \rangle}{G_{rec}} (MN) \right\}.$$

(4.34)

4.2.3.5   Calculation of Remaining Terms

The remaining terms are derived similarly to the aforementioned terms as:

$$P_{12} + P_{21} = 0$$

(4.35)

$$P_{24} + P_{42} = 0$$

(4.36)

$$P_{32} + P_{23} = 0$$

(4.37)

$$P_{34} + P_{43} = 2 \frac{\langle x_3 x_4 \rangle + \langle y_3 y_4 \rangle}{G_{rec}} (MN) \overline{\sqrt{G_{rec}}}^2$$

(4.38)

$$P_{13} + P_{31} = 0$$

(4.39)

$$P_{22} = 0$$

(4.40)

$$P_{33} = \frac{2\sigma_m^2}{G_{rec}} (MN) \overline{\sqrt{G_{rec}}}^2$$

(4.41)

4.2.3.6   Thematic Expression of the Correlation Power

Rearranging Equations (4.22), (4.32), and (4.34) to develop Equation (4.41) and using $\langle x_3 x_4 \rangle$, $\langle x_4^2 \rangle$, $\langle y_3 y_4 \rangle$, and $\langle y_4^2 \rangle$, we get P_C:

$$P_C = \left\{ (MN)^2 \cdot SCR \cdot a_d^2 \cdot D \cdot V + (MN)\frac{2\sigma^2}{G_{rec}} U \right\} \overline{\sqrt{G_{rec}}}^2$$

(4.42)

$$= (MN)^2 a_d^2 \cdot (SCR \cdot D \cdot V + Q \cdot U) \overline{\sqrt{G_{rec}}}^2$$

$$V - (1 + E)^? \tag{4.43}$$

$$U = \frac{1 + 2E}{1 + SNR} - \left(1 + \frac{1}{\eta^2}\right) \cdot E - \sqrt{\frac{2}{\pi}} \frac{1}{\eta} e^{-\frac{1}{2\eta^2}} \tag{4.44}$$

$$2\sigma^2 = a_d^2 (MN) G_{rec} Q \tag{4.45}$$

$$S_a = Erfc\left(\frac{1}{\sqrt{2}\eta}\right) \tag{4.46}$$

$$Q = 1 + SNR^{-1} \tag{4.47}$$

where S_a is the saturation rate (described later), and D is the processing efficiency that is 1.0 for the point target and 0.73 for the distributed target. There are three important parameters in Equation (4.42): V, U, and σ/C. The latter is the saturation-caused correlation gain loss for the correlative signal for the Cell; U is the correlation gain loss for the non-correlative signal outside the Cell, thermal noise, and ADC redundancy noise; and σ/C is the input power to ADC saturation level ratio, or "ISL." ISL depends on the signal-to-noise ratio (SNR), receiver gain (G_{rec}), and the target brightness ($a_d^2 MN$). SNR depends on G_{rec}, $a_d^2 MN$, thermal noise, ADC quantization interval, and so on. Thus, ISL can be assumed to depend on the receiver gain and the target brightness.

Equation (4.42) means that the SAR image power of each pixel is composed of the correlation power of all the received signals from the scatterers within the Cell, the non-correlation power for all the remaining scatterers within the IFOV but not in Cell, the thermal noise, and ADC redundancy noise; the saturation reduces correlation and non-correlation powers rated by V and U, respectively; the correlation gain loss depends on ISL but not on the pixel brightness itself (SCR).

4.2.3.7 Consistency with the ADC-Error-Free Case

Here, we consider a case in which ADC does not generate saturation or redundancy noise. The raw data then consist of signals from the resolution cell of interest, background signals, and thermal noise. Deleting the saturation-related terms in the previous equation, we have

$$P_C = (MN)^2 a_d^2 \cdot \left(SCR \cdot D + \frac{Q}{1 + SNR}\right) \sqrt{G_{rec}}^2 . \tag{4.48}$$

This is consistent with Freeman and Curlander's results (1989).

4.2.3.8 Standard Deviation of the Correlation Signal Power

The expectation of the standard deviation of the correlation power is derived (see Appendix 4A-3) as the normalized standard deviation (K_{pc}), which is the standard deviation (σ_{PC}) divided by an average (P_C) as

$$K_{PC} \equiv \frac{\sigma_{PC}}{P_C} = \frac{\left[Q^2 \cdot U^2 / 4 + 2 \cdot SCR \cdot U \cdot V \cdot Q\right]^{1/2}}{SCR \cdot D \cdot V + Q \cdot U}. \tag{4.49}$$

4.2.4 A Radar Equation for the SAR Correlation Power

The radar equation for the background power (a_d^2) is

$$a_d^2 = \frac{P_t G_0^2 G_{ele}^2 \lambda^2}{(4\pi)^3 R^4} \sigma^0 \frac{\delta_a \delta_r}{\sin \theta} L, \tag{4.50}$$

where P_t is the transmission power; G_0 is the antenna peak gain (one way); G_{ele} is the antenna elevation gain (one way); λ is the wavelength; R is the slant range; σ^0 is the normalized radar cross section; δ_r and δ_a are the slant range and azimuth pixel spacing; θ is the local incidence angle; and L is a system loss. Generally, the N_L independent data are incoherently summed to suppress the speckle noise. This is called a look summation. Combining this with Equation (4.42), the generalized radar equation for SAR correlation power becomes

$$P_C = (MN)^2 \frac{P_t G_0^2 G_{ele}^2 \lambda^2}{(4\pi)^3 R^4} N_L \sigma^0 \frac{\delta_a \delta_r}{\sin\theta} \cdot L \cdot (SCR \cdot D \cdot V \mid Q \cdot U) \overline{\sqrt{G_{rec}}}^2 , \tag{4.51}$$

which we can use as the base for the SAR calibration. Equation (4.51) governs the relationships among the correlated signal power, the target backscattering characteristics, and the radar parameters. When P_C is expressed by a short integer value, such as 16-bit data, it is converted to

$$DN_{PC} = C_f \cdot N_L (MN)^2 \frac{P_t G_0^2 G_{ele}^2 \lambda^2}{(4\pi)^3 R^4} \sigma^0 \frac{\delta_r \delta_a}{\sin\theta} L (SCR \cdot D \cdot V + Q \cdot U) \overline{\sqrt{G_{rec}}}^2 , \tag{4.52}$$

where DN_{PC} is the digital number for the correlation power and C_f is the conversion factor. In Equation (4.52), unknown parameters are constants and functions of slant range.

4.2.4.1   Constants
P_t, MN, δ_a, δ_r, λ, C_f, N_L, and L are radar and processor parameters. Some of these can be estimated from ground measurements acquired before launch. The possible gradual change of the sensor characteristics on orbit may, however, degrade the estimation.

4.2.4.2   Functions of Slant Range
The antenna elevation pattern (G_{ele}) and the slant range (R) change across the swath, so they must be accurately determined to calibrate SAR products. Determination of the slant range (R) requires the precise characterization of the time delay in the SAR transmitter and the receiver. The antenna elevation pattern can be determined in two ways. One is to deploy several reference targets with known radar cross sections at known locations over a swath and use them to derive the pattern. The other is to use a natural target with the same normalized radar cross section. The Amazon Rainforest data can be used in the latter method, which will be discussed in Chapter 7.

4.2.4.3   Calibration Method
After G_{ele} and R are determined precisely, the remaining problem is to determine the unknown parameters. If we set P_t, MN, λ, δ_r, δ_a, and L to the representative values (i.e., ground measurement values), C_f can be adjusted to make most of the DNs fit within the dynamic range of the SAR product and also to relate DNs to σ^0. This is how to convert DN to σ^0. This will be discussed in Chapter 9.

4.2.5   Several Expressions for the Real SAR Data
We summarize several expressions for the real SAR products based on Equation (4.51). For the SAR product that maintains the same azimuth resolution across the swath, the Doppler bandwidth should be constant. This implies that the azimuth correlation number (M) is proportional to the slant range (R) and is expressed by the satellite ground speed (V_g), pulse repetition frequency (PRF), theoretical azimuth resolution (ρ_a), and wave length (λ). The range correlation number (N) is determined by the radar parameters, pulse duration (τ), and sampling frequency (f_{sample}):

$$M = \frac{PRF}{2V_g \rho_a} \lambda R \tag{4.53}$$

$$N = \tau \cdot f_{sample} \tag{4.54}$$

If we use the scattering coefficient (γ^0) instead of σ^0, the final expression for the correlation signal power becomes:

$$P_C = \left(\tau f_{sample} \frac{PRF}{2V_g \rho_a} \lambda \right)^2 \frac{P_t G_0^2 G_{ele}^2 \lambda^2}{(4\pi)^3} N_L \frac{\gamma \cot\theta}{R^2} \cdot \delta_a \delta_r L \cdot (SCR \cdot D \cdot V) \overline{\sqrt{G_{rec}}}^2$$

$$+ \left(\tau f_{sample} \frac{PRF}{2V_g \rho_a} \lambda \right)^2 \frac{\overline{\sqrt{G_{rec}}}^2}{G_{rec}} R N_L 2\sigma_m^2 (1 + SNR) \cdot U \tag{4.55a}$$

In the case of a saturation-free signal, a more simplified equation is derived:

$$P_C = \left(\tau f_{sample} \frac{PRF}{2V_g \rho_a} \lambda \right)^2 \frac{P_t G_0^2 G_{ele}^2 \lambda^2}{(4\pi)^3} N_L \frac{\gamma \cot\theta}{R^2} \cdot \delta_a \delta_r L \cdot (SCR \cdot D) \overline{\sqrt{G_{rec}}}^2$$

$$+ \left(\tau f_{sample} \frac{PRF}{2V_g \rho_a} \lambda \right)^2 \frac{\overline{\sqrt{G_{rec}}}^2}{G_{rec}} R N_L 2\sigma_m^2 \tag{4.55b}$$

4.3 ANALYSIS OF THE PARAMETERS

In the correlation power model of Equation (4.42), U and V represent the power reduction quantities. In order to retrieve the saturation-free signal power, U and V should be evaluated under the possible saturation condition represented by (σ / C). This section discusses the saturation rate, a comparison of the ADC output power and correlated signal power, an analysis of these parameters under JERS-1/SAR conditions, a model comparison with JERS-1/SAR data, and the receiver gain mode. Only three-bit ADC is considered.

4.3.1 SATURATION RATE

To express the saturation quantitatively, "saturation rate," denoted by S_a, is introduced as the expected occurrence that the x or y component of $\mathbf{S}_1 + \mathbf{S}_2 + \mathbf{S}_3$ exists in all ADC conversion regions except IX (Figure 4-3). Because the distribution functions $P(x_1 + x_2)$ and $P(x_3)$ are Gaussian, the saturation rate of the x component obtained is

$$Saturation\ rate = \int_{-\infty}^{\infty} P(x_1 + x_2) \left\{ \int_{C_{AD}-x_1-x_2}^{\infty} + \int_{-\infty}^{-C_{AD}-x_1-x_2} P(x_3) dx_3 \right\} d(x_1 + x_2) = Erfc\left(\frac{1}{\sqrt{2}\eta} \right). \tag{4.56}$$

This parameter already has been introduced in Equation (4.42) as S_q. Because the two ADCs have the same characteristics, their saturation rates are the same. Figure 4-5 shows the relation between the saturation rate and σ / C. The saturation rate increases simply with a σ / C greater than 0.4. JERS-1/SAR data, which are always slightly saturated with S_a equal to 4% to 5%, has a σ / C of 0.5 (Shimada et al. 1993; Shimada and Nakai 1994).

4.3.2 COMPARISON OF CORRELATION AND NON-CORRELATION POWER RATIO

It is important to know the dependencies of the correlation power and the non-correlation power on the saturation and SNR. These power ratios, $(1 + SNR^{-1})U / (SCR \cdot D \cdot V)$, were calculated for SCR = 1, 10, and 100, and for SNR = 5 dB, 10 dB, and 20 dB. For these values, SCR = 1 corresponds

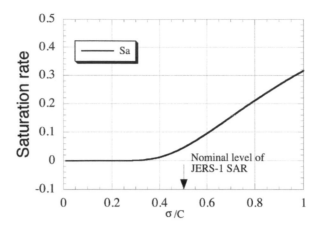

FIGURE 4-5   Saturation rate versus σ / C. Saturation rate increases with σ / C greater than 0.4. JERS-1 SAR data with a normal saturation rate of 4% to 5% correspond to 0.5 of σ / C.

to a wide area with similar brightness; SCR ≥ 100 corresponds to a bright point target (e.g., a corner reflector deployed on a dark target); and SNR = 20 dB is typical for the land (noise equivalent σ^0 for JERS-1/SAR is −20.5 dB, and σ^0 of land ranges from −10 dB to 0 dB) (Nemoto et al. 1991). Figure 4-6 shows that the correlation power always exceeds the non-correlation power in the given η range; the correlation power becomes much more dominant than the non-correlation power as the SNR increases. Representative power ratios are listed in Table 4-1. From this table, the non-correlation power is around 30% of correlation power at SCR = 1, SNR = 5 dB, and σ / C = 0.5; and the ratio at SNR = 20 dB and σ / C = 0.8 is 15%. The non-correlation power therefore cannot be neglected. The expectation of the ADC output power (P_{raw}) is obtained in the same way that P_C has been derived (see Appendix 4A-4):

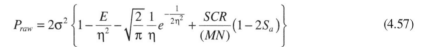

$$P_{raw} = 2\sigma^2 \left\{ 1 - \frac{E}{\eta^2} - \sqrt{\frac{2}{\pi}} \frac{1}{\eta} e^{-\frac{1}{2\eta^2}} + \frac{SCR}{(MN)}(1 - 2S_a) \right\} \qquad (4.57)$$

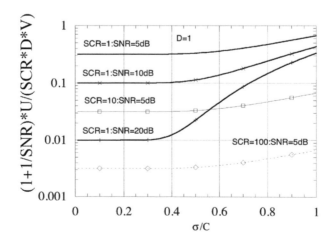

FIGURE 4-6   Dependence of the ratio of the non-correlation power to the correlation power on saturation is shown for five signal conditions where SCR = 1, 10, 100, and SNR = 5dB, 10dB, or 20dB. It shows that the correlation power exceeds the non-correlation power more in less saturation (σ / C < 0.5) and at less in more saturations (σ / C > 0.5).

TABLE 4-1

Representative Non-Correlation Power to Correlation Power Ratio

No.	SNR	0.5 of σ / C	0.8 of σ / C
1	5 dB	33%	47%
2	10 dB	11%	24%
3	20 dB	2%	15%

Dependence of the correlation gain loss in raw data power (P_{raw}) and the SAR image power (P_C) on the ISL is evaluated by using the parameters, $\Delta P_{raw} = P_{raw}/2\sigma^2$ and $\Delta P_C = P_C / \left\{ (MN)^2 \cdot a_d^2 \cdot SCR \cdot D \right\}$. The results in Figure 4-7 show that these two powers decrease as the ISL increases; the SAR image power decreases more than raw data power as the ISL increases (e.g., at σ / C of 0.8, ΔP_C is about −2 dB for SCR = 1 and ΔP_{raw} about −0.6 dB; at σ / C of 0.5, ΔP_C is −0.4 dB and ΔP_{raw} is −0.2 dB); as SCR increases, the SAR image power asymptotically approaches the final curve.

4.3.3 COMPARING S_a FOR MODEL AND DATA

Two saturated raw data areas, south of Mt. Fuji and west of Beppu City, both in Japan, (Table 4-2) were evaluated for saturation rate and raw data power. As shown in Figure 4-8, the measured saturation rate (Appendix 4A-5), averaged over all the segments in the full range, reaches 19% at worst, when the AGC monitoring window views the ocean. The AGC value ranges from 0 to 9 dB when the data are not saturated or slightly saturated and goes to 0 dB when the signal is heavily saturated. Note that AGC value is the attenuation in the receiver. Relationships between the saturation rate and P_{raw}/h^2 are plotted for the measured data and the model in Figure 4-9. The solid line represents the model for which the output of the intermediate amplifier is controlled to 4 dBm as AGC (Figure 4-2). The white circle is the measured value for Beppu, and the black triangle is that for Mt. Fuji. The models and measured data agree well for the saturation rate and the raw data power.

4.3.4 RECEIVER GAIN AND THE GAUSSIAN POWER $2\sigma^2$

Two gain-selecting modes are available in the JERS-1/SAR receiver: manual gain control (MGC) and automatic gain control (AGC). In the MGC mode, a constant gain is selected. In the AGC mode,

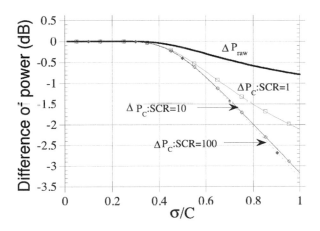

FIGURE 4-7 Dependence of the raw data power and the SAR image power on saturation. As saturation increases, the correlation power decreases more with more SCR and the raw data power decreases the least.

TABLE 4-2
Evaluated JERS-1 SAR Images

No.	Location	Lat. (deg)	Lon. (deg)	Path	Row	Acquisition	Remarks
1	Mt. Fuji	35.32	138.57	65	241	Sept. 7, 1995	325W of trans. power
2	Beppu	33.25	131.5	78	245	Nov. 3, 1995	325W ditto

Shimada et al. 1991 for path and row

the gain is selected automatically every 64 pulses to keep the ADC input power almost equal to $2\sigma_{cri}^2$ in 1 dB steps. Using the Gaussian notation, $[\cdot]$, the noise power is

$$2\sigma^2 = (MN)a_d^2 \cdot Q \cdot 10^{0.1 \cdot IG_{rec}} \tag{4.58}$$

$$IG_{rec} = \left[10\log_{10}\left\{ \left(2\sigma_{cri}^2 - \frac{h^2}{6}W(\sigma_{cri}) \right) \middle/ \left((MN)a_d^2 + 2\sigma_{rn}^2 \right) \right\} \right] \tag{4.59}$$

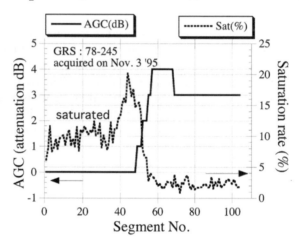

FIGURE 4-8   AGC and saturation patterns for the Mt. Fuji image. When AGC monitors the ocean, the AGC value drops to zero and the saturation goes up.

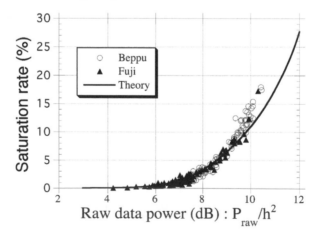

FIGURE 4-9   Comparison of the measured saturation rate and the model using the Peppu image (78-244) observed by JERS-1 SAR.

4.4 RADIOMETRIC CORRECTION OF JERS-1/SAR DATA

4.4.1 RADIOMETRIC CORRECTION OF THE SATURATED SIGNAL POWER

Since U and V behave differently depending on SNR and SCR, the received signal power cannot be estimated accurately by subtracting the non-correlation power from the SAR image power. Allowing an error of 15% in the seriously saturated case (σ / C of 0.8 and SNR = 20 dB in Table 4-1), we define the saturation-corrected SAR image power as

$$P'_C \equiv \frac{P_C}{V} = (MN)^2 \, a_d^2 SCR \cdot D \cdot \left(1 + \frac{1 + SNR^{-1}}{SCR \cdot D} \frac{U}{V}\right) \cong (MN)^2 \, a_d^2 SCR \cdot D, \qquad (4.60)$$

for which V requires the measured saturation rate. If $\sigma / C = 0.5$ and SNR = 20 dB, the final modification can be valid. If we simply apply the aforementioned correction to the SAR image power (produced by considering G_{rec} only), S_a may not be exactly co-registered with the SAR image power and artifacts may occur. This is because the location matching between the raw data and the SAR image depends on the SAR processor design, especially the azimuth reference function. Even if it is possible, the interpolation of S_a at each pixel location may cause artifacts.

Our proposal is to convolve the correction factor with the receiver gain correction during the SAR correlation. We call this the M-1 method (see Figure 4-10 for the correction flow):

(M-1) $$P'_C \propto \mathbf{f}'_C \cdot \mathbf{f}'^*_C \qquad (4.61)$$

$$\mathbf{f}'_C \propto \left\{\left(\sum_{m=1}^{4} \mathbf{f}_m \oplus \mathbf{g}^*_r\right) \cdot G_c\right\} \oplus \mathbf{g}^*_a \quad or \quad \left\{\left(\sum_{m=1}^{4} \mathbf{f}_m\right) \cdot G_c \oplus \mathbf{g}^*_r\right\} \oplus \mathbf{g}^*_a \qquad (4.62)$$

$$G_c = \frac{1}{\sqrt{G_{rec}} \cdot (1 - S_a)} \qquad (4.63)$$

where P'_C and $\mathbf{f}'_C$ are the saturation-corrected SAR image power and its complex expression, respectively. For comparison, we define two simple methods, M-2 and M-3:

(M-2) $$G_{cr} \cong \frac{1}{\sqrt{G_{rec}}} \qquad (4.64)$$

(M-3) $$G_{cr} \cong 1 \qquad (4.65)$$

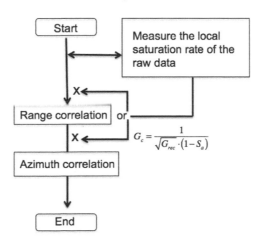

FIGURE 4-10 A correction flowchart for Method M-1.

TABLE 4-3
Simulation Condition

No.	Items	Description
1	ADC	3-bit ADC with h = 0.128 V and C = 0.512 V
2	$\overline{SNR}$	20 dB
3	MGC	Constant gain is used. $G_{rec} = 0$ dB
4	AGC	Selects the receiver gain in 1 dB steps to keep the ADC input power less than 0 dBm.
5	Impedance	50 ohm
6	SCR	1, 10, 100, 1,000

M-1 requires measurement of S_a and the receiver gain G_{rec}, and may require more execution time with better accuracy. M-2 only corrects the receiver gain, which does not provide sufficient correction for heavy saturation. M-3 does not correct receiver gain (AGC).

4.4.2 SIMULATION

A simulation study is conducted to estimate the saturation-free SAR image power for MGC and AGC modes and for the M-1 and M-2 methods (Table 4-3). Three cases (SCR = 1, 10, and 100) are examined in this simulation. Figure 4-11 shows the correction error in MGC and Figure 4-12 in AGC.

  i. MGC: In general, the error increases as the ISL increases. In M-2, the normal JERS-1 SAR conditions cause an error of −0.4 dB; the error becomes around −2 dB at a saturation rate of 20% ($\sigma / C = 0.8$). M-1 succeeds in saturation correction even for SCR = 1, yielding an error of 0.5 dB for a saturation rate of 20%.
  ii. AGC: There are several differences. First, M-2 reduces the error to less than 0.3 dB and produces a saw pattern, which is driven by the AGC's step-like operation. Second, M-1 succeeds almost perfectly in saturation correction regardless of SCR (the error is around 0.04 dB at SCR = 1).

The normalized standard deviations ($K_{pc}s$) for SAR image power saturation corrected by M-1 are shown in Figures 4-13 and 4-14. The MGC case (Figure 4-13) shows that the correction error increases as SCR decreases and as saturation increases. This is because the non-correlation powers

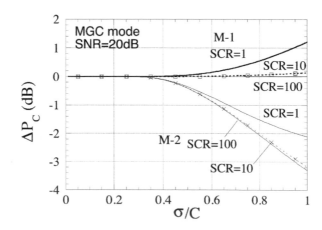

FIGURE 4-11   Saturation correction error for the data acquired in MGC mode. The error in M-1 in the high σ / C region is much smaller than that in M-2.

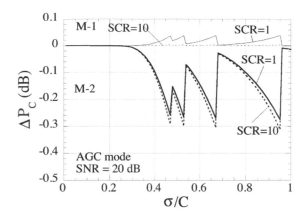

FIGURE 4-12 Saturation correction error for the data acquired in AGC mode. M-1 produces smaller errors than M-2; both M-1 and M-2 encounter smaller errors than those in MGC and do not vary at high σ / C.

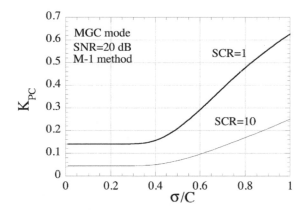

FIGURE 4-13 Accuracy of the power estimation by M-1 in MGC mode.

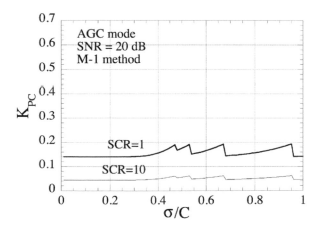

FIGURE 4-14 Accuracy of the power estimation by M-1 in AGC mode.

are produced by a random process and K_{PC} may gradually approach 1.0 for SCR = 1. The AGC case (Figure 4-14) shows that the correction error does not depend on the σ / C. The slight stepwise pattern is due to the AGC gain changes. Up to $\sigma / C = 0.5$, MGC and AGC yield the same K_{PC} value. Above that input level, MGC becomes less accurate, but AGC remains accurate.

From this, we conclude that M-1 corrects the saturated SAR data more accurately than M-2 for both AGC and MGC, that a combination of M-1 and AGC is best for saturation correction, and that M-1 attains a correction accuracy of 0.04 dB at SCR = 1, a saturation rate of 20%, and a smaller value for high SCR.

4.4.3 Comparison of Images Corrected by M-1, M-2, and M-3

A 40 km (south) by 6.1 km (east) image, which includes Mt. Fuji (latitude 33 19', longitude 131 34'), Japan, the north side of which is a non-saturated area and the south side of which is a saturated area, were corrected by the three methods, M-1, M-2, and M-3. Their related figure (Figure 4-15a, b, and c) was visually enhanced in the same way. As shown in Figure 4-8, the right half of the image is saturated at a maximum rate of 20%, and the left half is not so saturated (2.5% or so). Because both a highly saturated area and a normal intensity area are included simultaneously, this image may be a good example for evaluating the correction method. The original image looks fully saturated south of Mt. Fuji and

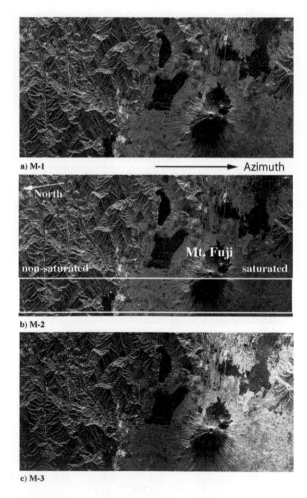

FIGURE 4-15   SAR slant range 45 km (south) by 22.5 km (east) image of the Mt. Fuji area; (a), (b), and (c), are corrected by M-1, M-2, and M-3, respectively.

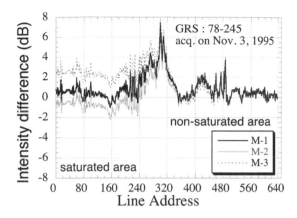

FIGURE 4-16   Averaged power profile of Beppu image.

appears bright in (c). M-2 makes the north side of the mountain brighter than the south side in (b). M-1 corrects the brightness uniformly over the full image in (a). The along-track profile was calculated by averaging the intensity over 8 pixels in azimuth and 320 pixels in range for the rectangular area (boxed in Figure 4-15b), and the two results are shown in Figure 4-16 for Beppu and Figure 4-17 for Mt. Fuji. The calibration instruments were not deployed in the saturated and non-saturated areas; thus, we cannot discuss the radiometric correction accuracy (as done in Gray et al. 1990; Ulander 1991). We may, however, assume the average normalized radar cross section of both saturated and non-saturated areas to be the same. We conclude that M-1 can recover the saturated images. In contrast, M-2 undercorrects the Beppu image by 1 dB and the Mt. Fuji image by 2 dB. ("Undercorrect" means that the corrected image for the saturated area is darker than the non-saturated area.) M-3 does not correct the saturation or AGC.

4.4.4   Merits and Demerits of the Methods

We will now summarize the merits and the demerits of the proposed method (M-1) and the current method (M-2).

Proposed method (M-1):

Merits   It can accurately correct the saturation-caused power loss for the bright target, depending on the SNR and SCR.

It can correct the clutter with an accuracy of 0.6 dB in bias and 1.7 dB in random at a saturation rate of 20%.

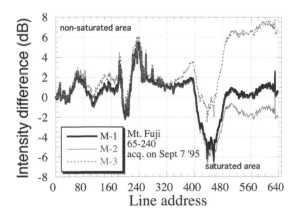

FIGURE 4-17   Averaged power profile of Mt. Fuji image.

Demerits It requires a saturation measurement over an image as well as more computational power.
It is not available for the extremely high saturation case (when the denominator becomes zero).

Current method (M-2):

Merits   Simple operation.
Demerits It cannot correct the saturation-caused power loss and leaves the gradual intensity change along the azimuth.

The estimated σ^0 has a bias error of 1.5 dB and a random error of 1.7 dB for clutter at a saturation rate of 20%.

The estimated σ^0 has a bias error of 2.0 dB and the smaller random error for the brighter target at a saturation rate of 20%.

4.4.5 APPLICATION TO RECENT SARs

In the 1990s, saturation was one of the critical issues in degradation of SAR image qualities. JERS-1 SAR sometimes recorded high saturation values in complex areas, with which islands, oceans, and AGC monitoring windows occasionally are associated, and the ERS-1 Active Microwave Instrument (AMI) with the MGC mode sometimes saturated in oceans with strong winds. In the 2000s, the saturation rate has decreased drastically because the data compression methods are well developed. In Japan's L-band SARs, the ALOS/PALSAR encountered several percentages of saturation on average because the MGC gain was selected optimally. ALOS-2/PALSAR-2 showed quite low saturation values, which are discussed in Chapter 9. It may be that future SARs may not need saturation corrections.

4.5 SUMMARY

In this chapter, we derived radar equations for SAR image intensities under realistic noise conditions, theoretical expressions of the NRCS for strip SAR, ScanSAR, and the raw data intensity of RAR and compared them. Based on these evaluations, we proposed a radiometric correction method for saturated SAR images. This method recovers the decreased correlation power that is due to the raw data saturation. Theoretical models are derived for the expected retrieved signal power and its normalized standard deviation as a function of the saturation rate. The distribution of the saturation rate over an image is then required. Correction can be made either for range correlation or azimuth correlation. Several simulations using the JERS-1 SAR saturated images indicated that the proposed method can correct the saturation.

APPENDIX 4A-1: ADC REDUNDANCY NOISE

ADC redundancy noise, σ^2_{ADC}, is given by

$$\sigma^2_{ADC} = \left\langle (X_{ADC} - X)^2 \right\rangle,   \tag{4A-1.1}$$

where

$$X_{ADC} = \left[\frac{X}{h} + 0.5 \right] \cdot h   \tag{4A-1.2}$$

is the ADC output; X is the input voltage to ADC; h is the ADC conversion interval; [ ] is the truncation operation; and $\langle\ \rangle$ is the expectation. A Gaussian distributed signal with $2\sigma^2$ of power is

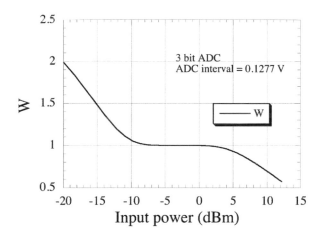

FIGURE 4A-1 Weighting factor (W) versus the ADC input power.

the input. The aforementioned expression can be calculated for each ADC interval. Integration over a 2^L interval gives the final expression for the ADC redundancy noise

$$\sigma^2_{ADC} = \frac{\sigma^2}{2} Erfc\left(-\frac{2^{L/2} \cdot h}{\sqrt{2}\sigma}\right) - \sqrt{\frac{2}{\pi}}\frac{h\sigma}{2}\sum_{i=-2^{L/2}}^{2^{L/2}-1}\left\{e^{-\frac{(i+1)^2 h^2}{2\sigma^2}} + e^{-\frac{i^2 h^2}{2\sigma^2}}\right\}$$

$$+ \frac{h^2}{2}\left(2^L + 0.5\right) Erfc\left(-\frac{2^{L/2} \cdot h}{\sqrt{2}\sigma}\right) + \frac{h^2}{2}\sum_{i=-2^{L/2}}^{2^{L/2}-1}\left\{i \cdot Erfc\left(\frac{i \cdot h}{\sqrt{2}\sigma}\right) + (i+1)Erfc\left(\frac{(i+1)\cdot h}{\sqrt{2}\sigma}\right)\right\}.$$

$$(4A-1.3)$$

We define a weighting factor

$$W = \sigma^2_{ADC} \cdot \frac{12}{h^2},\qquad (4A-1.4)$$

which is normalized by the variance of a signal that is normally distributed and not saturated; W versus $2\sigma^2$ is shown for the 3-bit ADC in Figure 4A-1. The figure shows that for a power less than 9 dBm, W exceeds 1.0, and when P_{in} exceeds 5 dBm, W is less than 1.0.

APPENDIX 4A-2: EXPECTATION OF x_4

An expectation of x_4 is given by:

$$\langle x_4 \rangle = \int_{-\infty}^{\infty} F(x_1, x_2)\cdot P(x_2)dx_2 \qquad (4A-2.1)$$

$$F(x_1, x_2) = \int_{C-x_1-x_2}^{\infty}(C - x_1 - x_2 - x_3)P(x_3)dx_3 + \int_{-\infty}^{-C-x_1-x_2}(-C - x_1 - x_2 - x_3)P(x_3)dx_3 \qquad (4A-2.2)$$

$$P(x_2) = \frac{1}{\sqrt{2\pi}\sigma_n}e^{-\frac{(x_2)^2}{2\sigma_n^2}} \qquad (4A-2.3)$$

$$P(x_3) = \frac{1}{\sqrt{2\pi}\sigma_m} e^{-\frac{(x_3)^2}{2\sigma_m^2}}$$

(4A-2.4)

$$2\sigma^2 = 2\sigma_n^2 + 2\sigma_{rn}^2$$

$$- 2\sigma_{in}^2 (SNR + 1)$$

(4A-2.5)

The term-by-term calculation gives

$$F(x_1, x_2) = \frac{C - x_1 - x_2}{2} Erfc\left(\frac{C - x_1 - x_2}{\sqrt{2}\sigma_{rn}}\right) - \frac{C + x_1 + x_2}{2} Erfc\left(\frac{C + x_1 + x_2}{\sqrt{2}\sigma_{rn}}\right)$$

$$- \frac{\sigma_{rn}}{\sqrt{2\pi}} e^{-\frac{(C - x_1 - x_2)^2}{2\sigma_{rn}^2}} - \frac{\sigma_{rn}}{\sqrt{2\pi}} e^{-\frac{(C + x_1 + x_2)^2}{2\sigma_{rn}^2}}.$$

(4A-2.6)

Equation (4A-2.1) means that $x_2 + x_3$ is a variable that distributes in the convoluted distribution function of x_2 and x_3. Then, Equation (4A-2.1) is expressed simply by

$$\langle x_4 \rangle = \frac{C - x_1}{2} Erfc\left(\frac{C - x_1}{\sqrt{2}\sigma}\right) - \frac{C + x_1}{2} Erfc\left(\frac{C + x_1}{\sqrt{2}\sigma}\right) - \frac{\sigma}{\sqrt{2\pi}} e^{-\frac{(C - x_1)^2}{2\sigma^2}} - \frac{\sigma}{\sqrt{2\pi}} e^{-\frac{(C + x_1)^2}{2\sigma^2}}.$$

(4A-2.7)

A similar expression is obtained as

$$\langle x_4^2 \rangle = \frac{1}{2}\{(C - x_1)^2 + \sigma^2\} Erfc\left(\frac{C - x_1}{\sqrt{2}\sigma}\right) + \frac{1}{2}\{(C + x_1)^2 + \sigma^2\} Erfc\left(\frac{C + x_1}{\sqrt{2}\sigma}\right)$$

$$- \frac{\sigma}{\sqrt{2\pi}}\left\{(C - x_1)e^{-\frac{(C - x_1)^2}{2\sigma^2}} + (C + x_1)e^{-\frac{(C + x_1)^2}{2\sigma^2}}\right\}$$

(4A-2.8)

$$\cong (C^2 + \sigma^2) \cdot Erfc\left(\frac{C}{\sqrt{2}\sigma}\right) - \sqrt{\frac{2}{\pi}}C\sigma e^{-\frac{C^2}{2\sigma^2}}.$$

$\langle x_3 x_4 \rangle$ is expressed by

$$\langle x_3 x_4 \rangle = \int_{-\infty}^{\infty} P(x_2) \cdot G(x_1, x_2) dx_2,$$

$$G(x_1, x_2) = -\frac{\sigma_{rn}^2}{2}\left\{Erfc\left(\frac{C - x_1 - x_2}{\sqrt{2}\sigma_{rn}}\right) + Erfc\left(\frac{C + x_1 + x_2}{\sqrt{2}\sigma_{rn}}\right)\right\}.$$

Numerical simulation solves the previous formula as

$$\langle x_3 x_4 \rangle \cong -\frac{\sigma^2}{1 + SNR} Erfc\left(\frac{C}{\sqrt{2}\sigma}\right).$$

(4A-2.9)

APPENDIX 4A-3:   NORMALIZED STANDARD DEVIATION

The normalized standard deviation K_{PC} is given by

$$K_{PC} \equiv \frac{\sqrt{V_{PC} - P_C^2}}{P_C}. \tag{4A-3.1}$$

The fourth-order moment of correlated signal f is expressed by V_{PC} and is given by

$$V_{PC} = \left\langle \left(\mathbf{S}_C \cdot \mathbf{S}_C^* \right)^2 \right\rangle$$

$$= \sum_{m=1}^{4} \sum_{n=1}^{4} \sum_{k=1}^{4} \sum_{l=1}^{4} \left\langle \mathbf{S}_{cra,m} \cdot \mathbf{S}_{cra,n}^* \cdot \mathbf{S}_{cra,k} \cdot \mathbf{S}_{cra,l}^* \right\rangle. \tag{4A-3.2}$$

Summation of 256 (=4*4*4*4) terms produces the expectation of squared correlated signal power. Using the following two conditions, K_{PC} can be calculated to be Equation (4.31):

$$x_1 \cong y_1 \cong 0 \tag{4A-3.3}$$

$$\langle x_1 x_2 x_3 x_4 \rangle = \langle x_1 \rangle \langle x_2 \rangle \langle x_3 \rangle \langle x_4 \rangle$$

$$+ \langle x_1 \rangle \langle x_2 x_3 x_4 \rangle \delta(t_2 - t_3) \delta(t_3 - t_4) + \ldots$$

$$+ \langle x_1 x_2 \rangle \langle x_3 x_4 \rangle \delta(t_2 - t_1) \delta(t_4 - t_3) + \ldots \tag{4A-3.4}$$

$$+ \langle x_1 x_2 x_3 x_4 \rangle \delta(t_1 - t_2) \delta(t_2 - t_3) \delta(t_3 - t_4)$$

$$+ \langle x_1 \rangle \langle x_2 \rangle \langle x_3 x_4 \rangle \delta(t_3 - t_4) + \ldots$$

APPENDIX 4A-4:   ADC OUTPUT POWER (RAW DATA POWER)

Raw data power, denoted by P_{raw}, is obtained in two steps in the same way as $\mathbf{R}_{mn}$ of Section 4.2:

$$P_{raw} \cong \left\langle \mathbf{S}_4 \cdot \mathbf{S}_4^* \right\rangle$$

$$= \left\langle \mathbf{S}_1 \cdot \mathbf{S}_1^* \right\rangle + \left\langle \mathbf{S}_2 \cdot \mathbf{S}_2^* \right\rangle + \left\langle \mathbf{S}_3 \cdot \mathbf{S}_3^* \right\rangle + \left\langle \left(\mathbf{S}_1 \cdot \mathbf{S}_2 \right)^* \right\rangle + \left\langle \left(\mathbf{S}_2 \cdot \mathbf{S}_3 \right)^* \right\rangle + \left\langle \left(\mathbf{S}_1 \cdot \mathbf{S}_3 \right)^* \right\rangle \tag{4A-4.1}$$

$$\left\langle \mathbf{S}_1 \cdot \mathbf{S}_1^* \right\rangle = \left\langle x_1^2 \right\rangle + \left\langle y_1^2 \right\rangle = G_{rec} a_d^2 \cdot SCR \tag{4A-4.2}$$

$$\left\langle \mathbf{S}_2 \cdot \mathbf{S}_2^* \right\rangle = \left\langle x_2^2 \right\rangle + \left\langle y_2^2 \right\rangle = 2\sigma^2 \tag{4A-4.3}$$

$$\left\langle \mathbf{S}_3 \cdot \mathbf{S}_3^* \right\rangle = \left\langle x_3^2 \right\rangle + \left\langle y_3^2 \right\rangle \tag{4A-4.4}$$

$$\left\langle \left(\mathbf{S}_1 \cdot \mathbf{S}_2 \right)^* \right\rangle = \left\langle \mathbf{S}_1^* \cdot \mathbf{S}_2 + \mathbf{S}_1 \cdot \mathbf{S}_2^* \right\rangle = \left\langle 2x_1 \langle x_2 \rangle + 2y_1 \langle y_2 \rangle \right\rangle = \left\langle 2x_1 0 + 2y_1 0 \right\rangle = 0 \tag{4A-4.5}$$

$$\left\langle \left(\mathbf{S}_2 \cdot \mathbf{S}_3 \right)^* \right\rangle = \left\langle \mathbf{S}_2^* \cdot \mathbf{S}_3 + \mathbf{S}_2 \cdot \mathbf{S}_3^* \right\rangle = 2 \langle x_2 x_3 \rangle + 2 \langle y_2 y_3 \rangle \tag{4A-4.6}$$

$$\left\langle \left(\mathbf{S}_1 \cdot \mathbf{S}_3 \right)^* \right\rangle = \left\langle \mathbf{S}_1^* \cdot \mathbf{S}_3 + \mathbf{S}_1 \cdot \mathbf{S}_3^* \right\rangle = \left\langle 2x_1 \langle x_3 \rangle + 2y_1 \langle y_3 \rangle \right\rangle \tag{4A-4.7}$$

On the right side of Equation (4A-4.1), the first, second, and third terms are positive. The fifth term is negative because x_3 is opposite to x_2. The sixth term is also negative because x_1 and $\langle x_3 \rangle$ are opposite. Therefore, the saturation reduces the non-saturated ADC output power (Shimada et al. 1993; Nemoto et al. 1991). Finally, we obtain

$$P_{raw} = 2\sigma^2 \left\{ 1 - \frac{E}{\eta^2} - \sqrt{\frac{2}{\pi}}\frac{1}{\eta}e^{-\frac{1}{2\eta^2}} + \frac{SCR}{(MN)}(1 - 2 \cdot S_a) \right\}. \tag{4A-4.8}$$

APPENDIX 4A-5: MEASURED SATURATION RATE, $\bar{S}_a$

The actual saturation rate, $\bar{S}_a$, can be measured from the raw data as follows. An image is segmented into small pieces, each of which is 512 pixels in range and 64 pixels in azimuth. A normalized histogram, $h_m[i]$ with $i = 0$ to 7, is measured for each segment. The histogram $h_m[i]$ for $i = 1$ to 6 is Gaussian distributed. Then, the following equation is obtained:

$$\int_{0.5}^{6.5} \frac{1}{\sqrt{2\pi}\sigma}e^{-\frac{(x-\mu)^2}{2\sigma^2}} dx = \sum_{i=1}^{6} h_m[i] \tag{4A-5.1}$$

where μ is calculated by the measurement, and then the unknown σ is determined by the iteration method. Finally, the measured saturation rate can be obtained by the following parameter:

$$\bar{S}_a = 1 - \int_{-0.5}^{7.5} \frac{1}{\sqrt{2\pi}\sigma}e^{-\frac{(x-\mu)^2}{2\sigma^2}}. \tag{4A-5.2}$$

REFERENCES

Freeman, A. and Curlander, J. C., 1989, "Radiometric Correction and Calibration of SAR Images," *Photogramm. Eng. Rem. S.*, Vol. 55, No. 9, pp. 1295–1301.
Gray, A. L., Vachon, P. W., Livingstone, E., and Lukowski, T. I., 1990, "Synthetic Aperture Radar Calibration Using Reference Reflectors," *IEEE Trans. Geosci. Rem. Sens.*, Vol. 28, No. 3, pp. 374–383.
Nemoto, Y., Nishino, H., Ono, M., Mizutamari, H., Nishikawa, K., and Tanaka, K., 1991, "Japanese Earth Resources Satellite-1 Synthetic Aperture Radar," *Proc. of the IEEE*, Vol. 79, No. 6, pp. 800–809.
Papoulis, A., 1977, "Part Three: Data Smoothing and Spectral Estimation," *Signal Analysis*, McGraw-Hill, New York.
Shimada, M., Nagai, T., and Yamamoto, S., 1991, "JERS-1 Operation Interface Specification," JAXA internal document, HE-89033, Revision-3, Tokyo, Japan.
Shimada, M., Nakai, M., and Kawase, S., 1993, "Inflight Evaluation of the L-band SAR of JERS-1," *Can. J. Remote Sens.*, Vol. 19, No. 3, pp. 247–258.
Shimada, M. and Nakai, M., 1994, "Inflight Evaluation of L Band SAR of Japanese Earth Resources Satellite-1," *Adv. Space Res.*, Vol. 14, No. 3, pp. 231–240.
Shimada, M., 1999a, "Radiometric Correction of Saturated SAR Data," *IEEE T. Geosci. Remote*, Vol. 37, No. 1, pp. 467–478.
Shimada, M., 1999b, "A Study on Measurement of Normalized Radar Cross Section of Earth Surfaces by Spaceborne Synthetic Aperture Radar," PhD Thesis, University of Tokyo.
Ulander, L. M. H., 1991, "Accuracy of Using Point Targets for SAR Calibration," *IEEE T. Aero. Elec. Sys.*, Vol. 27, No. 1, pp. 139–148.

5 ScanSAR Imaging

5.1 INTRODUCTION

Most of the recent SARs are equipped with three image modes—spotlight, strip, and scanning SAR (ScanSAR)—that fully utilize digital and radar technology for wide bandwidth, polarization, transmit and receive modules, and so on. Unlike the strip mode, ScanSAR disperses the pulses to several beams by activating each beam intermittently. As a result, ScanSAR is more advantageous. It is equipped with a wider image swath of 300 km to 500 km, resulting in a shorter revisit time to the Earth's surface compared to the strip mode and thereby compensating for reducing the imaging resolution and image qualities (Cumming and Wong 2005). Image quality issues with ScanSAR are represented by three types of artifacts: (1) periodic artifacts in the azimuth direction called scalloping, (2) truncation noise in the azimuth direction, and (3) banding between two neighboring scans.

The first artifact, residual scalloping, is caused by a mismatch of the real azimuth antenna pattern (AAP) and the modeled AAP, while the causes for the other phenomena are based on inaccurate knowledge of Doppler centroid frequency and modulation of noise by the AAP in regions of excessively low signal-to-noise ratio (SNR) (Bamler 1995; Leung et al. 1996). The scalloping can be suppressed if the noise floor level or the saturation rate of the SAR data is not too high and the Doppler frequency can be accurately estimated (Jin 1996). Several studies have investigated suppressing scalloping (Bamler 1995; Hawkins and Vachon 2002). Bamler (1995) proposed an excellent algorithm that generates an optimum weighting function by summing the different looks in such a way as to suppress the artifact for the given AAP and multiple looking intervals. Vigneron (1994) evaluated the inverse antenna pattern method and concluded that a higher SNR successfully suppressed the scalloping.

Amazon rain forest data have uniform backscattering characteristics independent of the incidence angle and are very good reference targets for SAR calibration (Shimada 2005). They are widely accepted as major calibration sources and are used for SAR calibrations (i.e., estimation of range antenna pattern [RAP] and monitoring radiometric calibration accuracy and sensor stabilities) (Shimada et al. 2009; Srivastava et al. 2001; Rosich et al. 2004; Shimada 1993). However, they have not been discussed specifically with regard to suppressing scalloping or AAP estimation. Although the causes of scalloping were clarified and a complex but sophisticated algorithm became available, a simpler algorithm could be possible either utilizing the Amazon data or creating a correction algorithm. Here, we describe a new method to estimate the AAP for the ScanSAR using only Amazon rain forest data (i.e., not using the antenna pattern measured on the ground or the one measured using the receiver on the ground during satellite passage) and an errorless multi-looking method to minimize scalloping.

The second artifact arises from signal truncation at the edge of the frequency spectrum. This can be solved by increasing the PRF so that the Doppler bandwidth of the illuminated area can be fully covered in order to satisfy the Nyquist theorem. However, the parameter selection of the PRF and the number of pulses within a burst for each beam are sometimes restricted by the SAR system. For example, some beams of Phased-Array L-band Synthetic Aperture Radar (PALSAR) onboard the Advanced Land-Observing Satellite (ALOS) suffered occasionally from prioritizing the 350-km imaging swath with five beams rather than the image quality (Shimada et al. 2009; Shimada et al. 2007). Thus, we propose to apply a band-limitation method.

For the third artifact (i.e., banding between scans), the representative correction method, which was developed for Radarsar-1 and ENVISAT, is to update the roll angle and the range-dependent gain corrections (RDGCs) mainly using the overlap region of the two neighboring sub-swaths under the condition that the range antenna pattern (RAP) for each of the multiple beams is given (Bast and Cumming 2002;

Dragosevic 1999; Hawkins et al. 2001). As an alternative to this approach, we adopt a dynamic balancing method that equalizes the intensity locally at the overlapped region and maintains the intensity in the high SNR region (i.e., the global center of the sub-swaths in the least square sense for range and azimuth directions). This method was validated by the JERS-1 SAR mosaicking approach (Shimada and Isoguchi 2002) and has been improved to suppress the intensity discontinuity at neighboring sub-swaths. In this chapter, we describe an innovative ScanSAR image-correction method (shimada 2009).

5.2 SCALLOPING CORRECTION

5.2.1 SCANSAR IMAGING

We consider the ScanSAR N scan observations, each burst of which consists of N_{az} pulses emitted at f_{PRF} frequency, a Doppler chirp rate of f_{DD}, the repeat interval of T_{SCAN}, and the satellite ground speed of vg. As introduced in Chapter 3, Figure 5-1 describes the geometric relationship of the repeatability of burst in time in (a) ground repeatability of the (b) burst (focused image), and (c) the ground range. Several relationships for the burst, burst separation, and the azimuth resolution are as follows:

1. Azimuth pixel spacing:

$$-\frac{f_{PRF}}{f_{DD} \cdot N_{az}} v_g \tag{5.1}$$

2. Azimuth image length:

$$-\frac{f_{PRF}}{f_{DD} \cdot N_{az}} N_{az} v_g - \frac{N_{az}}{f_{PRF}} v_g = -\left(\frac{f_{PRF}}{f_{DD}} + \frac{N_{az}}{f_{PRF}}\right) v_g \tag{5.2}$$

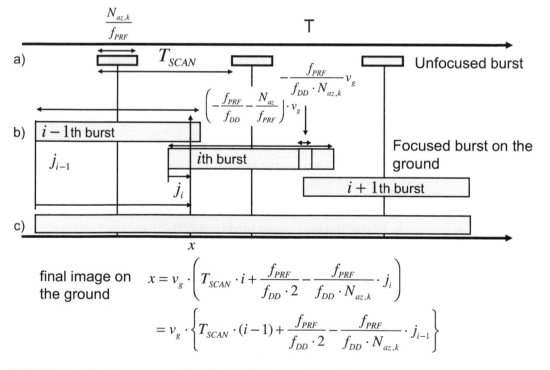

FIGURE 5-1 Coordinate system applied to ScanSAR processing. a) shows the repeatability of the burst location in time (or distance from the reference point), b) shows the relationship of the focused image location in time, and c) shows the relationship of the ScanSAR image in time.

3. Burst interval on the ground.

$$T_{SCAN} \cdot v_g \qquad (5.3)$$

Equation (5.1) is obtained as a per sample frequency resolution times the ground speed per Doppler rate. Equation (5.2) is the total ground length in azimuth subtracted from the synthetic aperture length, and Equation (5.3) is the azimuth length between two contiguous bursts.

When focusing on the same SCAN beam, the along-track properties of the range-correlated and spectral analysis (SPECAN)-processed ScanSAR image for the kth scan and ith burst can be expressed by the following equations. Hereafter, we use the word "burst" to mean "the focused burst."

First the, geometry of each pixel can be expressed by

$$X_{i,j}^k = i \cdot T_{SCAN} \cdot v_g + x_{i,j}^k \qquad (5.4)$$

$$x_{i,j}^k = \frac{f_{PRF,k}}{f_{DD}} \frac{v_g}{2} - \frac{f_{PRF,k} \cdot v_g}{f_{DD} \cdot N_{az,k}} j_{i,k} , \left(j_b \le j_{i,k} \le j_e \right) \qquad (5.5)$$

$$T_{SCAN} = \sum_{k=1}^{N_{SCAN}} \frac{N_{az,k}}{f_{PRF,k}}. \qquad (5.6)$$

Here, $x_{i,j}^k$ is the relative azimuth coordinate along the track within a burst; and $X_{i,j}^k$ is the azimuth coordinate along the track. Additionally, N_{oise} is the noise; i is the burst number either in the transmission group or the focused burst on the ground; $j_{i,k}$ is the relative image address in the ith burst (i.e., focused) image; T_{SCAN} is the time period for all the scans, which is also defined as "interburst time" or "cycle time"; N_{SCAN} is the number of scans or beams to cover the full ScanSAR swath; $f_{PRF,k}$ is the pulse repetition frequency and/or azimuth processing bandwidth for the kth scan; f_{DD} is the Doppler chirp rate (Hz/s); v_g is the satellite ground speed; and S_a is the saturation rate of the raw data (Shimada 1999).

As shown in Figure 5-1, each burst has its own azimuth image length of $-f_{PRF} v_g / f_{DD}$ and its azimuth resolution can be given by $-f_{PRF} v_g / N_{az,k} f_{DD}$. In Equation (5.5), j_b and j_e express the start and end of the good points given from the SPECAN processing, which is roughly 90% of the FFT points.

Next the radiometry of the SAR image can be expressed by Equation (5.7). This equation is similar to Equation (4.55) for the strip mode except the dependence of R is 4. This difference occurs such that the strip maintains the constant Doppler bandwidth across the track while SPECAN maintains the constant azimuth integration time across the track.

$$P_r\left(X_{i,j}^k\right) = G_p \frac{P_t G_a^2\left(X_{i,j}^k\right) G_r^2 \lambda^2}{(4\pi)^3 R^4} \frac{1}{1-S_a} \sigma^0 \frac{\rho_r \rho_a}{\sin\theta} N_{az,k}^2 N_{rg}^2 + G_p N_{oise} N_{az,k} N_{rg} \qquad (5.7)$$

In this case, P_r is the power of the ScanSAR processed image; P_t is the transmit power; $G_a\left(\phi_a\right)$ the one-way AAP, which depends on the azimuth angle ϕ_a or relative distance $x_{i,j}^k$ explained below; $G_r\left(\varphi_{off}\right)$ is the one-way RAP, which depends on the off-nadir angle φ_{off}; λ^2 is the wavelength; R is the slant range; σ^0 is the sigma-naught of the target; $N_{az,k}$ is the number of pulses transmitted for a burst at the kth scan; N_{rg} is the number of samples for the range reference signal; ρ_r is the slant-range resolution (i.e., the speed of light divided by two and the signal bandwidth in range); ρ_a is the azimuth resolution (i.e., antenna azimuth length times pulse repetition frequency divided by two and Doppler bandwidth); θ is the incidence angle; S_a is the saturation rate; and G_p is the processor gain independent of azimuth and range.

Equation (5.7) shows the power of the focused image. The squared term, $N_{az,k}^2 \cdot N_{rg}^2$, at the first term on right hand side is there because the returned signal from the target is correlated with a reference signal in voltage, and power becomes its squared value. On the other hand, the single appearance of $N_{az,k} \cdot N_{rg}$ as the second term on the right-hand side is because the noise is not correlated with the reference signal, but the mean noise power level will be amplified with the gain from $N_{az,k} \cdot N_{rg}$ (Shimada 1999; Freeman and Curlander 1989; Curlander and McDonough 1991).

In order to correct the gain loss due to raw data saturation, the saturation rate, which is defined as the ratio of the number of data with the least significant bit (LSB) and most significant bit (MSB) to the total number of data, appears once in Equation (5.7) because the SPECAN SAR imaging contains the correlation process for the range direction (Shimada 1999) (see Chapter 4). In this case, P_r is inversely proportional to R^4 because the integral time for deramping at the SPECAN is constant across the scan swath.

As shown in Equation (5.7), the SAR image is linearly linked with the NRCS of the target. Thus, we introduce a sigma-naught related parameter S_r, which is derived from P_r, to allow the discussion to be linked with calibration. Generally, the noise term cannot be ignored. When we discuss a low noise level SAR, such as PALSAR with a noise equivalent sigma-naught of −34 dB (Shimada et al. 2009), noise can be ignored and S_r is directly proportional to sigma-naught

$$S_r\left(X_{i,j}^k\right)\left(\approx \sigma^0\right) \equiv \frac{(4\pi)^3 R^4}{P_t G_a^2\left(X_{i,j}^k\right)G_r^2\lambda^2} \frac{(1-S_a)\sin\theta}{N_{az,k}^2 N_{rg}^2\rho_r\rho_a}\left\{\frac{P_r\left(X_{i,j}^k\right)}{G_p}\right\}. \tag{5.8}$$

Hereafter, we will focus on the kth scan for our discussion, and thus the suffix k will be omitted.

5.2.2 MULTI-LOOKING IN TIME DOMAIN AND SCALLOPING CORRECTION

Since the time series output shown in Equation (5.7) is truncated within the maximum number of pulses for a burst from each scan, multi-looking at the aforementioned data can be done by interpolating the data at the time (or space) domain and summing it up. The number of looks (NL) refers to the number of samples summed for a specific time or space and is given as

$$NL = \left(\frac{f_{PRF}}{-f_{DD}} - \frac{N_{az}}{f_{PRF}}\right) \cdot \frac{1}{T_{SCAN}}. \tag{5.9}$$

We use the full number of FFT output. In real processing, NL becomes the integer, and the better output points are summed. Using this expression, the multi-look power can be given by the following:

$$\bar{S}_r\left(X\right) = \sum_{j=1}^{NL} S_r\left(x_j\right). \tag{5.10}$$

Equation (5.10) behaves periodically because $G_a\left(x\right)$ is periodic and the data are summed after shifting. In this case, this periodicity causes scalloping, a well-known artifact in the SPECAN algorithm.

5.2.2.1 Method-1

Scalloping corrections have been discussed in Bamler (1995). If the SNR of the radar signal is high and the antenna pattern is accurate and stable, the power deviation due to this periodicity must be corrected simply by normalizing each power by the AAP shifted with time, as seen in Equation (5.11):

$$\bar{S}_{r,1}\left(X\right) = \frac{1}{NL}\sum_{j=1}^{NL} \frac{S_r\left(x_j\right)}{G_a^2\left(x_j\right)}. \tag{5.11}$$

In this case, we call this the inverse antenna pattern method or simply Method-1 from here on. However, this type of normalization may not be stable because the denominator (antenna pattern) may be incorrectly estimated in orbit or the image may be slightly shifted due to an incorrect Doppler estimation.

5.2.2.2 Method-2

The following normalization scheme suppresses the power deviation because integrals in the numerators and denominators may cancel out even though both have been affected by the antenna pattern errors:

$$\overline{S}_{r,2}(X) = \frac{\sum\limits_{j=1}^{NL} S_r(x_j)}{\sum\limits_{j=1}^{NL} G_a^2(x_j)}. \tag{5.12}$$

5.2.2.3 Simulation Study for Method-1 and Method-2

We propose a new correction method that we will call Method-2. We first compare the differences between the two methods through the simulation where the antenna pattern was deviated slightly. For a specific burst, we have the following ScanSAR image:

$$S_r(x) = A\tilde{G}_a^2(x-\delta,\varepsilon), \tag{5.13}$$

$$\tilde{G}_a^2(x-\delta,\varepsilon) = \left[1 - \left\{1 - G_a(x-\delta)\right\}(1-\varepsilon)\right]^2 \tag{5.14}$$

$$\approx (1-2\varepsilon)G_a^2(x) + 2\varepsilon G_a(x) + 2\delta\dot{G}_a(x)$$

where A is a constant, x the angular or along track distance in the ith burst, $\tilde{G}_a^2$ the true double-way AAP, and $\dot{G}_a$ its x derivative. The real antenna pattern could be shifted by δ in the azimuth direction and is deviated by an x dependent gain difference, which has 0 at the peak and ε at the edge.

Inserting Equation (5.13) into Equations (5.11) and (5.12) yields the following:

$$\overline{S}_{r,1}(X) \approx \frac{1}{NL} A \sum_{j=1}^{NL}\left(1 - 2\varepsilon + \frac{2\varepsilon}{G_a(x)} + \frac{2\delta\dot{G}_a}{G_a^2(x)}\right) \tag{5.15}$$

$$\overline{S}_{r,2}(X) \approx A\left\{1 - 2\varepsilon + 2\varepsilon\frac{\sum\limits_{j=1}^{NL} G_a(x)}{\sum\limits_{j=1}^{NL} G_a^2(x)} + 2\delta\frac{\sum\limits_{j=1}^{NL} \dot{G}_a(x)}{\sum\limits_{j=1}^{NL} G_a^2(x)}\right\}. \tag{5.16}$$

From these expressions, both models give the same accurate results only when $\varepsilon = 0$ and $\delta = 0$. Equation (5.15) is discontinuous at some x because $1/G_a$ varies with x but is always normalized by the same NL. The error shown in Equation (5.16) is lower because the NL is replaced by the summation of the antenna pattern and the summation of $\dot{G}_a$ over several looks in the sense that they cancel each other out. Thus, Equation (5.11) is easily affected by the AAP deviation and not recommended for operational use. Equation (5.12) has the function form of the minimized error by using the AAP as the weighting function. In a later section, we will study and confirm the error values using real data.

5.2.3 Azimuth Antenna Pattern (AAP)

As demonstrated previously, the AAP plays a very important role and must be calculated accurately. AAP can be obtained in one of two ways: (1) By measuring the history of power received by the ground-based receiver (Shimada et al. 2003) and converting the receiving time to the azimuth angle or (2) by estimating it using the SAR images of the Amazon Rainforest. Although the power history is intermittent, the first method provides a relatively easy means of obtaining AAP. However, its adaptation to the processor requires an accurate determination of the noise floor of the SAR system and processor. Uncertainty in estimating the noise floor may create another error source. In contrast, the second method always considers this noise floor.

The Amazon Rainforest is widely considered a time-independent uniform target and exhibits incidence-angle constancy at the gamma-naught. Thus, the Amazon Rainforest is often used to determine the SAR RAP (Shimada and Freeman 1995). We also employ Amazon Rainforest data to determine the AAP using the second method. The frequency spectrum of the deramped range-correlated signal indicates the azimuth response of the target in the presence of associated noise. An average of the responses over the range gives the AAP, G_a, as follows:

$$G_a(x) = \overline{g'}(x)$$

$$= \sum_{i=0}^{M-1} a_i x^i \tag{5.17}$$

where $\overline{g'}(x)$ represents the averaged output; the second expression is its polynomial approximation of the azimuth coordinate, x; and M is the maximum number of the exponent.

5.2.4 WINDOW FUNCTION

A weighting or window function is sometimes required to limit the signal bandwidth and suppress side lobes. If the selected sampling frequency is not high enough, a window function must be provided. Of several possible functions, the Kaiser window is selected for its superior in-band characteristics:

$$W_i = \frac{I_0\left(\pi\alpha\sqrt{1-(2k/N_{az}-1)^2}\right)}{I_0(\pi\alpha)} \quad 0 \le K \le N_{az}-1 \tag{5.18}$$

In this case, I_0 is a first-order Bessel function; α is a parameter for expressing the steepness of the in-band characteristics; and k is the integer address within the window size, N_{az}.

Applying Equation (5.18) to image generation in the range-migration process, Equation (5.12) suppresses scalloping more effectively. This window function may reduce the signal truncation noise due to the PRF being lower than the requested SAR Doppler bandwidth.

5.2.5 CALIBRATION OF THE GAIN OFFSET AMONG THE SCANS

The antenna peak gain and the attenuators of the SAR receiver were calibrated on the ground before the satellite launch but have to be updated before being applied to the operation because the SAR is sensitive to outside circumstances and most of the spaceborne SARs were calibrated for these terms during the in-flight calibration phase. The gain difference between the beams can be determined based on the assumption that the representative gamma-naught of each scan (i.e., at the center of the beam) should be the same for the Amazon Rainforest. Thus, the gain offset for each scan, ΔA_k, can be determined by the following:

$$\Delta A_k = \frac{\dfrac{\overline{S}_k}{\cos\theta_k}}{\displaystyle\sum_{k=1}^{N_{SCAN}} \dfrac{\overline{S}_k}{\cos\theta_k}}. \tag{5.19}$$

5.3 SUPPRESSION OF THE INTER-SCAN BANDING

Inter-scan banding (ISB) often appears as an intensity difference at the border of two neighboring scans. This may occur anywhere—even if the RAP is calculated exactly—and may be due to the temporal change of the RAP and the variation in the noise floor and the background intensities. The most reasonable answer is that an SAR image is composed of the backscattering of the target that is amplified by the antenna pattern associated with the noise floor and that the RAP calculated from the Amazon differs from the noise floor values. Those two cannot be perfectly corrected, even if their characteristics are well known.

The method we describe is a dynamic balancing method composed of the following three-step correction: (1) Building the accumulated multiplication factors to eliminate the intensity difference of the two neighboring scans, (2) re-correcting under- or over-correction in the range direction in reference to the calibrated ScanSAR data (before banding extraction), and (3) smoothing polynomial coefficients obtained in Step 2 in the azimuth direction by assuming that all the RAPs and the gain offsets among the scans are well determined using the Amazon Rainforest and strip data.

5.3.1 STEP 1: GAINS ACCUMULATED

Figure 5-2 illustrates the processing scenario. The calculated power ratio, $g_l(R)$, between the overlapped region of the neighboring scan and its accumulated power ratio, $G_k(R)$, to the scan of interest in the kth scan can be expressed by

$$g_l(R) = \frac{s_{l+1,near}(R)}{s_{l,far}(R)}, \text{ and} \tag{5.20}$$

$$G_k(R) = \prod_{l=1}^{k} \langle g_l(R) \rangle \tag{5.21}$$

where k is the scan number of interest, l is the scan number within the kth scan, the suffix near (far) refers to the region overlapped in the near (far) range, and $\langle \ \rangle$ indicates the moving average over several small regions in the azimuth direction. The $\prod_{l=1}^{k}$ function works as a multiplication of any component satisfying the integer conditions.

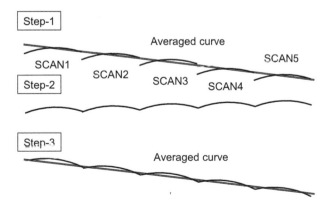

FIGURE 5-2 Schematic view of the scan-to-scan correction. At Step-1, the ScanSAR data intensity is modeled by a quadratic equation of the slant range. In Step-2, the near range of SCAN2 is made continuous to the far range of SCAN1 with a multiplication factor. The near range of SCAN3 is made continuous to the far range of SCAN2 with a multiplication factor. Further steps will be implemented to SCAN5. In Step-3, the continuously connected line is rotated so that its center axis is aligned with that of Step-1.

Using the previous function, each pixel value can then be corrected to eliminate banding as follows:

$$S_{k,j} = G_k(R)s_{k,j} \tag{5.22}$$

where $S_{k,j}$ is the intermediate corrected pixel value and $s_{k,j}$ is the pixel value before correction. This correction includes the error propagation from near range to far range and in the azimuth direction.

5.3.2 STEP 2: CORRECTION OF THE OVER/UNDER ESTIMATION

Assuming that the gain offsets among the scans are well calibrated from Equation (5.19), the power average at the mid-region of the scan (i.e., where the SNR is sufficiently high) should be independent of the scan number. The range-dependent correction factor $G_k(R)$ should then be nearly equal to unity for any of the scans, and its deviation from unity can be suppressed by using a self-range continuous function, $g_c(R)$, an mth-order polynomial of the range function, as follows:

$$g_c(R) = \sum_{l=0}^{m} a_l \cdot R^l, \tag{5.23}$$

where a_l is the coefficient of the polynomial, and R is the slant range. The final correction function can be expressed by the following equation:

$$\tilde{g}_C^k(R) = \frac{G_k(R)}{g_c(R)} \tag{5.24}$$

Using these functions, the conversion can be expressed as follows:

$$S_{k,j} = \tilde{g}_C^k(R) \cdot s_{k,j} \tag{5.25}$$

5.3.3 STEP 3: SMOOTHING IN AZIMUTH

In this third step, we apply the smoothing process for the coefficients acquired in Step 2. We selected the moving average method based on the median filter.

Finally, the error (ε_{rror}) associated in this dynamic balancing can be expressed by the least-squared error as follows:

$$\varepsilon_{rror} = 10 \cdot \log_{10} \sqrt{\left(\frac{G_k(R)}{g_c(R)}\right)^2} \tag{5.26}$$

In this case, the calculation will be conducted over all the ranges to obtain the error in Equation (5.26).

5.4 EXPERIMENTS

We used the PALSAR/ScanSAR data to validate the method that is proposed in this paper. A brief description of PALSAR/ScanSAR follows.

5.4.1 PALSAR/SCANSAR

PALSAR/ScanSAR was implemented with bandwidths of 14 MHz or 28 MHz and HH (horizontal transmit and receive) or VV (vertical transmit and receive) polarization. The 14-MHz mode, which is called WB1, is often used for standard observation with an imaging swath of 350 km and three or more looks, whereas the 28-MHz mode, which is called WB2, is designed for the ScanSAR-ScanSAR interferometry. A summary of the ScanSAR specifications is given in Table 2-3. The number of looks varies with the scan number as indicated in Table 5-1.

TABLE 5-1

Look-Number Distribution of PALSAR/ScanSAR

No. of Scans	Long/Short Burst Mode	Number of Bursts Number of Looks
3	Short	247, 356, 274
		6.54, 9.73, 7.42
4	Short	247, 356, 274, 355
		4.82, 7.13, 5.44, 7.12
5	Short	247, 356, 274, 355, 327
		3.6, 5.35, 4.08, 5.34, 5.03
3	Long	480, 698, 534
		3.21, 3.94, 3.01
4	Long	480, 698, 534, 696
		2.35, 2.89, 2.20, 2.82
5	Long	480, 698, 534, 696, 640
		1.85, 2.27, 1.73, 2.21, 2.13

Note: All burst numbers have a zero data reception window of 12 to 13.

PALSAR was designed to be able to operate continuously for up to 70 min, which is about 70% of one orbit period. Due to the antenna length (8.9 m), which is shorter than that of the JERS-1 SAR, and the satellite height (i.e., 691.5 km), the PRF can change up to seven times during one-half of the orbit (i.e., from the North Pole to South Pole or vice versa) and the sampling window start time (SWST) can vary at least every 30 seconds. The changes in PRF and SWST are non-synchronized among all the beams. Thus, the processing of the ScanSAR data for long-path products must undergo a synchronization of the data among the different PRFs as well as the image framings for all the beams. ALOS is operated in the yaw-steering mode, and the Doppler frequency will be accommodated independently of the latitude. During the last three years, calibration and validation have been conducted successfully (Shimada et al. 2009).

5.4.2 CALIBRATIONS AND DATA SETS

In this evaluation study, we used several ScanSAR data sets that were acquired from Amazon Rainforest regions. Table 5-2 presents the data sets from the Amazon Rainforest data used for this calibration. These data were collected at three different times in 2006 and 2007 from a uniform forest area. Because the Amazon has two seasons, the wet season from November to April and the dry season from May to October, efforts were made to collect these data sets from both seasons almost equally.

The gain offset of each beam must be determined for preparing the banding correction proposed in this paper. Using Equation (5.19) and the Amazon data sets listed in Table 5-2, the gain offset for all the

TABLE 5-2

Data Set Used for ScanSAR Analysis

No.	Observation Date	Latitude (degrees)	Longitude (degrees)
1	Nov. 24, 2006	−5.35868	−67.37804
2	Nov. 24, 2006	−2.87441	−66.84801
3	Jan. 9, 2007	−2.87888	−66.87921
4	Jan. 9, 2007	−5.35855	−67.41248
5	July 12, 2007	−2.87004	−66.86024
6	July 12, 2007	−5.36289	−6.739107
7	July 12, 2007	−7.84948	−67.94334

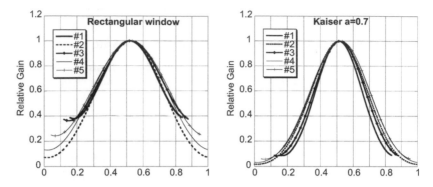

FIGURE 5-3 Azimuth antenna patterns calculated from the Amazon data using (a) the rectangular window (left), and (b) the Kaiser window ($a = 0.7$; right).

beams was calculated as shown in Table 5-4. The two lines in this table refer to the fact that the gain characteristics of ScanSAR might have changed slightly with the attenuator changes applied to the SAR receiver on August 7, 2006, to reduce the saturation rate, which was higher before that date. These two lines indicate the change in the values, especially for the No. 2, No. 3, No. 4, and No. 5 beams.

5.4.3 WINDOW FUNCTION AND AAP

The AAP is dependent on the window function applied in Equation (5.18). We present two patterns: one obtained by using the rectangular window and the other obtained by using a Kaiser window of $a = 0.7$. These patterns are depicted in Figure 5-3. The x-axis was normalized with the maximum number of pulses for each burst, $N_{az,k}$. Beams 1 and 3 exhibit small rises in the curve at the end in Figure 5-3. This indicates that the signal is truncated.

The PRF changes with the satellite altitude obtained from the real-time GPS receiver onboard ALOS. The typical PRF patterns for PALSAR/ScanSAR over Hokkaido, Japan, on April 18, 2006, and over the Amazon and the United States are presented in Table 5-3. Two beams, No. 1 and No. 3,

TABLE 5-3
Typical PRF (Hz) Observed at WB1 5 SCAN

Scan No.	1	2	3	4	5
Hokkaido	1,694	2,375	1,718	2,164	1,923
Amazon	1,677	2,352	1,700	2,141	1,901
Louisiana	1,686	2,358	1,709	2,150	1,912
Toyama	1,689	2,364	1,712	2,155	1,916

Note: Doppler bandwidth for all scans is 1,700 Hz.

TABLE 5-4
Gain Offset by SCAN

SCAN	1	2	3	4	5
ΔAa (dB)	0.21	1.18	0.89	1.59	1.10X
ΔAb (dB)	0.16	0.34	−0.61	0.33	−0.19

Note: Suffix "a" refers to data collected after August 7, 2006, and suffix "b" refers to data collected before that date. On August 7, 2006, the attenuator of the receiver was changed slightly for each scan, thereby slightly changing the data saturation rate.

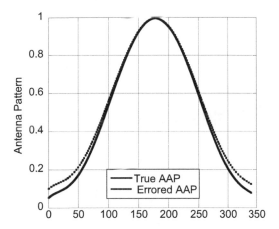

FIGURE 5-4   True azimuth antenna pattern (solid line) and deviated azimuth antenna pattern (broken line). The multiplicative error in the vertical antenna pattern is 5%.

are slightly under-sampled, and three beams, No. 2, No. 4, and No. 5, are over-sampled. The two under-sampled beams are degraded by the azimuth ambiguity caused by the signal truncation. The best solution for suppressing this azimuth artifact is to increase the PRF, but applying the bandwidth limitation using the aforementioned window function in Equation (5.18) is another option.

5.4.4   VALIDATION OF THE PROPOSED METHOD USING REAL SCANSAR DATA

5.4.4.1   Scalloping Correction

First, we conducted a simulation study to investigate how sensitive both methods (Method-1 and Method-2) are to the error parameters. Because dislocation due to incorrect Doppler estimation is not an issue investigated in this paper, we considered only a case in which the AAP differs slightly (i.e., with $\varepsilon = 0.05$ and $\delta = 0.0$) from the real one to determine how Equations (5.15) and (5.16) behave. The simulation was conducted using the real AAP of the PALSAR/ScanSAR, which is shown in the solid line in Figure 5-4 and the deviated pattern (broken line). The results, which are shown in Figure 5-5, suggest that the proposed method (i.e., Method-2, solid line) is error-free and

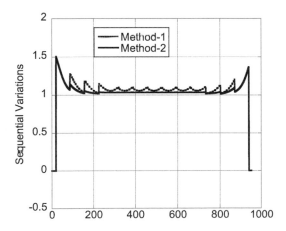

FIGURE 5-5   Comparison of the two scalloping correction methods. Method-1 (broken line) does not correct the variation of the data over time; Method-2 (solid line) suppresses the variation of the antenna pattern. In this case, we adopted the simulation with the following parameters: $f_{DD} = -490.2$ Hz/s; $f_{PRF} = 1,923$ Hz; $T_{SCAN} = 0.7891$; $N_{az} = 342$; $\varepsilon = 0.05$; and $\delta = 0.0$.

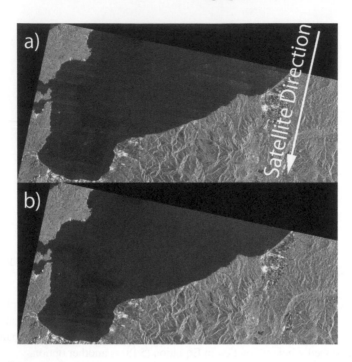

FIGURE 5-6   Scalloping observed in the image of the Gulf of Toyama, Japan, when processed by (a) Method-1 and removed by (b) Method-2.

performs well in reducing scalloping; whereas Method-1 does not correct the scalloping and still exhibits periodic perturbation of the intensity in the broken line.

Next, we conducted a test using real data from the PALSAR/ScanSAR, which was acquired over a heterogeneous area, the Gulf of Toyama, Japan. This area was observed by the No. 5 beam, which is operated with a sufficiently satisfactory PRF to cover the needed Doppler bandwidth (Table 5-3). Figure 5-6 presents the comparative study result for correcting scalloping using the two methods. According to the result, Method-1 suffered from severe scalloping in the gulf along the azimuth direction, while Method-2 does not present this artifact at all. Clearly, Method-2 significantly improves the image quality.

5.4.4.2   Reduction of the Signal Truncation and Azimuth Error in the Image

We generated two images using the two window functions and compared the image quality relating to how the window function works regarding the signal truncation issue. The Amazon Rainforest was selected as a uniform target, and the coastal region of Hokkaido, which is often affected by azimuth ambiguities, was selected as the nonuniform target.

   a. Amazon: As depicted in Figure 5-7a and b, both windows exhibit ideal responses; thus, no irregularities were observed. The power reduction due to the bandwidth suppression should be corrected at the NRCS calculation stage.
   b. Coastal region of Hokkaido: Figure 5-8 presents the image of Hokkaido. The image from the rectangular window retains high ambiguity in the coastal region, especially in beam No. 3, which is shown in the red ellipse. This is because beam No. 3 is often operated with an unsatisfactory PRF. Therefore, the resultant edge response in the higher spectrum is associated with the noise floor, and the truncated signal appears as azimuth ambiguity. The image from the Kaiser window has no azimuth ambiguity in beam No. 3 because the weighting of the window suppresses the signal truncation; although the resolution in the azimuth direction may be slightly reduced.

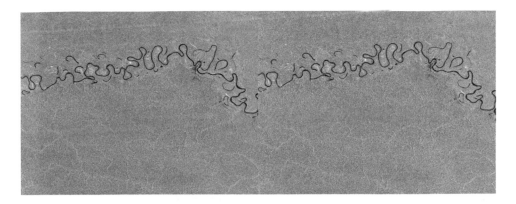

FIGURE 5-7   ScanSAR images of an area in the Amazon with a horizontal width of 130 km do not exhibit vertical stripes for either window function.

5.4.4.3   Inter-Scan Banding Correction

Two examples of the range dependence of the gain difference between the accumulated gain corrections and the normalization factors are presented in Figure 5-9; the Amazon Rainforest is the uniform case, and Japan is the nonuniform case. The Amazon case indicates that the gain difference is a triangular function with five peaks; the Japan case exhibits a nonuniform distribution in the gain pattern across the track. Errors for the averaged gain corrections of Equation (5.26) are summarized in Table 5-5 and Figure 5-10. The average error over four samples was 0.0054, and its standard deviation was 0.00034. Thus, this correction method provides sufficiently accurate gain corrections.

TABLE 5-5
Errors of the Scan-to-Scan Disbanding Processes

Scene	Average Error	Standard Deviation
Japan	0.0040	0.00039
Amazon	0.0040	0.00022
Amazon2	0.0060	0.00019
Sea ice	0.0075	0.00047
Average	0.00540	0.00034

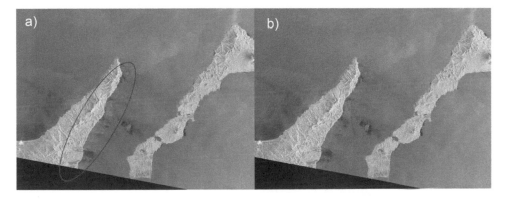

FIGURE 5-8   ScanSAR image of an area east of Hokkaido, Japan, with a horizontal width of 130 km is depicted in two windows. The azimuth ambiguity south of the Shiretoko peninsula appearing in the red ellipse of (a) is corrected by the proposed window function and thus cannot be seen in (b).

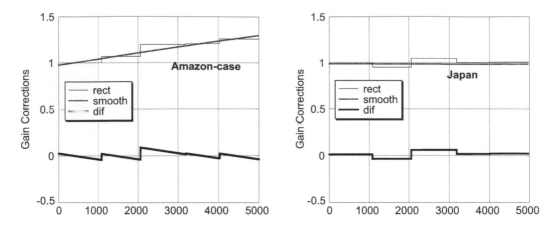

FIGURE 5-9   Samples of the inter-scan destriping process for two cases: the Amazon on the left and Japan on the right. Each has three correction curves: a thin solid line for the accumulated curve, a thin dotted curve for the calibrated curve, and a bold solid line for the final correction curve in the range direction.

Figure 5-11 compares an image of northern Europe that was corrected using the proposed banding method to an uncorrected image. The image using the proposed banding method was significantly improved in terms of quality; in fact, no minor stripes can be seen along the track.

Figure 5-12 shows representative ScanSAR images acquired for four regions: the Sahara Desert in Africa, the Amazon Rainforest, Hokkaido and the ocean, and Antarctica, each of which has an image size of 350 km × 350 km. All images were processed using the SPECAN and the image correction algorithms. We can confirm that these heterogeneous and homogeneous images were corrected free of artifacts. As introduced in Shimada et al. (2009), the radiometric and geometric accuracies meet the specifications.

5.4.4.4   Simpler Implementation of the Scalloping Correction Algorithm

As a summary, Sections 5.2 and 5.3 represent the main part of this scalloping correction algorithm. The determination of the AAP in Sections 5.3 and 5.4 using the Amazon data and applying it to multi-looking using Equation (5.12) makes the scalloping correction simple and stable. The implementation of this algorithm is easy. Notably, the calibration of the gain offsets among the scans in Section 5.5 is necessary for the preparation process.

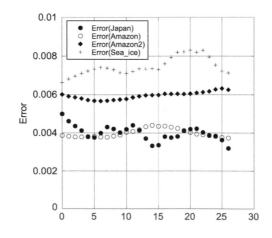

FIGURE 5-10   Averaged error associated with the scan-to-scan normalization for four different images.

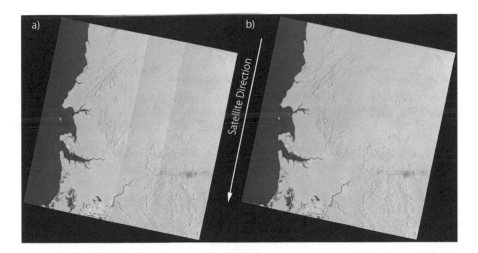

FIGURE 5-11 Comparison of inter-scan banding for a PALSAR/ScanSAR image of northern Europe. Image (a) is before correction; image (b) is after correction.

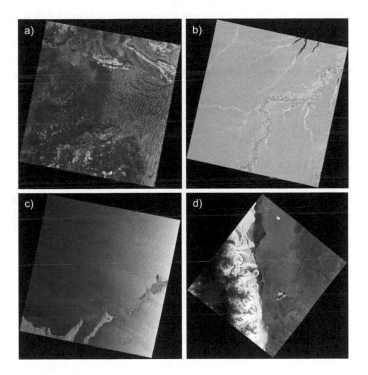

FIGURE 5-12 Sample PALSAR/ScanSAR images processed by using the proposed algorithm: (a) Sahara Desert, b) Amazon Rainforest, (c) Hokkaido and the ocean, and (d) Antarctica.

5.5   SUMMARY

In this paper, we proposed a new method to correct the scalloping, the azimuth ambiguity, and the scan-to-scan banding artifacts that typically appear in ScanSAR images. Specifically, the method to correct scalloping is based on the AAP derived from data from the Amazon Rainforest and its adaptation to the weighted summation of the multi-looking process. This theory was validated using a numerical simulation with real data obtained from the PALSAR ScanSAR.

REFERENCES

Bamler, R., 1995, "Optimum Look Weighting for Burst-Mode and ScanSAR Processing," *IEEE T. Geosci. Remote*, Vol. 33, No. 3, pp. 722–725.

Bast, D. C. and Cumming, I. G., 2002, "RADARSAT ScanSAR Roll Angle Estimation," *Proc. IGARSS*, Vol. 1, Toronto, ON, Canada, June 24–28, 2002, pp. 152–154.

Cumming, I. G. and Wong, F. G., 2005, *Digital Processing of Synthetic Aperture Radar Data*, Artech House, Norwood, MA, pp. 369–423.

Curlander, J. C. and McDonough, R., 1991, *Synthetic Aperture Radar, Systems and Signal Processing*, Wiley, Hoboken, NJ, 1991.

Dragosevic, M. V. and Davidson, G. E., 2000, "Roll Angle Measurement and Compensation Strategy for RADARSAT ScanSAR," ESA SP-450, *Proc. of the CEOS SAR Workshop*, Toulouse, France, October 26–29, 1999, pp. 545–549.

Freeman, A. and Curlander, J. C., 1989, "Radiometric Correction and Calibration of SAR Images," *Photogramm. Eng. Rem. S.*, Vol. 55, No. 9, pp. 1295–1301.

Hawkins, R. K. and Vachon, P., 2002, "Modeling SAR Scalloping in Burst Mode Products from RADARSAT-1 and ENVISAT," ESA SP-526, *Proc. of CEOS Working Group on Calibration and Validation*, London, September 24–26, 2002.

Hawkins, R. K., Wolfe, J., Murnaghan, K., and Jefferies, W. C., 2001, "Exploring the Elevation Beam Overlap Region in RADARSAT-1 ScanSAR," *Proc. of the CEOS Calibration Working Group on Calibration and Validation*, Tokyo, April 2–5, 2001, pp. 77-83.

Jin, M., 1996, "Optimal Range and Doppler Centroid Estimation for a ScanSAR System," *IEEE T. Geosci. Remote*, Vol. 34, No. 2, pp. 479–488.

Leung, L., Chen, M., Shimada, J., and Chu, A., 1996, "RADARSAT Processing System at ASF," *Proc. of IGARSS'96, International Geoscience and Remove Sensing Symposium*, Lincoln, NE, May 27–31, 1996.

Rosich, B., Meadows, P. J., and Monti-Guarnieri, A., 2004, "ENVISAT ASAR Product Calibration and Product Quality Status," *Proc. CEOS SAR Workshop*, Ulm, Germany, May 27–28, 2004.

Shimada, M., 1993, "An Estimation of JERS-1's SAR Antenna Pattern Using Amazon Rainforest Images," *Proc. 1993 SAR Calibration Workshop* (CEOS SAR CAL/VAL), Noordwijk, The Netherlands, September 20–24, 1993, pp. 185–208.

Shimada, M., 1999, "Radiometric Correction of Saturated SAR Data," *IEEE T. Geosci. Remote*, Vol. 37, No. 1, pp. 467–478.

Shimada, M., 2005, "Long-Term Stability of L-band Normalized Radar Cross Section of Amazon Rainforest Using the JERS-1 SAR," *Can. J. Remote Sensing*, Vol. 31, No. 1, pp. 132–137.

Shimada, M., 2009, "A new method for correcting SCANSAR scalloping using forest and inter SCAN banding employing dynamic filtering," *IEEE Trans. GRS*, Vol. 47, No. 12, pp. 3933–3942.

Shimada, M. and Freeman, A., 1995, "A Technique for Measurement of Spaceborne SAR Antenna Patterns Using Distributed Targets," *IEEE T. Geosci. Remote*, Vol. 33, No. 1, pp. 100–114.

Shimada, M. and Isoguchi, O., 2002, "JERS-1 SAR Mosaics of Southeast Asia Using Calibrated Path Images," *Int. J. Remote Sensing*, Vol. 23, No. 7, pp. 1507–1526.

Shimada, M., Isoguchi, O., Tadono, T., and Isono, K., 2009, "PALSAR Radiometric and Geometric Calibration," *IEEE T. Geosci. Remote*, Vol. 47, No. 12, pp. 3915–3932.

Shimada, M., Tanaka, H., Tadono, T., and Watanabe, M., 2003, "Calibration and Validation of PALSAR (II) Use of Polarimetric Active Radar Calibrator and the Amazon Rainforest Data," *Proc. IGARSS 2003, International Geoscience and Remote Sensing Symposium*, Vol. 2, Toulouse, France, July 21–25, 2003, pp. 1842–1844.

Shimada, M., Watanabe, M., Moriyama, T., Tadono, T., Minamisawa, M., and Higuchi, R., 2007, "PALSAR Radiometric and Geometric Calibration (in Japanese)," *J. Remote Sens. Soc. Japan*, Vol. 27, No. 4, pp. 308–328.

Srivastava, S. K., Hawkins, R. K., Banik, B. T., Adamovic, M., Gray, R., Murnaghan, K., Lukowski, T. I., and Jefferies, W. C., 2001, "RADARSAT-1 Image Quality and Calibration—A Continuing Success," *Adv. Space Res.*, Vol. 28, No. 1, pp. 99–108.

Vigneron, C., 1994, "Radiometric Image Quality Improvement of ScanSAR Data," Thesis of B. Eng. (High Distinction), Carleton University.

6 Polarimetric Calibration

6.1 INTRODUCTION

PolSAR consists of two orthogonally polarized transmitters and two receivers. It measures a target's scattering property through scattering data. The transmitters and receivers differ from each other, and the data are affected by the differences. In order to fully utilize the (uncalibrated) data, polarimetric calibration, PolCAL, which corrects the difference, is mandatory.

Figure 6-1 shows the conceptual signal flow in PolSAR. From the left, the original signal is divided into two paths by the splitter and circulator: the upper H transmit path and the lower V transmit path. This example represents the linear H-V PolSAR, and the right-left circular PolSAR simply can be imagined. In reality, the simultaneous transmission of the dual pol signals is quite difficult due to signal contamination. Instead, the transmitter generates two pol signals alternatively (such as H-V-H-V and so on), and the signal routing paths are switched above-down-above-down and so on. The line length between the transmitter and antenna differs in H and V, resulting in different attenuations. Finally, the channel imbalance, f_1, and the cross talks, δ_1 (δ_2), from H to V (and vice versa) differ. As a result, the antenna creates the electric field E_H and E_V, which is slightly different in amplitude and phase between polarizations.

The receiving process from the target shows that difference in the receivers' properties also creates a crosstalk and a channel imbalance. The four measured components contain these differences, and their correction should be applied in order to attain the correct scattering matrix.

Study of the PolCAL algorithm begin in the United States and Canada in the late 1980s, when the PolSAR first appeared on airplanes and spacecraft (Freeman 1992; Freeman et al. 1992). For various reasons, the properties of the transmission and reception have ten unknowns, which include crosstalk, channel-imbalances, and noise levels; their values have different orders of magnitude; they are complex values; solutions become more unstable as more unknowns are included; even with multiple calibration instruments or a natural target whose backscattering properties are well known, PolCAL is difficult to attain. It should be noted that the absolute SAR calibration that determines the radar cross section of the target is a completely different topic from PolCAL.

PolCAL has two purposes. The first is to determine the two 2×2 polarimetric distortion matrices (PDMs), which express the polarimetric transformation between transmission and reception using channel imbalances and crosstalk. The second is to convert uncalibrated PolSAR data to calibrated data using the PDMs. This chapter aims to determine the PDMs, and we imply PolCAL as the first meaning. The transmission or reception PDM consists of one channel imbalance and two crosstalk terms.

Table 6-1 summarizes all the PolCAL methods developed to date. The following assumptions and calibration instruments are associated with these methods.

a. Radar reciprocity: When the SAR uses a passive SAR antenna, such as JERS-1 SAR and Pi-SAR-L/L2, transmit and receive properties are the same as are the distortion matrices. Recent SARs can quickly change the antenna beam direction in elevation and azimuth, enabling wide image observation (ScanSAR mode) and very high-resolution observation (spotlight mode). These high performances are realized by the transmit-receive modules (TRMs) and phase shifter, each of which is composed of many TR-independent and phase change units each with low transmission power, which avoids discharging in space. Thus, recent SARs (launched after 2010) are invalid under this assumption.

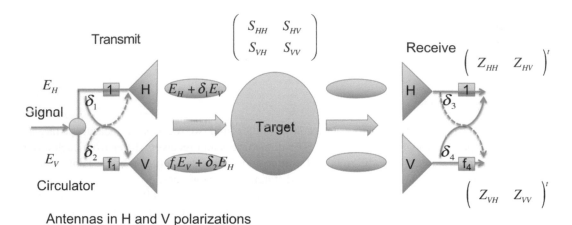

FIGURE 6-1   Polarization mixtures in transmit and receive modes in the PolSAR and why PolCAL is necessary. The notations will be clarified later in this chapter.

    b. Reflection symmetry: When considering the microwave backscattering from a uniform target, the like-scattering component is not correlative with the cross-scattering component—for example, $<S_{HH}S_{HV}^*> = <S_{HH}S_{VH}^*> = <S_{VV}S_{HV}^*> = <S_{VV}S_{VH}^*> = 0$. Here, S_{ij} is the (complex) scattering coefficient at the ijth polarization; $<>$ is the average; the suffix i is the receive polarization; and j is the transmit polarization. This polarization order is similar in ease to the matrix calculation. It also should be noted that "like" means the transmit component and the receive component have the same polarization: HH and VV, while the "cross" means different polarizations, such as HV and VH. By using this assumption, the covariance calculation can be simplified. Other symmetries include rotation symmetry and azimuth symmetry (Nghiem et al. 1992; Yueh et al. 1994). The reflection symmetry assumption has often been used in many PolCAL methods.

TABLE 6-1
PolCAL Methods and Assumptions

Methods	Data Expression	Instruments	Assumptions
Freeman 1992	Scattering matrix	PARC	None (importance of the PARC deployment angle)
Freeman et al. 1992	Scattering matrix	Trihedral CR Dihedral CR	Symmetrization of the cross pol
Van Zyl 1990	Covariance matrix	Rough surface + T-CR	Reflection symmetry + scattering reciprocity + radar reciprocity
Quegan 1994	Covariance matrix	Rough surface + T-CR	Reflection symmetry + scattering reciprocity
Kimura 2009	Scattering matrix Orientation angle	Urban area	Reflection symmetry + scattering reciprocity
Ainsworth et al. 2006	Covariance matrix Orientation angle	Natural target	Scattering reciprocity
Fujita and Murakami 2005; Takeshiro et al. 2009	Scattering matrix	Phase preserving CR Phase variable CR	Radar reciprocity
Touzi and Shimada 2009	Scattering matrix	CR	Scattering reciprocity Symmetrization of the cross pol Azimuthal symmetry
Shimada 2011	Covariance matrix	Forest + trihedral-CR	Reflection symmetry Scattering reciprocity

c. Scattering reciprocity: In scattering reciprocity, the assumption is made that all cross-polarization scattering from natural targets is the same ($S_{HV} = S_{VH}$) and is used for all PolCAL that uses a natural target. This assumption makes the equation simple.

d. Calibration instruments: Reflectors consist of two or three square or triangular metallic planes, each sized larger than ten wavelengths. Their radar cross sections are well known. Thus, they are often used for the absolute calibration and PolCAL. There are several artificial targets used for generating the specific scattering matrix; these are active radar calibrators.

Table 6-1 summarizes all the PolCAL methods with regard to authors, conditions, and contents. From this, it becomes apparent that most PolCAL methods make two assumptions, reflection symmetry and scattering reciprocity, and that they also use the corner reflector (CR)—in particular, the trihedral (CR), which determines the channel imbalance and is one of the unknowns in PolCAL. These methods reflect the features of the SARs in each era. For example, Van Zyl (1990) simplified the radar system using the reciprocity assumption (i.e., passive-antenna-based polarimetric SAR) that the transmission and reception distortions have equal properties. The single PDM is determined through iterative calculations by using depolarized targets satisfying the reflection symmetry and reciprocal conditions (e.g., rough surface or target slightly covered with vegetation) and a trihedral CR.

The other methods are as follows:

1. Freeman et al. (1992) proposed two methods, one of which adopted several different scattering matrices intensively generated by the polarimetric active radar calibrator (PARC), and another that used symmetrized cross components supported by the trihedral CR (Klein 1992).
2. Van Zyl (1990) adopted radar reciprocity, scattering reciprocity, reflection symmetry, and one trihedral CR.
3. Quegan (1994) utilized only the reflection symmetry for a natural target as a minimum requirement and more intensively supported by a trihedral CR. He proposed an iterative method to determine the PDMs of non-reciprocal radar systems, such as PALSAR, TerraSAR-X, RADARSAT-2, and SIR-C (Spaceborne Imaging Radar), which consisted of a large number of transmit-receive modules (TRMs) using depolarized targets for satisfying the reflection symmetry and a CR.
4. Kimura (2009) developed a new PolCAL method that uses preservation of the polarization orientation induced in built-up areas and poses the reflection symmetry on the target.
5. Ainsworth et al. (2006) also proposed a new PolCAL method that iteratively solves the PDM and only ingests the scattering reciprocity as the weakest constraints with which the polarimetric system has been associating.
6. Fujita and Murakami (2005) and Takeshiro et al. (2009) used a trihedral CR and polarization selective CRs.
7. Ridha and Shimada (2009) enhanced the Van Zyl and Freeman methods for accurate estimation of the antenna crosstalk.
8. Shimada (2011) used the reflection symmetry and scattering reciprocity of the Amazon forest and one CR deployed there in considering the polarization-dependent signal penetration in the forest.

In this way, the PolCAL method reflects PolSAR characteristics and its progress. ALOS/PALSAR, launched in 2006, was the world's first operational L-band polarimetric SAR, and its PolCAL was performed using the Quegan method (Shimada et al. 2009; Moriyama et al. 2007).

Most PolCAL methods using natural targets pose an assumption of reflection symmetry, which implies the ensemble averages of the co-pol (hereafter Cpol) and cross-pol (hereafter Xpol) correlations are zero and simplifies the equation manipulation afterward. Reflection symmetry and other symmetries (i.e., rotation and azimuthal symmetries) were investigated theoretically by Nghiem et al. (1992), and it was clarified that the covariance matrix can be represented by three ensemble averages (i.e., $HH \times VV$, $HV \times HV$, and $VV \times VV$, each being normalized by $HH \times HH$), and their forms depend on the type of symmetry. Covariance matrices on real targets were investigated by Van Zyl (1990) using the Jet Propulsion Laboratory's (JPL) multi-frequency Airborne Synthetic Aperture Radar (AIRSAR) over ocean surfaces, clear-cut forests, and forested areas. For the Shasta-Trinity National Forest in California, he also stated that the forest crown is approximated by a huge number of randomly oriented thin cylinders, and $HH \times VV$ and $HV \times HV$ tend to be equal.

Rainforests over flat terrain (e.g., in the Amazon and the Congo basin) behave as uniform distributed targets (Lambertian targets) in the radar backscatter and are often used as stable calibration targets for spaceborne SARs. They provide incidence-angle-independent backscattering coefficients (i.e. gamma-naught) and depolarization sources, which may be supported by a huge number of randomly distributed small scatterers (i.e., crown canopies, leaves, branches, low vegetation above ground, and trunks). They have been adopted as standard targets by the Committee on Earth Observation Satellites (CEOS) SAR working community and are often used to determine the antenna elevation pattern of spaceborne SARs (Shimada and Freeman 1995; Zink and Rosich 2002; Shimada et al. 2009; Lukowski et al. 2003; Attema et al. 1997; Cote et al. 2005). However, data acquired over rainforests have not often been used as a PolCal source except in a preparative study for the C-band SAR of RADARSAT-2 (Luscomb 2001). C-band or higher frequency data have very little or no signal penetration through forest canopies and equalizing the Cpol signals simplifies the PolCAL process. At the L-band or lower frequency, a problem may arise with the polarization-dependent signal penetration through the forest. One possible disadvantage of forest data is that the volume scattering target could behave as a source of cross talk.

Several polarimetric classification techniques were developed in coherent and/or incoherent ways (Krogager 1990; Cloude and Pottier 1996; Yamaguchi et al. 2005; Freeman 2007). Krogager (1990) decomposes polarimetric data in odd-number reflection, even-number reflection, and helix reflection targets as a coherent method. Freeman and Durden (1998) developed an incoherent target decomposition method that classifies land use by splitting the signal power into three scattering components (i.e., volume, double-bounce, and surface scattering) quantitatively. They also showed that L-band backscatter from dense forest is mainly composed of volume and double-bounce scattering with negligible surface scattering. Freeman (2007) also developed a two-scattering component model that reduces the singularity of the solution finding. Neumann et al. (2009, 2010) modeled the forest property by using spherical and ellipsoidal scatterers with orientation randomness.

The approach introduced in this chapter is to apply incoherent target decomposition to the rainforest PolSAR covariance data and build up the PolCAL equation, which is explicitly associated with the PDM. We also show a procedure to solve the nonlinear equation for obtaining the PDM. Here, we use the rainforest PolSAR data as a polarization-dependent signal penetration subject. The method requires two targets in an image: one is the volume scattering target and the other is the surface scattering target. Ideally, the latter needs to be a trihedral CR with a known complex backscattering coefficient for each polarization. As a second opportunity, we experimentally seek possible calibration sources from natural targets (e.g., global rough surfaces and wider forested regions) that can substitute surface scattering from trihedral targets to determine channel imbalances, in case of the unavailability of artificial targets (Shimada 2011).

In this chapter, the proposed polarimetric calibration method using forest PolSAR data is derived and solved in Sections 6-2 and 6-3, respectively. Section 6-4 introduces a summary of the calibration experiment. The assumptions made by the proposed method are validated, and the stability of its solution is evaluated, in Section 6-5. Then, the calibration is carried out and the results are compared with Quegan's in Section 6-6. We end with a discussion and conclusion in Sections 6-7 and 6-8.

6.2 THEORETICAL EXPRESSION OF THE POLARIMETRIC CALIBRATION METHOD

6.2.1 Assumptions

In the following sections, we discuss the assumptions regarding the PolSAR and the target scattering property.

6.2.1.1 Nonreciprocity of the PolSAR

The polarimetric property of the PolSAR can be approximated by two independent nonsymmetrical distortion matrices: This assumption is general and realistic because the most recent PolSAR consist of a large number of TRM and are no longer reciprocal.

6.2.1.2 Backscattering from the Forest

The Amazon Rainforest consists of various types of tree species with different heights and volumes. These tree groups are composed of crown canopies (as a group of leaves and branches) and trunks; their heights reach several tens of meters. Penetration, attenuation, and backscattering properties of L-band signals through or from the forest show a complex behavior depending on polarizations (Figure 6-2a shows the representative signal return passes). Incoherent signal decomposition using a polarimetric covariance leads to the quantitative interpretation of the backscattering property from the target even when the signal penetrates the target. To interpret the contribution from the forest, we adopt the Freeman-Durden method (1998) as a base and modify it for the purpose of sensitivity analysis.

6.2.1.3 Reflection Symmetry and Reciprocity

Polarimetric clutter from random or semi-random media theoretically implies the symmetry property of the covariance matrix and the perfect decorrelation of the Cpol and Xpol covariances (Nghiem et al. 1992; Borgeaud et al. 1987). Under the symmetric condition, the covariance matrix is represented by the three components C_{HVHV}, C_{VVVV}, and C_{HHVV}, whose ranges of values depend on the type of symmetry. Here, the notation C_{HVHV} is a covariance of HV and HV normalized by that for HHHH. Some studies were conducted to analyze and model the polarimetric backscattering property from the real forest assuming the azimuthal symmetry target (Van Zyl 1992; Freeman 2007; Freeman and Durden 1998). Since the Amazon Rainforest is very dense and seems to be very uniform (except for the clear-cut regions), we assume the Amazon Rainforest as a reflection symmetry target and as a reflection reciprocity target (i.e., $<HH \cdot HV^*> = <HH \cdot VH^*> = <VV \cdot HV^*> = <VV \cdot VH^*> = 0$ and $HV = VH$).

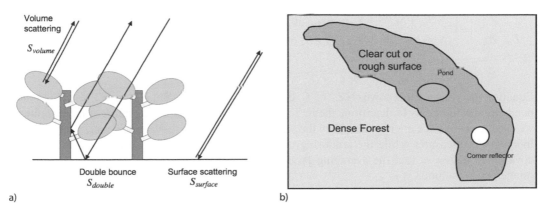

FIGURE 6-2 Expression of (a) scattering components and (b) simplified forest distribution.

6.2.1.4   Faraday Rotation

When the electromagnetic wave propagates through an external magnetic field such as the geomagnetic field, the polarization plane rotates around the radar line of sight with the Faraday rotation angle (FRA) proportionate to the external magnetic intensity at the radar line of sight and the Total Electron Content (TEC). The FRA can be estimated from the calibrated PolSAR data (Bickel and Bates 1965). Although the FRA can be estimated as an additional unknown during the PolCAL process, the other PolCAL parameters may reduce the confidence. The calibration site used in this study is located in Rio Branco (in western Brazil, bordering Peru), where the geomagnetic line is almost parallel to the ascending orbit of the satellite so that the FRA is negligible. This calibration site also contains bare surfaces, forested areas, and a CR.

6.2.1.5   Forest Cover Status

Although the Amazon has been a world reservoir of natural dense rainforest, deforestation is reducing the current forest coverage. Even within one PolSAR image from a 35-km observation swath, both dense forested and clear-cut areas normally are included, as schematically illustrated in Figure 6-2b). An image classification could be necessary to discriminate non-forested areas before the newly developed algorithm is performed.

6.2.2   Expression of the Uncalibrated Scattering Matrix

The uncalibrated scattering matrix can be expressed by

$$
\begin{pmatrix} Z_{hh} & Z_{hv} \\ Z_{vh} & Z_{vv} \end{pmatrix} = A \cdot e^{\frac{-4\pi r}{\lambda}} \begin{pmatrix} 1 & \delta_3 \\ \delta_4 & f_2 \end{pmatrix} \begin{pmatrix} \cos\Omega & \sin\Omega \\ -\sin\Omega & \cos\Omega \end{pmatrix} \begin{pmatrix} S_{hh} & S_{hv} \\ S_{vh} & S_{vv} \end{pmatrix} \begin{pmatrix} \cos\Omega & \sin\Omega \\ -\sin\Omega & \cos\Omega \end{pmatrix} \begin{pmatrix} 1 & \delta_1 \\ \delta_2 & f_1 \end{pmatrix}
$$

$$
+ \begin{pmatrix} N_{hh} & N_{hv} \\ N_{vh} & N_{vv} \end{pmatrix}, \tag{6.1a}
$$

where Z_{pq} is the measurement matrix; q is the transmission polarization; p is the reception polarization; A is the amplitude; r is the slant range; S_{pq} is the true scattering matrix of the target; f_1 is the channel imbalance of the transmission distortion matrix; f_2 is that for the reception matrix; $\delta_1(\delta_2)$ is the transmission cross talk; $\delta_3(\delta_4)$ is the reception cross talk; N_{pq} is the noise component; and Ω is the FRA. Based on the aforementioned fourth assumption, we can set $\Omega = 0$, and Equations (6.1a) and (6.1b) can be expanded to

$$
\begin{pmatrix} Z_{hh} & Z_{hv} \\ Z_{vh} & Z_{vv} \end{pmatrix} = \begin{pmatrix} S_{hh} + S_{hv}\delta_2 + S_{vh}\delta_3 + S_{vv}\delta_2\delta_3 + N_{hh} & S_{hh}\delta_1 + S_{hv}f_1 + S_{vh}\delta_1\delta_3 + S_{vv}f_1\delta_3 + N_{hv} \\ S_{hh}\delta_4 + S_{hv}\delta_2\delta_4 + S_{vh}f_2 + S_{vv}\delta_2 f_2 + N_{vh} & S_{hh}\delta_1\delta_4 + S_{hv}f_1\delta_4 + S_{vh}\delta_1 f_2 + S_{vv}f_1 f_2 + N_{vv} \end{pmatrix}
$$

$$\tag{6.1b}$$

Equation (6.1) has ten unknowns (i.e., f_1, f_2, δ_1, δ_2, δ_3, δ_4, N_{hh}, N_{hv}, N_{vh}, N_{vv}), and it needs ten independent measurements to determine them. To increase the number of independent measurements, we use the covariance matrix rather than the scattering matrix. The covariance matrix generates ten independent measures while the scattering matrix gives us only four. Another advantage of the covariance matrix is that the averaging process reduces speckle noise, although it also reduces image pixel resolution.

Recent SARs have been built using several hundred TRMs for signal transmission and reception (e.g., 80 for PALSAR, 180 for PALSAR-2/ALOS-2, 512 for RADARSAT-2, 320 for ASAR, and 384 for TerraSAR-X); SARs become nonreciprocal systems. Channel imbalances are on the order of one, cross talks and noise levels are generally about -20 dB and -25 dB, respectively. PALSAR has recorded the smallest noise equivalent sigma naught with a value equal to -34 dB (Shimada et al. 2009). The signal-to-noise ratio (SNR) depends on the target of interest. The SNR of the forest (whose backscattering coefficient or sigma naught, σ^0, is approximately -7 dB) may be as large as 27 dB, and noise may be negligibly small. For rough surfaces (σ^0 approximately -13 dB), the noise can be relatively small since the SNR is about 20 dB. The existence of the noise can be fully ignored in the case of CRs. Thus, how the noise component is handled really depends on the backscattering coefficient of the target. Even if the noise may not be an issue when forested areas or brighter targets are selected as polarimetric calibration sources, this method includes and assesses the noise component.

6.2.3 Covariance Matrix

A covariance of the measured scattering matrix is given by

$$
\langle \mathbf{Z} \cdot \mathbf{Z}^* \rangle = \left\langle
\begin{array}{cccc}
z_{hh}z_{hh}^* & z_{hv}z_{hh}^* & z_{vh}z_{hh}^* & z_{vv}z_{hh}^* \\
z_{hh}z_{hv}^* & z_{hv}z_{hv}^* & z_{vh}z_{hv}^* & z_{vv}z_{hv}^* \\
z_{hh}z_{vh}^* & z_{hv}z_{vh}^* & z_{vh}z_{vh}^* & z_{vv}z_{vh}^* \\
z_{hh}z_{vv}^* & z_{hv}z_{vv}^* & z_{vh}z_{vv}^* & z_{vv}z_{vv}^*
\end{array}
\right\rangle ,
\tag{6.2}
$$

where the asterisk (*) indicates the complex conjugate and the angular bracket, $\langle\ \rangle$, stands for the ensemble average. Referring to a relationship between the covariance matrix and the scattering matrix for volume and double bounce, we can expand Equation (6.2) for our purposes where the covariance matrix (see Cloude and Pottier 1996; Freeman and Durden 1998) is expanded from 3×3 to 4×4 for enabling the PolCAL parameter estimation.

Equation (6.2) is a Hermitian matrix, and only its upper triangular ten components are independent and meaningful. Inserting Equations (6.1a) and (6.1b) into Equation (6.2), we can express all sixteen components by the elements of the scattering matrix and the distortion matrices. Expansion and simplification of all ten equations are similar. As an example, $\langle z_{hh}z_{hh}^* \rangle$ can be expressed as

$$
\begin{aligned}
\langle z_{hh}z_{hh}^* \rangle &= \left\langle \left((S_{hh} + S_{hv}\delta_2 + S_{vh}\delta_3 + S_{vv}\delta_2\delta_3)(S_{hh} + S_{hv}\delta_2 + S_{vh}\delta_3 + S_{vv}\delta_2\delta_3)^* + N_{hh}N_{hh}^* \right\rangle \\
&= \langle S_{hh}S_{hh}^* \rangle + \langle S_{vv}S_{hh}^* \rangle \delta_2\delta_3 + \langle S_{hv}S_{hv}^* \rangle \delta_2\delta_2^* + \langle S_{vh}S_{hv}^* \rangle \delta_3\delta_2^* + \langle S_{hv}S_{vh}^* \rangle \delta_2\delta_3^* \\
&\quad + \langle S_{vh}S_{vh}^* \rangle \delta_3\delta_3^* + \langle S_{hh}S_{vv}^* \rangle \delta_2^*\delta_3^* + \langle S_{vv}S_{vv}^* \rangle \delta_2\delta_2^*\delta_3\delta_3^* + \langle N_{hh}N_{hh}^* \rangle ,
\end{aligned}
\tag{6.3}
$$

where the first line of the right-side terms was expanded and simplified to the following using reflection symmetry:

$$
\langle S_{hh}S_{hv}^* \rangle = \langle S_{hh}S_{vh}^* \rangle = 0 .
\tag{6.4}
$$

6.2.4 Covariance Matrix for the Three Scattering Components

Based on Freeman and Durden (1998), the covariance of forest data can be expressed by a summation of three covariances on volume, double-bounce, and surface scattering as expressed by Equations (6.5) to (6.8), assuming that the volume scatterings are fully depolarized and that the surface and double-bounce scatterings are partially depolarized;

$$
\langle \mathbf{S} \cdot \mathbf{S}^* \rangle_{total} = \langle \mathbf{S} \cdot \mathbf{S}^* \rangle_{volume} + \langle \mathbf{S} \cdot \mathbf{S}^* \rangle_{double} + \langle \mathbf{S} \cdot \mathbf{S}^* \rangle_{surface}
\tag{6.5}
$$

$$\langle \mathbf{S} \cdot \mathbf{S}^* \rangle_{volume} = \begin{bmatrix} 1 & 0 & 0 & \rho \\ 0 & \gamma & \gamma' & 0 \\ 0 & \gamma^* & \gamma & 0 \\ \rho^* & 0 & 0 & 1 \end{bmatrix} f_v, \left(0 \leq \gamma, |\gamma'|, |\rho| \leq 1 \right), \left(-\pi \leq \arg(\gamma'), \arg(\rho) \leq \pi \right) \quad (6.6)$$

$$\langle \mathbf{S} \cdot \mathbf{S}^* \rangle_{double} = \begin{bmatrix} |\mathbf{x}|^2 & 0 & 0 & \mathbf{x} \\ 0 & 0 & 0 & 0 \\ 0 & 0 & 0 & 0 \\ \mathbf{x} & 0 & 0 & 1 \end{bmatrix} f_d, \left(|\mathbf{x}| \geq 1.0 \right), \left(-\pi \leq \arg(\mathbf{x}) \leq \pi \right) \quad (6.7)$$

$$\langle \mathbf{S} \cdot \mathbf{S}^* \rangle_{surface} = \begin{bmatrix} |\mathbf{y}|^2 & 0 & 0 & \mathbf{y} \\ 0 & 0 & 0 & 0 \\ 0 & 0 & 0 & 0 \\ \mathbf{y} & 0 & 0 & 1 \end{bmatrix} f_s \quad (6.8)$$

where f_v is the volume scattering component within the total power; f_d is the double-bounce component of vertical polarization (VV); f_s is the surface scattering component for vertical polarization; ρ is $\langle S_{hh} S_{vv}^* \rangle / \langle S_{hh} S_{hh}^* \rangle$, γ is $\langle S_{hv} S_{hv}^* \rangle / \langle S_{hh} S_{hh}^* \rangle$; and γ' is $\langle S_{hv} S_{vh}^* \rangle / \langle S_{hh} S_{hh}^* \rangle$. Based on the reciprocity of the target, γ is equal to γ'; $\mathbf{x}$ is a polarized ratio (HH/VV) of the signal components that penetrate through the canopy and are double-bounced off the ground and the trunk.

A detailed description (Freeman et al. 1992; Freeman and Durden 1998) results in:

$$\mathbf{x} = e^{j2(\kappa_H - \kappa_V)} \left(R_{gH} R_{tH} / R_{gV} R_{tV} \right) \quad (6.9)$$

where κ is a complex number and represents any attenuation and phase change of the vertically and horizontally polarized waves as the signal propagates from the radar to the ground and back again; R_g is the Fresnel reflection coefficient of the ground, and R_t is the reflection coefficient of the trunks; and $\mathbf{y}$ is the polarized ratio (HH/VV) of the surface scattering components. The expression for $\mathbf{y}$ is obtained under the first-order statistics of the Bragg model, and it becomes a real value.

It should be noted that our decomposition model has two degrees of freedom for the volume covariance (γ, ρ) while the Freeman-Durden model sets a constant value of 1/3 for both γ and ρ. This variable parameter can allow investigation into how volume covariance contributes to the determination of the forest parameter. Volume scattering is highly depolarized and can be assumed as an azimuthal symmetry target (and thus, the VV and HH components are equal to 1.0). The terms in Equation (6.7) show a contribution from the double bounce.

6.2.5 Consideration of the Scattering Model for a Dense Forest

While the three-component covariance model is commonly used to express polarimetric backscattering from dense forest, the main part of the proposed PolCAL method is to explicitly separate the volume scattering part, which is uniform for any polarization, from the remaining parts: the double-bounce and surface scatterings, which may exhibit polarization dependency for signal scattering from forests. The double-bounce and surface scattering components have similar functional forms. In this PolCAL study, the purpose is not to separate the three components quantitatively, but to find a stable solution for the PDM. The surface scattering component is much smaller than the double-bounce component over dense forest (Freeman and Durden 1998). Thus, we can set an assumption for the magnitude order of the double-bounce and surface scatterings in Equation (6.10). This allows us to ignore f_s when dealing with the polarimetric backscatter from a dense forest.

$$f_s \ll f_d \quad (6.10)$$

On the other hand, non-forested regions illustrated by the brown area in Figure 6-2b induce a single existing component, f_s.

Thus, the covariance matrix can be expressed by the following equation, where we normalized all the terms by the HH component:

$$C = \left\langle \mathbf{S} \cdot \mathbf{S}^* \right\rangle / \left\langle S_{hh} \cdot S_{hh}^* \right\rangle$$

$$= \begin{pmatrix} f_v + |\mathbf{x}|^2 f_d & 0 & 0 & \rho f_v + \mathbf{x} f_d \\ 0 & \gamma f_v & \gamma f_v & 0 \\ 0 & \gamma f_v & \gamma f_v & 0 \\ \rho^* f_v + \mathbf{x}^* f_d & 0 & 0 & f_v + f_d \end{pmatrix} / \left(f_v + |\mathbf{x}|^2 f_d \right) \qquad (6.11)$$

$$= \begin{pmatrix} 1 & 0 & 0 & \dfrac{\rho f_v + \mathbf{x} f_d}{f_v + |\mathbf{x}|^2 f_d} \\ 0 & \dfrac{\gamma f_v}{f_v + |\mathbf{x}|^2 f_d} & \dfrac{\gamma f_v}{f_v + |\mathbf{x}|^2 f_d} & 0 \\ 0 & \dfrac{\gamma f_v}{f_v + |\mathbf{x}|^2 f_d} & \dfrac{\gamma f_v}{f_v + |\mathbf{x}|^2 f_d} & 0 \\ \dfrac{\rho^* f_v + \mathbf{x}^* f_d}{f_v + |\mathbf{x}|^2 f_d} & 0 & 0 & \dfrac{f_v + f_d}{f_v + |\mathbf{x}|^2 f_d} \end{pmatrix}$$

6.2.6 EXPRESSION OF THE COVARIANCE MATRIX COMPONENTS

Inserting Equations (6.6) and (6.7) into Equation (6.2), expanding all the terms, and applying the reflection symmetry of Equation (6.4), we have the following ten equations, which are grouped to the larger magnitude components in Equation (6.12) and the smaller magnitude components in Equation (6.13). Here, we do not ignore any higher order terms of cross talk multiplicative.

$$\left\langle z_{hh} z_{hh}^* \right\rangle = f_v + |\mathbf{x}|^2 f_d + D_1$$

$$D_1 = f_v \left\{ \rho \left(\delta_2 \delta_3 + \delta_2^* \delta_3^* \right) + \gamma \left(\delta_2 + \delta_3 \right) \left(\delta_2^* + \delta_3^* \right) + |\delta_2|^2 \cdot |\delta_3|^2 \right\}$$

$$\left\langle z_{vv} z_{hh}^* \right\rangle = f_1 f_2 \cdot \left(\rho f_v + \mathbf{x} f_d \right) + D_2$$

$$D_2 = f_v \left\{ \delta_1 \delta_4 + f_1 f_2 \delta_2^* \delta_3^* + \gamma \left(f_1 \delta_4 + f_2 \delta_1 \right) \left(\delta_2^* + \delta_3^* \right) + \rho \delta_1 \delta_4 \delta_2^* \delta_3^* \right\}$$

$$\left\langle z_{hv} z_{hv}^* \right\rangle = \gamma f_v |f_1|^2 + D_3$$

$$D_3 = f_v \left\{ |\delta_1|^2 + \left(\gamma |\delta_1|^2 + |f_1|^2 \right) |\delta_3|^2 + f_1 \delta_1^* \left(\rho \delta_3 + \gamma \delta_3^* \right) + f_1^* \delta_1 \left(\gamma \delta_3 + \rho \delta_3^* \right) \right\} \qquad (6.12)$$

$$\left\langle z_{vh} z_{hv}^* \right\rangle = \gamma f_v f_1^* f_2 + D_4$$

$$D_4 = f_v \left\{ \delta_1^* \delta_4 + \delta_2 \delta_3^* \left(\gamma \delta_1^* \delta_4 + f_1^* f_2 \right) + f_2 \delta_1^* \left(\rho \delta_2 + \gamma \delta_3^* \right) + f_1^* \delta_4 \left(\gamma \delta_2 + \rho \delta_3^* \right) \right\}$$

$$\left\langle z_{vh} z_{vh}^* \right\rangle = \gamma f_v |f_2|^2 + D_5$$

$$D_5 = f_v \left\{ |\delta_4|^2 + \left(\gamma |\delta_4|^2 + |f_2|^2 \right) |\delta_2|^2 + f_2 \delta_4^* \left(\rho \delta_2 + \gamma \delta_2^* \right) + f_2^* \delta_4 \left(\gamma \delta_2 + \rho \delta_2^* \right) \right\}$$

$$\left\langle z_{vv} z_{vv}^* \right\rangle = \left(f_v + f_d \right) |f_1|^2 |f_2|^2 + D_6$$

$$D_6 = f_v \left\{ |\delta_1|^2 |\delta_4|^2 + \gamma |f_1|^2 |\delta_4|^2 + \gamma |f_2|^2 |\delta_1|^2 + f_1 \delta_1^* \left(\rho f_2 \delta_4^* + \gamma f_2^* \delta_4 \right) + f_1^* \delta_1 \left(\gamma f_2 \delta_4^* + \rho f_2^* \delta_4 \right) \right\}$$

$$z_{hv}z_{hh}^* = \left(f_v + x^2 f_d\right)\delta_1 + \gamma f_v f_1 \delta_2^* + f_1\left(\rho f_v + \mathbf{x}f_d\right)\delta_3 + \gamma f_v f_1 \delta_3^* + \Delta_1$$

$$\Delta_1 = f_v\left\{\left(\gamma\delta_1 + f_1\delta_2^*\right)|\delta_3|^2 + \left(\gamma\delta_3 + \rho\delta_3^*\right)\delta_1\delta_2^*\right\}$$

$$z_{vh}z_{hh}^* = f_2\left(\rho f_v + \mathbf{x}f_d\right)\delta_2 + \gamma f_v f_2 \delta_2^* + \gamma f_1 f_2 \delta_3^* + \left(f_v + x^2 f_d\right)\delta_4 + \Delta_2$$

$$\Delta_2 = f_v\left\{\left(\gamma\delta_4 + f_2\delta_3^*\right)|\delta_2|^2 + \left(\gamma\delta_2 + \rho\delta_2^*\right)\delta_4\delta_3^*\right\}$$

$$z_{vv}z_{hv}^* = \gamma f_v f_2 f_1^* \delta_1 + f_1 f_2\left(\rho f_v + \mathbf{x}f_d\right)\delta_1^* + |f_1|^2 f_2\left(f_v + f_d\right)\delta_3^* + \gamma f_v |f_1|^2 \delta_4 + \Delta_3$$

$$\Delta_3 = f_v\left\{\left(\delta_4 + \gamma f_2\delta_3^*\right)|\delta_1|^2 + \left(\gamma f_1\delta_1^* + \rho f_1^*\delta_1\right)\delta_4\delta_3^*\right\}$$

$$z_{vv}z_{vh}^* = \gamma f_v |f_2|^2 \delta_1 + |f_2|^2 f_1\left(f_v + f_d\right)\delta_2^* + \gamma f_v f_1 f_2^*\delta_4 + f_1 f_2\left(\rho f_v + \mathbf{x}f_d\right)\delta_4^* + \Delta_4$$

$$\Delta_4 = f_v\left\{\left(\delta_1 + \gamma f_1\delta_2^*\right)|\delta_4|^2 + \left(\gamma f_2\delta_4^* + \rho f_2^*\delta_4\right)\delta_1\delta_2^*\right\}$$

(6.13)

These ten equations are the keys to this PolCAL method and have the following features:

1. These equations are valid for targets inducing volume and double-bounce components such as the dense forest in tropical regions (e.g., the Amazon).
2. Polarization dependence of signal penetration to the volume scattering target, which occurs at the L-band or lower frequencies, is expressed by $\mathbf{x}$.
3. Contributions from volume and double-bounce scatterings are expressed by f_v and f_d, respectively.

6.3 SOLUTIONS

To determine the ten unknowns (i.e., f_1, f_2, δ_1, δ_2, δ_3, δ_4, f_v, f_d, $\mathbf{x}$, and noise) from the aforementioned equations, we propose three individual processes, each of which deals with the different magnitude terms. They are mixed and iterated until convergence is obtained.

6.3.1 CHANNEL IMBALANCES AND OTHER FIRST-ORDER TERMS

We introduce the following abbreviations for convenience:

$$a = z_{hh}z_{hh}^*$$
$$\mathbf{b} = z_{vv}z_{hh}^*$$
$$c = z_{hv}z_{hv}^*$$
$$\mathbf{d} = z_{vh}z_{hv}^*$$
$$e = z_{vh}z_{vh}^*$$
$$f = z_{vv}z_{vv}^*$$

(6.14)

and

$$F_i = |f_i|$$
$$\theta_i = \arg\left(f_i\right), (i = 1, 2)$$

(6.15)

Here, bold type indicates the complex value, and normal type, the real value.

In Equation (6.1), there are two types of unknown parameters with regard to their magnitudes. For the larger quantities, channel imbalances, we have six equations in Equation (6.12). By subtracting the noise from a, c, e, and f and, leaving $u' = a - n_0$, $c' = c - n_0$, $e' = e - n_0$, $f' = f - n_0$ (see Section 6.3.3), we obtain the following six equations (two complex and four real):

$$f_v + |\mathbf{x}|^2 f_d = a' - D_1$$

$$F_1 F_2 \left(\rho f_v + f_d x \cos\theta_x + f_d x \sin\theta_x j \right) \left\{ \cos(\theta_1 + \theta_2) + j\sin(\theta_1 + \theta_2) \right\} = \mathbf{b} - D_2$$

$$\gamma f_v F_1^2 = c' - D_3$$

$$\gamma f_v f_2 f_1^* = \mathbf{d} - D_4 \qquad (6.16)$$

$$\gamma f_v F_2^2 = e' - D_5$$

$$\left(f_v + f_d \right) F_1^2 F_2^2 = f' - D_6$$

Here, we express $\mathbf{x}$ in the polar coordinate of $xe^{j\theta_x}$. There is a total of eight unknowns, that is, $\mathbf{x}(x,\theta_x), f_d, f_v, F_1, F_2, \theta_1, \theta_2$ in Equation (6.16). In Equation (6.16-3), the absolute values of Equation (6.16-4) and Equation (6.16-5) are cyclic, and these eight equations are not independent. Here, Equation (6.16-3) means the third equation in Equations (6.16). Thus, an additional equation is necessary. We find one more equation for the surface scattering target, which is out of the forested region but within the SAR measurement scheme, as follows:

$$F_1 F_2 \left\{ \cos(\theta_1 + \theta_2) + j\sin(\theta_1 + \theta_2) \right\} = \mathbf{b}_s \qquad (6.17)$$

Here, the suffix "s" stands for the surface scattering component. With some reference scatterers in the image, such as trihedral CRs or ideal surfaces, the eight unknowns can be analytically determined. At each iteration step, the measured covariance is renewed by its being subtracted by a newly calculated D_i (e.g., $a' - D_1 \rightarrow a''$). In order to reduce the complexity of the expression, we use the same notation (i.e., $a' - D_1 \rightarrow a'$).

From Equations (6.17) and (6.16-4), we have

$$\theta_2 = \frac{1}{2} \left(\arg(\mathbf{d}) + \theta_s \right)$$

$$\theta_1 = \frac{1}{2} \left(\theta_s - \arg(\mathbf{d}) \right) \qquad (6.18)$$

where θ_s is the phase difference of the surface scatterer between $\mathbf{S}_{s,hh}$ and $\mathbf{S}_{s,vv}$. Selecting the trihedral CR as the reference target gives $\theta_s = 0$. Equations (6.16-1), (6.16-2), and (6.16-6) become

$$f_v + |\mathbf{x}|^2 f_d = a' \qquad (6.19\text{-}1)$$

$$\left(\rho + \frac{f_d}{f_v} x \cos\theta_x + \frac{f_d}{f_v} x \sin\theta_x j \right) e^{j\theta_s} = \frac{\mathbf{b}}{g} \qquad (6.19\text{-}2)$$

$$\left(f_v + f_d \right) = \frac{f_v^2}{g^2} f' \qquad (6.19\text{-}3)$$

$$g \equiv \frac{\sqrt[3]{c'de'}}{\gamma} \qquad (6.19\text{-}4)$$

Here, $d = |\mathbf{d}|$. Although these equations have two additional parameters (i.e., γ and ρ), the proposed method cannot estimate them. When they are given, we can solve for the eight unknowns.

The phase component of Equation (6.19-2) determines θ_x and the arrangement of Equation (6.19-1, −2, and −3) gives a relationship between f_v and f_d as follows:

$$\theta_x = \tan^{-1}\left\{ \frac{I_m\left(\dfrac{\mathbf{b}}{g}e^{-j\theta s} - \rho\right)}{R_e\left(\dfrac{\mathbf{b}}{g}e^{-j\theta s} - \rho\right)} \right\} \tag{6.20}$$

$$x = h\frac{f_v}{f_d} \tag{6.21}$$

and

$$h = \sqrt{\left\{R_e\left(\frac{\mathbf{b}}{g}e^{-j\theta s} - \rho\right)\right\}^2 + \left\{I_m\left(\frac{\mathbf{b}}{g}e^{-j\theta s} - \rho\right)\right\}^2} . \tag{6.22}$$

From Equation (6.19-3) we have

$$f_d = f_v \cdot \left(\frac{f'}{g^2}f_v - 1\right). \tag{6.23}$$

By combining Equations (6.19-1), (6.19-3), and (6.21), the following quadratic equation in f_v is retrieved:

$$f_v \cdot \left\{ \frac{f'}{g^2}f_v^2 - \left(\frac{f'}{g^2}a' + 1 - h^2\right)f_v + a'\right\} = 0. \tag{6.24}$$

Equation (6.24) has real solutions when

$$\left(\frac{f'}{g^2}a' + 1 - h^2\right)^2 - 4a'\frac{f'}{g^2} \geq 0. \tag{6.25}$$

The validity of Equation (6.25) has been evaluated using real PALSAR data, and it appears that PALSAR data can satisfy the previous condition for f_v (see Section 6.4). The smaller root of Equation (6.24) is retained to ensure that $f_v + f_d < 1, 0 \leq f_d, f_v \leq 1$.

$$f_v = \frac{\dfrac{f'}{g^2}a' + 1 - h^2 - \sqrt{\left(\dfrac{f'}{g^2}a' + 1 - h^2\right)^2 - 4a'\dfrac{f'}{g^2}}}{2\dfrac{f'}{g^2}} \tag{6.26}$$

Finally, $\mathbf{x}$, f_d, F_1, and F_2 are obtained from the forest parameters by

$$F_1 = \sqrt{\gamma\frac{f'}{e'}\frac{f_v}{f_v + f_d}}$$

$$F_2 = \sqrt{\gamma\frac{f'}{c'}\frac{f_v}{f_v + f_d}} . \tag{6.27}$$

When the CR is available, the following expression provides the amplitude of channel imbalances:

$$F_1 = \sqrt{\frac{c'}{d'}} f_{CR}^{1/4}$$

$$F_2 = \sqrt{\frac{e'}{d'}} f_{CR}^{1/4}$$

(6.28)

where f_{CR} is a covariance component of f' measured for the CR response.

6.3.2 CROSS-TALK OF DISTORTION MATRICES

In this section, the four complex cross talk terms are expressed by the real unknowns (A, B, C, D, E, F, G, and H) as follows:

$$\delta_1 = A + Bj$$
$$\delta_2 = C + Dj$$
$$\delta_3 = E + Fj$$
$$\delta_4 = G + Hj$$

(6.29)

Rearrangement of Equation (6.13) gives some linear expressions for the four cross talk terms, which can be summarized by a matrix formulation as:

$$\mathbf{MY} = \mathbf{N},$$ (6.30)

where

$$\mathbf{M} = \begin{pmatrix} m_{1r} & -m_{1i} & m_{2r} & m_{2i} & m_{3r}+m_{4r} & -m_{3i}+m_{4i} & 0 & 0 \\ m_{1i} & m_{1r} & m_{2i} & -m_{2r} & m_{3i}+m_{4i} & m_{3r}-m_{4r} & 0 & 0 \\ 0 & 0 & m_{5r}+m_{6r} & -m_{5i}+m_{6i} & m_{7r} & m_{7i} & m_{8r} & -m_{8i} \\ 0 & 0 & m_{5i}+m_{6i} & m_{5r}-m_{6r} & m_{7i} & -m_{7r} & m_{8i} & m_{8r} \\ m_{9r}+m_{10r} & -m_{9i}+m_{10i} & 0 & 0 & m_{11r} & m_{11i} & m_{12r} & -m_{12i} \\ m_{9i}+m_{10i} & m_{9r}-m_{10r} & 0 & 0 & m_{11i} & -m_{11r} & m_{12i} & m_{12r} \\ m_{13r} & -m_{13i} & m_{14r} & m_{14i} & 0 & 0 & m_{15r}+m_{16r} & -m_{15i}+m_{16i} \\ m_{13i} & m_{13r} & m_{14i} & -m_{14r} & 0 & 0 & m_{15i}+m_{16i} & m_{15r}-m_{16r} \end{pmatrix}$$

(6.31)

$$\mathbf{Y} = \begin{pmatrix} A \\ B \\ C \\ D \\ E \\ F \\ G \\ H \end{pmatrix} \qquad \mathbf{N} = \begin{pmatrix} \mathrm{Re}\left(\langle z_{hv} z_{hh}^* \rangle - \Delta_1\right) \\ \mathrm{Im}\left(\langle z_{hv} z_{hh}^* \rangle - \Delta_1\right) \\ \mathrm{Re}\left(\langle z_{vh} z_{hh}^* \rangle - \Delta_2\right) \\ \mathrm{Im}\left(\langle z_{vh} z_{hh}^* \rangle - \Delta_2\right) \\ \mathrm{Re}\left(\langle z_{vv} z_{hv}^* \rangle - \Delta_3\right) \\ \mathrm{Im}\left(\langle z_{vv} z_{hv}^* \rangle - \Delta_3\right) \\ \mathrm{Re}\left(\langle z_{vv} z_{vh}^* \rangle - \Delta_4\right) \\ \mathrm{Im}\left(\langle z_{vv} z_{vh}^* \rangle - \Delta_4\right) \end{pmatrix}$$

(6.32)

with

$$m_1 = \left(f_v + x^2 f_d\right) \quad m_2 = \gamma f_v f_1 \qquad\qquad m_3 = f_1\left(\rho f_v + x f_d\right) \qquad m_4 = \gamma f_v f_1$$

$$m_5 = f_2\left(\rho f_v + x f_d\right) \quad m_6 = \gamma f_v f_2 \qquad\qquad m_7 = \gamma f_v f_2 \qquad\qquad m_8 = \left(f_v + x^2 f_d\right)$$

$$m_9 = \gamma f_v f_2 f_1^* \qquad m_{10} = f_1 f_2\left(\rho f_v + x f_d\right) \quad m_{11} = |f_1|^2 f_2\left(f_v + f_d\right) \quad m_{12} = \gamma f_v |f_1|^2$$

$$m_{13} = \gamma f_v |f_2|^2 \qquad m_{14} = |f_2|^2 f_1\left(f_v + f_d\right) \quad m_{15} = \gamma f_v f_1 f_2^* \qquad m_{16} = f_1 f_2\left(\rho f_v + x f_d\right)$$

$$(6.33)$$

In Equation (6.31), the suffixes "r" and "i" stand for the real and imaginary parts, respectively. Finally, the aforementioned equation can be solved by using a linear algebraic method.

$$\mathbf{Y} = \mathbf{M}^{-1}\mathbf{N} \tag{6.34}$$

In this way, the cross talk terms are obtained as the solutions of four linear equations, and the higher order terms can be corrected in Equation (6.32) by iterative processes.

6.3.3 Noise Estimation

The covariance of Cpol-Cpol and Xpol-Xpol include some noise components, which is not the case for the others. Thus, Equation (6.2) has the following expressions for noise:

$$\left\langle \left(S_{hh} + N_{hh}\right)\left(S_{hh} + N_{hh}\right)^*\right\rangle = \left\langle S_{hh}S_{hh}^* + N_{hh}N_{hh}^*\right\rangle = \left\langle S_{hh}S_{hh}^*\right\rangle + n_0$$

$$\left\langle S_{hv}S_{hv}^* + N_{hv}N_{hv}^*\right\rangle = \left\langle S_{hv}S_{hv}^*\right\rangle + n_0$$

$$\left\langle S_{vh}S_{vh}^* + N_{vh}N_{vh}^*\right\rangle = \left\langle S_{vh}S_{vh}^*\right\rangle + n_0$$

$$\left\langle S_{vv}S_{vv}^* + N_{vv}N_{vv}^*\right\rangle = \left\langle S_{vv}S_{vv}^*\right\rangle + n_0 \tag{6.35}$$

$$\left\langle S_{hh}S_{vv}^* + N_{hh}N_{vv}^*\right\rangle = \left\langle S_{hh}S_{hh}^*\right\rangle$$

$$\left\langle S_{hv}S_{vh}^* + N_{hv}N_{vh}^*\right\rangle = \left\langle S_{hv}S_{vh}^*\right\rangle$$

$$n_0 \equiv \left\langle N_{hh}N_{hh}^*\right\rangle = \left\langle N_{hv}N_{hv}^*\right\rangle = \left\langle N_{vh}N_{vh}^*\right\rangle = \left\langle N_{vv}N_{vv}^*\right\rangle$$

Here, n_0 is defined before Equation (6.16). Using the abbreviation for the aforementioned equations, we have the following:

$$\left(\gamma f_v + n_0\right) F_1^2 = c - D_3 \left(\equiv c''\right)$$

$$\left(\gamma f_v + n_0\right) F_2^2 = e - D_5 \left(\equiv e''\right) \tag{6.36}$$

$$\gamma f_v F_1 F_2 = d - D_4 \left(\equiv d''\right)$$

From them, we have

$$\left(c'' - n_0 F_1^2\right)\left(e'' - n_0 F_2^2\right) = \left(d''\right)^2. \tag{6.37}$$

We have noise level n_0 as a solution of the squared equation

$$n_0 = \frac{e'' F_1^2 + c'' F_2^2 - \sqrt{\left(e'' F_1^2 + c'' F_2^2\right)^2 - 4 F_1^2 F_2^2 \left(c'' e'' - d''^2\right)}}{2 F_1^2 F_2^2}. \tag{6.38}$$

Finally, the noise can be determined iteratively.

6.3.4 ITERATIONS

The previous sections introduced the solutions of Equations (6.12) and (6.13), which are Equations (6.16) and (6.30), respectively. The higher order multiplicative terms of the cross talks in Equations (6.12) and (6.13) can be corrected iteratively until the channel imbalances and cross talks converge.

6.3.5 FEATURES OF THE PROPOSED METHOD

The features of the proposed method are summarized as follows:

1. Out of the two estimations of the channel imbalance amplitude, the first method— Equation (6.27)—uses forest decomposition and is affected by volume/double-bounce decomposition accuracy and the Xpol covariance (γ). Accuracy evaluation is necessary before it can be used as a second opportunity. The second method of Equation (6.28) provides accurate results relying on the CR.
2. The phase of the channel imbalances (c.f., Equation [6.18]) depends only on the accuracy of the VV-HH phase difference, which the reference target can preserve. The artificial target (e.g., CR) preserves the phase difference accurately and can be used as a reference target. However, a natural target needs to be fully evaluated before being used as a reference target.
3. The accuracy of the cross talk estimation from Equation (6.34) possibly is affected by volume/double-bounce components and the Xpol covariance. Evaluation of this accuracy issue is necessary.

6.4 EXPERIMENT DESCRIPTION

The Japan Aerospace Exploration Agency (JAXA) launched the Advanced Land Observing Satellite (ALOS) on January 24, 2006, to an altitude of 691.5 km in a sun-synchronous orbit with a 46-day repeat cycle, carrying two high-resolution optical sensors and a Phased-Array L-Band Synthetic-Aperture Radar (PALSAR). After the initial calibration, ALOS started operation on October 24, 2006 (Shimada et al. 2009). PALSAR has the following features: (I) Use of L-band frequency with superior signal penetration in forested areas and the advantage of higher interferometric coherence, (II) a wide bandwidth of 28 MHz (higher spatial resolution), (III) 80 TRMs for quick off-nadir angle changes, and (IV) operational spaceborne full polarimetry. PALSAR has five operation modes: fine beam single (FBS), a single-polarization, high-resolution strip with a 70-km swath; fine beam dual (FBD), a dual-polarization reception (HH + VH, where VH indicates transmitted in horizontal polarization and received in vertical polarization) enabling deforestation monitoring; ScanSAR, with a 350-km swath for quick deforestation and sea-ice monitoring; polarimetry for scattering mechanism clarification; and a reduced-resolution strip mode. A dual-frequency GPS receiver, yaw steering, and an orbital tube less than 500 m in diameter maintain the quality of PALSAR images and interferometric coherence. The total system meets the requirement of accurate deformation detection.

We applied the proposed PolCAL method to calibrate and evaluate the PALSAR polarimetric data using the time series data sets acquired in the dense rainforest of Rio Branco, Brazil, located at 9.7599 S latitude and 68.073 W longitude, over 2 years and 5 months from July 20, 2006, to December 11, 2008. This test site was selected because the FRA (Wright et al. 2003; Meyer et al. 2006) is small or almost negligible in these latitudes especially in the ascending orbit, and the calibration unknowns could only be for the distortion matrix. The site was established through collaboration among JAXA, the Alaska Satellite Facility (ASF), and the Brazilian Geophysical Agency, *Instituto Brasileiro de Geografia e Estatística* (IBGE), as a joint calibration of the PALSAR. Two 2.5-m, leaf-sized trihedral CRs were deployed, one directed toward the ALOS descending

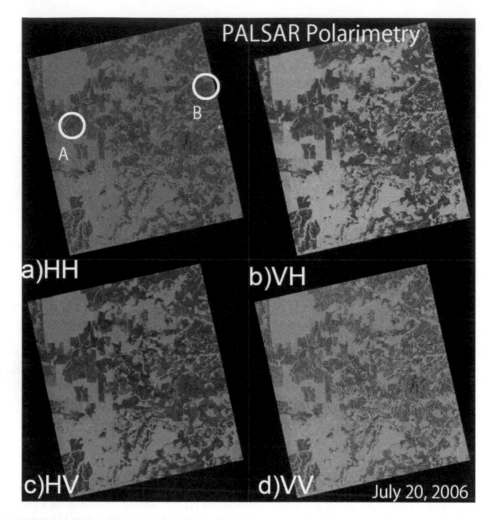

FIGURE 6-3   Uncalibrated PALSAR polarimetric data observed for Rio Branco, Brazil, on July 20, 2006, in ascending orbit. The images are (a) HH, (b) VH, (c) HV, and (d) VV. All the images are geo-coded on an equal latitude and longitude coordinate system. Here, light gray indicates dense forest and the dark areas are surface scattering areas. The area extends 35 km in the cross-track direction and 40 km in the along-track direction.

orbits (orbit number is 431) and the other for ascending orbits (orbit number is 090). Among the PALSAR polarimetric modes, which serve in a total of 12 off-nadir angles, an off-nadir angle of 21.5 degrees was selected.

Figure 6-3 presents the geo-coded and uncalibrated PLR21.5 image. In this image, two types of surface scatterers were selected: the trihedral CRs in circles A and B with their close-up in Figure 6-4, and the rough surface area that is shown as darker. Table 6-2 lists the polarimetric PALSAR time series data. A total of 26 data sets is available.

6.5   VALIDATION OF THE ASSUMPTIONS AND THE STABILITY OF THE DECOMPOSITION MODEL

6.5.1   EVALUATION OF COVARIANCE MATRIX MEASURED FOR THE FOREST AND THE SURFACE

In this subsection, the characteristics of the covariance matrix are measured using PALSAR for forested and non-forested areas. To increase measurement credibility, the distortion matrix obtained

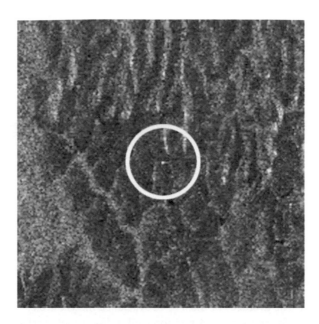

FIGURE 6-4 Enlarged parts of the targets in the Rio Branco test site at a CR site with the centered bright point.

by Quegan's PolCAL method (not the proposed method) is used to calibrate all 26 PALSAR data. Figure 6-5a plots the representative component of a 4×4 covariance matrix for all the data (i.e., C_{HHHV}, C_{HHVH}, C_{HHVV}, C_{HVHV}, C_{HVVH}, C_{HVVV}, C_{VHVH}, C_{VHVV}, and C_{VVVV}), and Figure 6-5b shows the phase component of the representative two components (i.e., C_{HHVV} and C_{HVVH}). Phases of C_{HVVH} exist around zero degrees, and those for C_{HHVV} are slightly offset at 2.5 degrees. Both figures are drawn as a function of the time after the ALOS launch. A slight annual variation of the amplitudes

TABLE 6-2
PALSAR Polarimetry Data Acquired for Rio Branco, Brazil

No.	Acquisition Date	A/D	Path-No.	No.	Acquisition Date	A/D	Path-No.
1	July 20, 2006	A	090	1	July 21, 2006	D	431
2	Sept. 4, 2006	A	090	2	Sept. 5, 2006	D	431
3	Oct. 20, 2006	A	090	3	Oct. 21, 2006	D	431
4	March 7, 2007	A	090	4	Jan. 21, 2007	D	431
5	April 22, 2007	A	090	5	July 24, 2007	D	431
6	Sept. 7, 2007	A	090	6	Sept. 8, 2007	D	431
7	Oct. 23, 2007	A	090	7	Oct. 24, 2007	D	431
8	Dec. 8, 2007	A	090	8	Dec. 9, 2007	D	431
9	Jan. 23, 2008	A	090	9	Jan. 24, 2008	D	431
10	Oct. 25, 2008	A	090	10	March 10, 2008	D	431
				11	April 25, 2008	D	431
				12	June 10, 2008	D	431
				13	July 26, 2008	D	431
				14	Sept. 10, 2008	D	431
				15	Oct. 26, 2008	D	431
				16	Dec. 11, 2008	D	431

Note: A = ascending and D = descending.

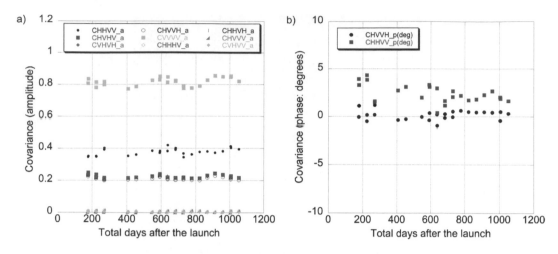

FIGURE 6-5 Stability of the representative components of the covariance matrix, (a) the amplitude and (b) phase difference. Here, the horizontal axis is the total days after the launch of the ALOS.

appears in Figure 6-5a. The averages of the 4×4 covariance matrix are calculated for the screened forested areas and also for the screened non-forested areas (rough surfaces). They are listed in Tables 6-3 and 6-4, respectively.

From these figures and tables (especially Table 6-3 for the forested area), we observe:

1. Cpol-Xpol covariances (i.e., C_{HHHV}, C_{HHVH}, C_{VVHV}, and C_{VVVH}) are decorrelated.
2. Xpol-Xpol covariances (i.e., C_{HVHV} and C_{VHVH}) are equal to 0.222 and smaller than 1/3, the theoretical γ of the Freeman-Durden model, and this means that either of γ less than 1/3 or fv is less than 1.0 or both are valid.
3. C_{HVVH} is almost equal to C_{HVHV}, zero-phase, and reciprocity is valid.
4. C_{HHVV} is slightly larger than 1/3.
5. C_{VVVV} is equal to 0.813 and slightly smaller than C_{HHHH} (1.0).

TABLE 6-3
Covariance Matrix of the Forested Area in the Amazon

Polarization	HH	HV	VH	VV
HH	(1.0, 0)	(0.00452, 17.334)	(0.00643, −54.943)	(0.377, 2.491)
HV	(0.00452, −17.334)	(0.222, 0)	(0.211, 0.0611)	(0.00600, −3.380)
VH	(0.00643, 54.943)	(0.211, −0.0611)	(0.222, 0)	(0.00620, 59.225)
VV	(0.377, −2.491)	(0.00600, 3.380)	(0.00620, −59.225)	(0.813, 0)

Note: The two values in the brackets are the amplitude of the covariance and the phase of the covariance in degrees.

TABLE 6-4
Covariance Matrix of the Non-Forested Area in the Amazon

Polarization	HH	HV	VH	VV
HH	(1.0, 0)	(0.00643, 8.6685)	(0.00703, −42.11)	(0.55433, −9.4777)
HV	(0.00643, −8.6685)	(0.12747, 0)	(0.096824, −0.70376)	(0.00766, −3.380)
VH	(0.00703, 42.11)	(0.096824, 0.70376)	(0.13175, 0)	(0.007155, 28.561)
VV	(0.55433, 9.4777)	(0.00766, 3.380)	(0.007155, −28.561)	(0.85668, 0)

Concerning the observations (2) and (5), this can be due to the fact that the double bounce occurs at the bottom of the volume and the trunks. Since fv is real in Equation (6.11), the volume scattering target maintains that the reciprocity of $\gamma = \gamma'$ and γ is real, and either or both of the double bounce or ρ has little imaginary value (real values).

With regard to the covariance of the non-forested area, we find some features (C_{HVHV}, C_{VHVH}, and C_{HVVH}) are almost the same and are considerably smaller than C_{HHHH}; that C_{VVVV} is smaller than 1.0; and that C_{HHVV} is larger than C_{HVHV}, which is different from the forested area. C_{HHHV}, C_{HHVH}, C_{HVVV}, and C_{VHVV} are almost equal to 0.0, although they are slightly larger than that for the forested area.

6.5.2 Reflection Symmetry of the Forest Data

There have been several theoretical investigations on the properties of the symmetrical scattering target (Nghiem et al. 1992; Borgeaud et al. 1987; Yueh et al. 1994). Since the symmetric assumption on the scattering target simplifies the model-based analysis, it is often adopted. Similar to the other methods, the proposed method adopts reflection symmetry for the dense Amazon Rainforest. The question is "Does the dense Amazon Rainforest maintain the reflection symmetry?" As introduced by Yueh et al. (1994), perfect theoretical proof on this question would be quite difficult. Here, we evaluate the forest data by simply averaging the Cpol-Xpol with different averaging numbers and measuring it if it converges to zero. We select the PALSAR data of July 20, 2006, shown in Figures 6-3 and 6-6, and Figure 6-7a and b shows the relationship between the averaging number and the covariances for four Cpol-Xpol covariances (i.e., correlation coefficient) over the screened forest and the surface-like scattering targets, respectively. From these figures, for the averaging number larger than 10,000, the four components become smaller than 0.01, and the outcome of the reflection symmetry is valid.

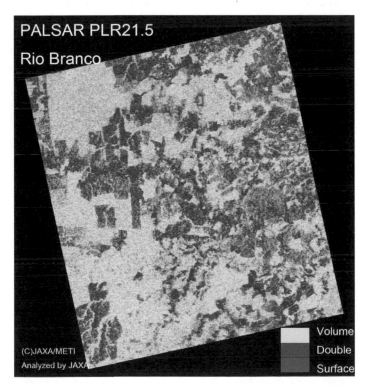

FIGURE 6-6   Freeman-Durden decomposition results. Green: Volume scattering component. Red: Double-bounce component. Blue: Surface scattering component.

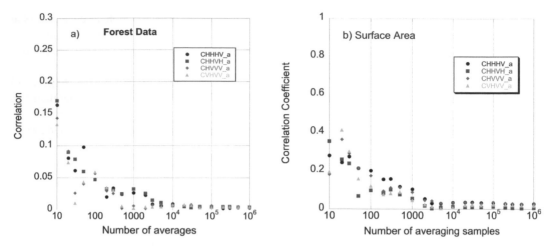

FIGURE 6-7 Dependences of the four covariance components on the number of samples are shown for (a) the forested region and (b) the surface scattering region.

6.5.3 APPLICATION OF FREEMAN-DURDEN DECOMPOSITION

The Freeman-Durden decomposition method is applied to the PALSAR data to check if the assumption that dense forest has a smaller surface scattering component than a double-bounce component is valid. Figure 6-6 depicts the three component decomposition results. Figure 6-8 plots the incidence angle dependence of the three components calculated for the same area where the data are averaged in the azimuth direction to obtain the incidence angle dependence. The representative values for the three components are 90.54% for volume, 8.41% for double bounce, and 1.85% for surface scattering. The surface scattering components are much smaller than the volume scattering components, being less than a quarter of the double-bounce scattering components. Thus, the assumption of Equation (6.10) is valid in this forest.

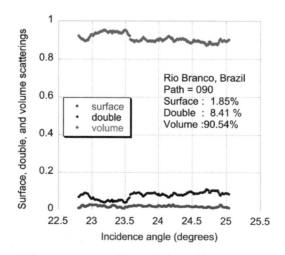

FIGURE 6-8 Incidence angle dependence of the three components derived from the Freeman-Durden method, where the blue circle is surface scattering, the red circle is double-bounce scattering, and the green diamond is volume scattering.

TABLE 6-5
Representative σ^0 for the Forest and the Surface Target

Area	σ^0-HH (dB)	σ^0-HV (dB)	σ^0-VV (dB)	Area-ratio
Forest	−7.15	−13.59	−8.04	0.76
Surface	−10.58	−19.68	−11.43	0.24

6.5.4 IGNORANCE OF SURFACE SCATTERING IN HV UNDERLYING THE VOLUME AND ITS EFFECT: EVALUATION

The Freeman-Durden model assumed the surface scattering in HV and VH are zero. The proposed PolCAL method also ignores all the surface scatterings. In this subsection, assessments are carried out to estimate how this aspect induces errors in the covariance measurement and how it affects the polarimetric parameter estimation. Using the 26 available Amazon scenes, we measured the representative σ^0 of HH, HV, and VV for the forest and surface regions as shown in Table 6-5. Assigning −6 dB as a criterion for the span (i.e., $\sigma_{HH}^0 + 2\sigma_{HV}^0 + \sigma_{VV}^0$), the forested and surface regions are classified. Here, the area-ratio "Forest" is the ratio of number of pixels classified as "Forest" to total number of pixels in the image.

Assuming that σ^0 of the contaminated signal is composed of $\alpha\%$ of the surface scattering and $100-\alpha\%$ of the forest scattering, dependence of C_{HVHV}/C_{HHHH} on α can be calculated as follows (Table 6-6): C_{HVHV}/C_{HHHH} is shifted only by 0.0024 if the surface scattering component varies from 0 (as the Freeman-Durden model does) to 5%, and the distortion matrix (i.e., cross talks) may not be estimated differently (refer to Section 6.5.5).

6.5.5 STABILITY OF THE PROPOSED METHOD ON SOLUTION EXISTENCE

The proposed method relies on the incoherent decomposition model and the covariance measurement to determine the distortion matrix. While several decomposition models were developed (i.e., Freeman-Durden, Freeman-2, and Neumann), their differences appear at the volume covariance in Equation (6.6), that is, γ and ρ. This aspect may impact the retrieved distortion matrices. The Freeman-Durden model has 1/3 for ρ and γ; Freeman-2 is adapted to the measurement; and Neumann has two parameters for the expression of the forest structures, that is, the orientation randomness (τ) and the particle shape (δ), both of which can be converted to ρ and γ. The incoherent decomposition method derives the scattering contribution from the target quantitatively and is often used for interpretation and classification of the PolSAR data. However, the Freeman-Durden model sometimes overestimates the volume scattering component. Thus, the proposed method may be affected by this issue. In this subsection, the stability of the proposed method is investigated for a large range of the volume covariances.

TABLE 6-6
Simulated Variation of Covariance versus Surface Contamination in the Forested Region

α	C_{HVHV}/C_{HHHH}	Ratio	Difference
0%	0.2270	100	0.0
1%	0.2265	99.78%	−0.0005
5%	0.2246	98.94%	−0.0024

Equation (6.25) describes the necessary condition that must be satisfied for the proposed method to retrieve meaningful solutions for the volume scattering component f_v as rewritten as follows:

$$F(\gamma, \rho) \geq 0. \tag{6.39}$$

$$F(\gamma, \rho) \equiv \left(\frac{f'}{g^2} a' + 1 - h^2 \right)^2 - 4a' \frac{f'}{g^2}. \tag{6.40}$$

Here, $F(\gamma, \rho)$ is defined so that we can investigate the dependence of the volume covariance on F. Beyond the fact that F depends on γ and ρ through the variables h and g (see Equations [6.19-4] and [6.22]), F also depends on the variation of the measurements (i.e., g', f', a', b', c', d'). We will discuss this measurement dependency in the latter part of this section. The equality condition of Equation (6.39) gives the borders for solution regions, and γ is a two-value function of ρ as follows:

$$\gamma = \frac{\sqrt[3]{c'd'e'}}{\sqrt{a'f'}} \left\{ 1 \pm \sqrt{ \left\{ R_e \left(\frac{b\gamma}{\sqrt[3]{c'd'e'}} e^{-j\theta_s} - \rho \right) \right\}^2 + \left\{ I_m \left(\frac{b\gamma}{\sqrt[3]{c'd'e'}} e^{-j\theta_s} - \rho \right) \right\}^2 } \right\}. \tag{6.41}$$

Although the regions with positive F (we call them "positive F regions") are candidates for successful decomposition, the following conditions limit the solution regions:

$$\begin{aligned} f_v + f_d &< 1 \\ 0 \leq f_v &\leq 1 \\ 0 \leq f_d &\leq 1 \end{aligned} \tag{6.42}$$

Using the Amazon data set of July 20, 2006, the stability of the proposed model is evaluated and the possible solution regions provided by the proposed method are calculated for γ and ρ ranging from 0.0 to 0.5. In addition to this, the γ and ρ values are computed for the four different decomposition models as listed in Table 6-7. The Neumann model can use a range of volume covariances, and the statistical evaluation of the Amazon data determines the representative values (refer to Appendix 6A-1). One empirical volume covariance is determined by applying the least square method to the Amazon data (refer to Appendix 6A-2).

Figure 6-9a shows a perspective view of F values depending on the γ and ρ. It shows that there are two positive F regions. In parallel, Figure 6-9b shows the distribution of f_v depending on γ–ρ. Whereas one of two positive F regions has f_v smaller than 1.0, which is correct, the other region is incorrect ($f_v > 1.0$). Figure 6-9c shows the ρ–γ dependence of the solution area, where REGION-1

TABLE 6-7
Representative γ and ρ for Four Case Studies

Model	γ	ρ
Freeman-Durden	1/3	1/3
Freeman-2	0.3282	0.3436
Neumann	0.3135	0.3495
Empirical parameter	0.295	0.325

Note: Neumann model can use a value range for the volume covariance. Here, we determined these volume covariances analyzing the 26 Amazon data assuming $fv = 0.9$. The Freeman-2 parameter was calculated referring to Freeman (2007).

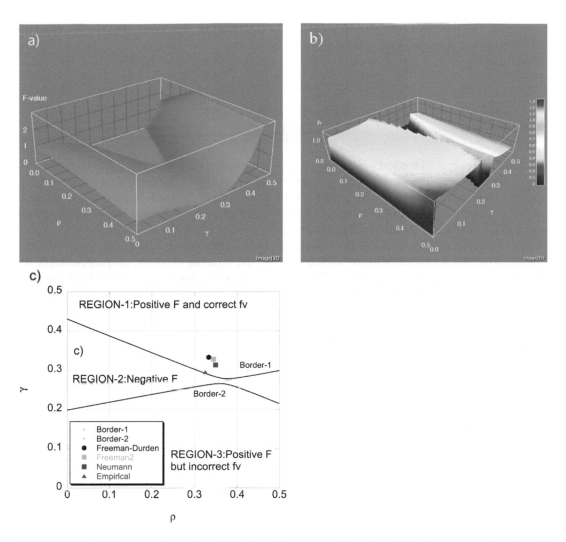

FIGURE 6-9 (a) A perspective view of F-values, (b) a perspective view of the forest parameter, fv, and (c) descriptions of the three regions. The regions depend on the positiveness of F and the value range of fv. REGION-1 produces the correct forest parameters. In this region, Freeman-Durden, Freeman-2, Neumann, and the empirical parameters are included.

has positive F values and a correct forest parameter (i.e., $0 \leq f_v < 1.0$), REGION-2 has negative F and no solution, and REGION-3 has positive F values but incorrect f_v larger than 1.0. In Figure 6-9c, the four points derived from the decomposition models and the empirical methodology are indicated. They confirm that all the models are in the REGION-1 and succeed the decomposition of the forest parameters. There are two borders, Border-1 and Border-2. The first can be obtained by the plus sign in Equation (6.41) and Border-2 can be obtained by the minus sign in Equation (6.41). It is noted that REGION-2 has very little negative value after ρ is larger than 0.3.

Next, the variations of the covariance and the F value are evaluated depending on the averaging number on the Amazon data of July 20, 2006, for a given ρ–γ pair. Here, the Freeman-Durden model is selected with $\rho = \gamma = 1/3$. Both F value and fv converge to constant values after the averaging number reaches 10,000. This suggests that an averaging number should be selected that is greater than 10,000 for the parameter estimation in this method (Shimada et al. 2009).

Finally, the stability of the proposed method for different γ and ρ parameters ranging from 0.3 to 0.4 mentions that the cross talk values of δ_1, δ_2, δ_3, and δ_4 are estimated between −30 dB and approximately −40 dB and appear to be relatively stable.

6.6 POLARIMETRIC CALIBRATION RESULTS

6.6.1 POLARIMETRIC CALIBRATION PROCEDURE

Although the proposed method is valid only for a combination of fully distributed target areas, which are dominated by volume scattering, the double-bounce component (the green area in Figure 6-2b), and the surface scattering component (the wood-colored area in Figure 6-2b), recent deforestation in the Amazon makes it difficult to select a fully forested area (Shimada et al. 2009). To effectively apply the SAR data to the proposed method, area filtering is necessary to extract the volume scattering. We selected a forested area such that the σ^0 for the HH channel is larger than −10 dB and a rough surface region such that the σ^0 is lower than −13 dB. After this process, the proposed PolCAL method is applied to the screened data. Figure 6-10 presents a flowchart of the PolCAL method. Selection of the surface scattering target (reference target) is essential to determine θ_s, which is the angular difference of the surface scatterer between HH and VV polarizations. Here, we prepare three targets for the surface reference scatterers: the trihedral CR, the rough surface with the σ^0 lower than −13 dB, and the forested region with the σ^0 larger than −10 dB. Although the clear-cut region is one possibility, it is easily affected by surface roughness and the unknown difference of the backscattering properties in HH and VV polarizations. However, the theoretical expression for their difference, given later, also describes the unknown but not-so-small errors in θ_s. Thus, the trihedral CR may be the only reference for PolCAL.

6.6.2 COMPARATIVE STUDY FOR SELECTING THE SURFACE SCATTERING TARGET

To statistically evaluate the proposed method and its variability depending on the reference surface scatterer, we prepared the following seven case studies as summarized in Table 6-8. Methods-1 through 1''', Method-2, and Method-3 examine the proposed PolCAL method. Methods-1 through 1''' differ only in volume covariances, with Method-4 being the reference.

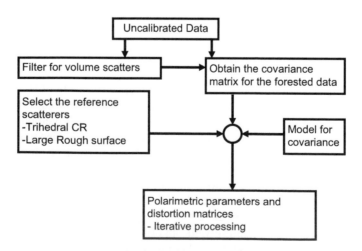

FIGURE 6-10 Flow chart of the PolCAL method proposed in this chapter.

TABLE 6-8
Definition of the Case Studies in This Research

Name	Description
Method-1	Proposed method + trihedral CR as the surface scatterer where the 7 pixels in azimuth and 7 pixels in range are averaged, noise subtracted, and CR is used for channel imbalance determination. Here, $\gamma = 1/3$, $\rho = 1/3$ (Freeman-Durden).
Method-1'	Proposed method + trihedral CR as the surface scatterer where the 7 pixels in azimuth and 7 pixels in range are averaged, noise subtracted, and CR is used for channel imbalance determination. Here, $\gamma = 0.3282$, $\rho = 0.34362$ (Freeman-2).
Method-1"	Proposed method + trihedral CR as the surface scatterer where the 7 pixels in azimuth and 7 pixels in range are averaged, noise subtracted, and CR is used for channel imbalance determination. Here, $\gamma = 0.3135$, $\rho = 0.3495$ (Neumann parameter).
Method-1'''	Proposed method + trihedral CR as the surface scatterer where the 7 pixels in azimuth and 7 pixels in range are averaged, noise subtracted, and CR is used for channel imbalance determination. Here, $\gamma = 0.295$, $\rho = 0.325$ (empirical parameter).
Method-2	Proposed method + larger surface regions as the surface scatterers, where all the pixels with σ^0 of lower than -13 dB are averaged, noise is subtracted, and the surface is used for channel imbalance determination. Here, $\gamma = 0.295$, $\rho = 0.325$.
Method-3	Proposed method + large forest regions as the surface scatterers, where all the pixels with σ^0 of larger than -10 dB are averaged, noise is subtracted, and the forest is used for channel imbalance determination. Here, $\gamma = 0.295$, $\rho = 0.325$.
Method-4	Quegan's method derived for PALSAR using CR (Quegan 1994, Shimada et al. 2009, and Moriyama et al. 2007)

6.6.3 PolCAL Parameters and Time Series Analysis

The polarimetric calibration parameters are calculated for Methods-1 through 3 and their accuracies are evaluated using the CR responses. Results of Method-4 are obtained using the parameters in Quegan (1994), Shimada et al. (2009), and Moriyama et al. (2007). We also established the time series properties of the distortion matrices obtained from Method-1''' using the data set presented in Table 6-2. Figure 6-11a and b depicts, respectively, the amplitude and the phase of the channel

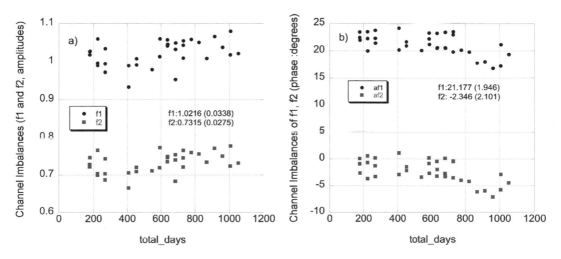

FIGURE 6-11 Temporal variation of the channel imbalances: (a) amplitude of the channel imbalances and (b) phase variation in degrees. Here, fi is $|f_i|$ and afi is arg (f_i).

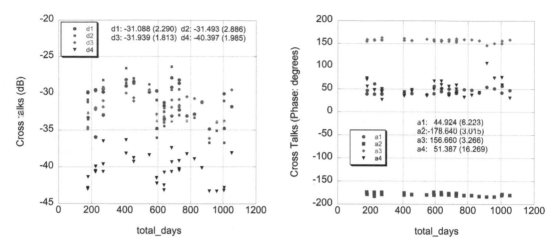

FIGURE 6-12 Temporal variations of the four cross talk signals (δ_1, δ_2, δ_3, and δ_4) as a function of the total days after ALOS launch: (a) power in decibels and (b) phase in degrees; d1 is $10 \cdot \log_{10}|\delta_i|^2$, and a1 is $\arg(\delta_i)$.

imbalance of f_1 and f_2. In a similar manner, Figure 6-12 illustrates the temporal variation of the four cross talks, both for their amplitudes and their phases. Figure 6-13 shows the temporal variation of the volume scattering component and the double-bounce components in HH and VV.

These figures illustrate that the polarimetric calibration parameters deviate slightly in time but are almost constant on average. This test site has been observed by the PALSAR in ascending and descending orbits and sometimes only 1 day apart (e.g., July 20, 2006, in ascending and July 21, 2006, in descending). Thus, the plotted time series of Figures 6-11, 6-12, and 6-13 shows data separated by either 1 or 45 days. Although ionospheric irregularities have tended to appear over the Amazon since 2006 at about 4 hours after local sunsets, and they exhibit seasonality (Shimada et al. 2008), these 26 data sets are not affected (confirmed by manual inspections).

From these figures, the parameters appear to be very stable and have no clear deviation between the ascending and descending orbits. Thus, the averaged PolCAL parameters can be calculated over

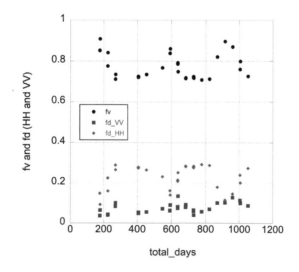

FIGURE 6-13 Temporal variations of the volume scattering component and the double-bounce scattering components of VV and HH. The double-bounce component of HH is nine times larger than that of VV. This calculation is obtained by Method-1'''.

all the orbit data, as presented in Table 6-9 for the distortion parameters issued from the different methods. Although Method-1''' shows the lowest cross talks among the methods, the difference is only 1 dB. From this, we see that there is no clear difference in PDM among the decomposition models (the Method-1s). However, Method-1''' has the best agreement between the forest covariance and the model covariance as shown in the bottom row of Table 6-9. And, the signal penetration issue needs to rely on Method-1''' and its derived PDM.

6.6.4 Comparison of the Calibration Parameters

6.6.4.1 Point-Target Analysis

Using the distortion matrices listed in Table 6-9, we calibrated all 26 PALSAR polarimetric data sets of Table 6-2 first, derived the following five key parameters using the responses from the CRs and natural target, and evaluated the performance of the proposed methods: (1) amplitude ratio of VV/HH, (2) phase difference of the VV and HH, (3) cross talk between VH and HH measured for the impulse response function of the CR, (4) as well as those for HV and VV, and (5) those derived from natural targets by their cross correlation process. Parameters (3) and (4) can be obtained by finding the CR in the HH image, reading pixel values in HH and the corresponding VH images, and calculating their power ratio after subtracting the background level values where the CR is deployed (here, 50×50 pixels are used for this calculation). Parameter (5) can be obtained by calculating the cross correlation of 200×200 pixel areas between the HH and VH images.

The results are summarized in Table 6-10. Temporal variations of the amplitude ratio, VV-HH phase difference, and the cross talks obtained from Method-1''' are shown in Figure 6-14a and b and in Figure 6-15, respectively. From these figures and this table, the Method-1s yield the best values among the methods, yielding the best for VV/HH with an amplitude ratio of 1.0036 (Method-1) and phase difference of 0.03025 degrees (Method-1) while the cross talks are as low as -32.412 dB for VH/HH, -31.866 for HV/VV, and -42.853 for natural targets (Method-1'''). Method-4, the reference method, is similar to the Method-1s. Method-2 shows an almost acceptable amplitude but with a slightly large phase difference. Method-3 shows acceptable phase difference but the amplitude difference is relatively large.

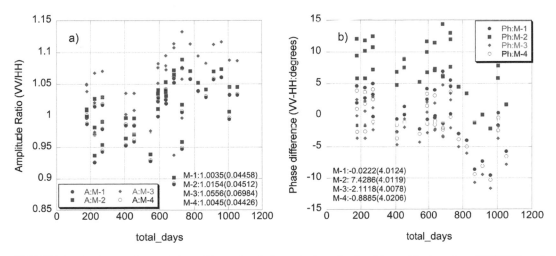

FIGURE 6-14 Comparison of the calibrated polarimetric data for the (a) amplitude ratio of VV/HH and (b) phase difference between HH and VV for the four methods.

TABLE 6-9
Comparison of the PDM Parameters Derived from Different Conditions and Covariance Agreement

Parameters	Method-1 (Freeman-Durden)	Method-1' (Freeman-2)	Method-1" (Neumann)	Method-1''' (Empirical)	Method-2	Method-3	Method-4
f_1	1.022(0.034)/21.172(1.956)	1.022(0.034)/21.172(1.956)	1.022(0.034)/21.172(1.956)	1.022(0.032)/21.116(1.760)	1.009(0.015)/17.500(0.862)	0.971(0.008)/22.262(0.449)	1.031/21.8050
f_2	0.73145(0.026)/-2.351(2.110)	0.73143(0.028)/-2.351(2.110)	0.73147(0.028)/-2.351(2.110)	0.73155(0.026)/-2.407(1.916)	0.72252(0.011)/-6.023(0.912)	0.69511(0.009)/-1.264(0.483)	0.72210/-1.8790
δ_1	-29.406(1.611)/39.873(3.305)	-29.498(1.641)/40.114(3.413)	-29.959(1.790)/41.367(3.995)	-31.072(2.245)/44.853(6.028)	-30.993(1.718)/44.201(3.871)	-29.711(1.241)/40.534(2.216)	-37.6175/79.3693
δ_2	-27.713(2.184)/-178.140(2.943)	-27.920(2.205)/-178.150(2.946)	-28.948(2.316)/-178.200(2.967)	-31.477(2.812)/-178.360(2.921)	-31.563(1.861)/-181.910(1.859)	-28.980(1.397)/-177.060(1.417)	-37.6851/-151.5032
δ_3	-28.652(1.601)/158.460(2.483)	-28.836(1.605)/158.380(2.487)	-29.744(1.601)/157.970(2.531)	-31.928(1.754)/156.630(3.003)	-32.033(1.904)/153.160(2.769)	-29.829(1.564)/159.330(2.645)	-40.4867/131.4864
δ_4	-37.556(1.783)/34.282(7.488)	-37.725(1.802)/35.007(7.778)	-38.558(1.869)/38.922(9.314)	-40.390(1.938)/51.199(15.977)	-40.538(2.078)/51.994(14.992)	-38.285(1.670)/37.359(8.007)	-39.8263/-1.8790
$\langle\delta\rangle$	-30.832	-30.995	-31.802	-33.717	-33.782	-31.701	-38.9039
E	0.115	0.113	0.0975	0.062	-	-	-

Note: Values are expressed in polar coordinates (i.e., the first value is amplitude and the second value after "/" is the angle in degrees). The values in brackets are the standard deviation; $\langle\delta\rangle$ is the averaged cross talks; E is the squared summation of the difference between measured forest covariance and the model calculated using Equation (6-A-6).

TABLE 6-10
Comparison of the Amplitude Ratio and Phase Difference between VV and HH Cross Talk Using the CR Response and Cross Talk as the Cross Correlation of the Like-Polarized and Cross-Polarized Data

	Method-1	Method-1'	Method-1"	Method-1'''	Method-2	Method-3	Method-4
Amplitude ratio (VV/HH)	1.0034(0.04458)	1.0034(0.4457)	1.0034(0.4454)	1.0036(0.04446)	1.0154(0.04498)	1.0558(0.04686)	1.0045(0.04426)
Phase difference in degrees (VV-HH)	0.03025(4.0123)	0.03270(4.0126)	0.04453(4.0141)	0.0608(4.0169)	7.4733(4.0171)	-2.0721(4.0138)	-0.8885(4.0206)
Cross talk 1 (VH/HH: dB)	-32.708(6.2428)	-31.135(6.2262)	-32.440(6.1672)	-32.412(5.5587)	-32.220(5.4365)	-32.011(6.1813)	-31.815(5.1476)
Cross talk 2 (HV/VV: dB)	-32.067(4.8685)	-32.036(4.8606)	-32.013(5.1227)	-31.866(5.4226)	-31.959(5.3883)	-32.421(4.9415)	-31.501(5.2084)
Cross talk 3 (natural: dB)	-41.961(6.0499)	-42.130(6.0243)	-42.651(5.7093)	-42.853(5.7266)	-42.881(5.785)	-42.625(5.7436)	-42.191(6.1546)

Note: Amplitude ratio and phase differences are measured by averaging the CR response for several pixels surrounding the center of the bright targets.

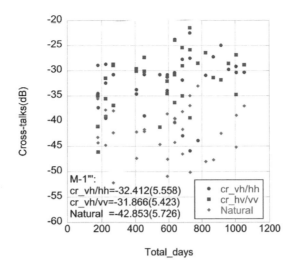

FIGURE 6-15 Cross talks calculated for the responses from the CRs and the cross correlation of the natural targets using Method-1'''. Here, "cr_vh/hh" corresponds to Cross Talk-1, "cr_hv/vv" to Cross Talk-2, and "cross" to Cross Talk-3 in Table 6-10.

6.6.4.2 Polarimetric Signatures

Polarimetric signatures are often used to visually evaluate the polarimetric calibration results. In Figure 6-16, we compare the eight cross- and like-polarimetric signatures calculated for the impulse responses from a CR deployed at the Rio Branco site in Brazil, observed on July 20, 2006, and calibrated by the seven polarimetric distortion matrices: uncalibrated scattering matrix, Methods-1 through 4. From these figures, the Method-1s and Method-4 provide the best polarimetric signatures for both like and cross polarizations. Method-2 provides acceptable signatures. However, Method-3 is the worst among the calibration methods.

6.6.5 SIGNAL DECOMPOSITION FOR THE FOREST

In addition to providing the PDM, the proposed method decomposes the signal backscattering from the forest and provides quantitative information on the forest parameters (e.g., volume scattering, double bounce, and signal penetration, and its polarization dependence). As an example, the incidence angle dependence of the retrieved forest parameters is plotted in Figure 6-17. The volume scattering for the vertical and horizontal polarizations (f_v), the double bounce in vertical and horizontal polarizations (f_{d_vv}, f_{d_hh}), and the HH/VV ratio of the double bounce $(|\mathbf{x}|)$ are illustrated; $f_v + f_d$ is always less than 1.0 because the horizontal component exceeds the vertical component. The HH/VV (amplitude) ratio of the double bounce ranges from 1.5 to 2.5 and is almost independent from the incidence angle. The double-bounce component lies around 0.050 (VV) and 0.232 (HH), meaning that about 5% (VV) and 23% (HH) of the incoming signals to the forest may penetrate the forest canopy and double-bounce back to the radar.

Next, we compare two covariances calculated for the calibrated and uncalibrated PALSAR data for the forest. Figure 6-18 depicts the six covariances of the calibrated data using the PDM based on Method-1'''. Amplitudes after calibration (Figure 6-18b) indicate that the VV power (C_{VVVV}) is 80% of the HH power (denoted C_{HHHH}). The 20% difference is due to VV having less signal penetration and thus fewer double-bounce components. The three cross terms (i.e., C_{HVHV}, C_{HVVH}, and C_{VHVH}) have almost the same value (0.23), suggesting that the polarimetric calibration was performed properly. The C_{HHVV} value is equal to 0.33, which is the same as the Freeman-Durden value. Comparing these values with the uncalibrated covariance (Figure 6-18a) confirms that the amplitudes of the covariances are very well balanced.

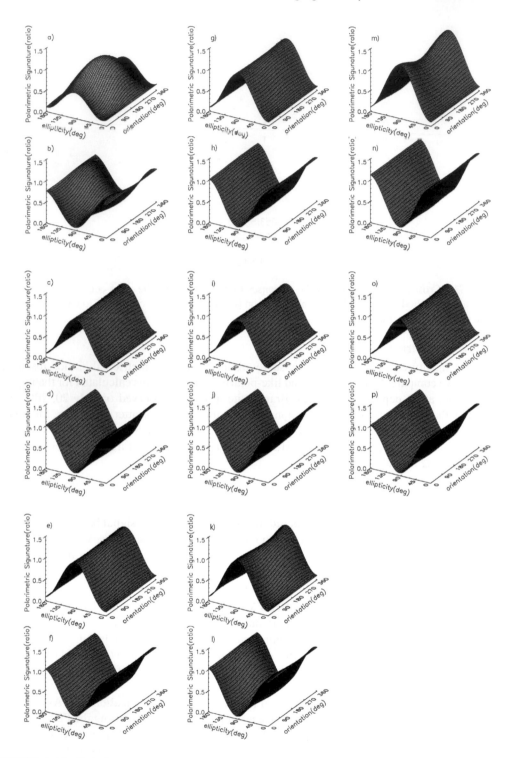

FIGURE 6-16 Polarimetric signatures calculated for the uncalibrated data and the calibrated data. The upper two rows (a) - (h) are the polarimetric signatures (PolSIG) for the CR in circle A of Figure 6-3.; the lower two rows (i) - (p) are for the CR in circle B of Figure 6-3. (a), (b): Like-polarimetric PolSIG and cross-polarimetric PolSIG for the uncalibrated data. (c), (d): Method-1. (e), (f): Method-1'. (g), (h): Method-1". (i), (j): Method-1'''. (k), (l): Method-2. (m), (n): Method-3. (o), (p): Method-4.

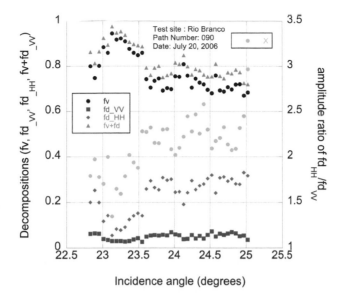

FIGURE 6-17 Incidence angle dependence of the forest parameters estimated from the proposed PolCAL method. Within a 3-degree incidence angle range, two components—the volume scattering component and the double-bounce component—are almost independent of the incidence angle. The amplitude ratio of the double-bounce component in HH to that in VV is nearly equal to 2.1.

The phase differences of C_{HHVV} and C_{HVVH} are also plotted in Figure 6-19 where (a) is the uncalibrated data and (b) is the calibrated data. The phase difference of ±25 degrees observed in the uncalibrated data converges to almost zero degrees (0.03 degrees), especially in C_{HVVH}. This means that the phase was well calibrated. The arg (C_{HHVV}) value remains around 4.52 degrees. This can be explained by HH polarization containing a larger double-bounce component than VV polarization. Most volume scattering contributes randomly to HH and VV polarizations.

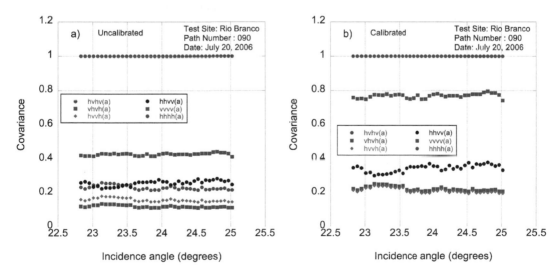

FIGURE 6-18 Incidence angle dependence of the amplitude of covariances at C_{HHHH}, C_{VVVV}, C_{HHVV}, C_{HVHV}, C_{VHVH}, and C_{HVVH}. (a): Uncalibrated data. (b): Calibrated data. Here, "hvhv" stands for C_{HVHV}. Other notations follow similarly.

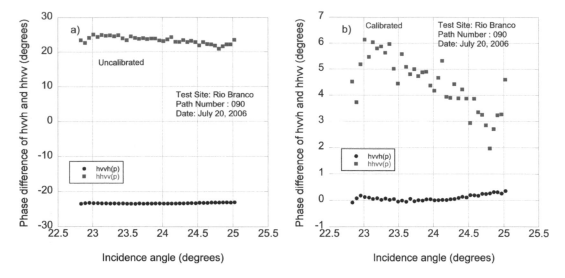

FIGURE 6-19 Incidence angle dependence of the phase difference of C_{HHVV} and C_{HVVH}. (a): Uncalibrated data. (b): Calibrated data.

6.7 DISCUSSIONS

We developed a new polarimetric calibration method that uses the dense forest, surface scatterer, and incoherent decomposition model to determine the polarimetric distortion matrices and backscattering property of forest data and that allows consideration of signal penetration by the lower frequency SAR. The largest advantage in using the dense rainforest in western Brazil is that its number of scatterers is tremendously large, that the reflection symmetry is valid (when the appropriate number of pixels is averaged), and that the backscattering of HH and VV can be balanced when the polarization dependence of the penetration is clarified. The disadvantage appears to be the complexity of decomposing the polarization dependence of the scattering mechanism. Here, we discuss the following points to interpret the results gained in this analysis: the evaluation summary of the proposed method, reference sources for the channel imbalances, forest characteristics, and future improvement.

6.7.1 COMPARISON OF THE PROPOSED METHODS

As described in the previous section, we compared the accuracy of the proposed methods (six plus one reference) using four evaluation parameters (i.e., amplitude gain ratio of VV/HH at the CR, phase difference of VV and HH at the CR, cross talks measured in two ways, and polarimetric signatures at the deployed CR). The Method-1s and Method-4 are ranked as highly usable, Method-2 ranked second, and Method-3 ranked third. Method-2 is weak in the phase difference between the HH-VV (e.g., 7 degrees, while the allowance of the CEOS SAR Working Group on Calibration and Validation [CALVAL] is 5 degrees), although the other parameters are acceptable. Method-3 is weak at the amplitude ratio of VV-HH (e.g., 1.05, which is 0.21 dB, while the allowance of the CEOS SAR CALVAL is 0.2 dB). In addition to this, we evaluated the stability of the proposed method in a variety of volume covariances (i.e., γ and ρ both ranging from approximately 0.3 to 0.4), simulating the cases for the Freeman-Durden (Method-1), Freeman-2 (Method-1'), and Neumann (Method-1'') models, and the empirically observed volume covariance (Method-1'''), and confirmed that the solutions are very stable and the derived PDM is similar among the models.

The biggest difference in the six methods (Methods-1 through 3) appears at the selection of the reference target for surface scattering (e.g., phase referencing) and amplitude balancing. In the past, determination of the channel imbalance was troublesome. Most research relied on the CR because its backscattering property is confident and equalized for HH and VV. In our research, we tried to find such a signal source that is a natural target. Because both Methods-2 and -3 deviated from the specification requirement of the CEOS SAR, currently CR is the only realizable reference source. In Method-2, we selected an "assumed rough surface" because the σ^0 is less than -13 dB. Visually, the selected area seems to be a real rough surface but the phase angle seems to differ from zero (probably about 7 degrees). If the phase difference between the HH-VV is well known for rough surfaces, it could be usable as a reference surface target.

6.7.2 REFERENCE SURFACE SCATTERER

In this experiment, we tested a rough surface as a reference surface scattering target. The test was not successful because the rough surface may have had a different Fresnel reflection coefficient depending on the incidence angle and polarization. If we assume that the dielectric constant of the rough surface is $\varepsilon = 28.0 - j8.5$ and the incidence angle is 25 degrees, the Fresnel reflection coefficient in HH is $R_{HH} = 0.71514e^{177.08j}$ and the coefficient in VV is $R_{VV} = 0.95664e^{176.49j}$ (see Appendix 6A-3). Their ratio is $R_{HH} / R_{VV} = 0.7476e^{0.59j}$ where the unit in the exponent is degrees. Thus, the rough surface could be useful as a reference surface scatterer. However, the results in Table 6-10 (Figures 6-13 and 6-14) do not support this (measured data is 7 degrees), possibly because the rough surface has some deviation from the normal condition. In this experiment, we could not achieve more promising results but hope to do so in the future. On the other hand, in using the dense rainforest for the surface scatter, the phase difference of the HH/VV is 2 degrees and an acceptable value. This also needs further investigation.

6.7.3 CROSS TALK

Cross talks of the radar system are required to be as small as possible. The PolSAR mode of ALOS/PALSAR specified the cross talk -25 dB as the minimum at the manufacture phase. In order to estimate the cross talks accurately as the PolCAL output, the proposed method maintained all the higher order terms (e.g., cross talks) in expressing the calibration equations and determined the PDM in two steps, one for channel imbalance and the other cross talks, iteratively. As a result, the proposed method exhibited -33.717dB of cross talks on average (Table 6-9). However, this value is almost 5 dB lower than that of Method-4, which linearizes the covariance matrix and obtains the PDM iteratively for the approximated covariance. If we compare the evaluated cross talks in Table 6-10, which were obtained from CR and the natural targets in PALSAR images, however, all methods show similar values.

The agreement in Table 6-10 and disagreement in Table 6-9 can be explained as follows: Method-4 (Quegan's approach) explicitly sets all co-pol/x-pol correlations to zero and then estimates the x-talk parameters. Methods 1-3 all fit the measured co-pol/x-pol correlations in the data when solving for the x-talk parameters. These measured Cpol/Xpol correlations are approximately -27dB for averages of 105 to 106 pixels (from Figure 6-7). Hence, in Table 6-9, Methods-1 through -3 have larger mean cross talk values (between -30dB and -33dB) than Method-4 (-39dB). In Table 6-10, the cross-talk assessments are derived from the calibrated PALSAR imagery. Therefore, the calibrated imagery reintroduces the small (but non-zero) Cpol/Xpol correlations in all methods. Imagery calibrated using Method-4 is different from, but quantitatively no better than, imagery calibrated using Method-1. Thus, Quegan's approach underpredicts the magnitudes of the polarimetric cross-talk parameters by explicitly setting the Cpol/Xpol correlations to zero. Thus, the evaluated cross talks are satisfactory.

6.7.4 FOREST CHARACTERISTICS

From the analysis of the dense forest data, we can interpret the signal penetration of the L-band SAR data through the forest canopy. In Equation (6.9), **x** is the amplitude ratio of the double-bounce component (or penetrated signal) of the HH polarization to that of VV polarization. Figure 6-17 graphs the incidence angle dependence of **x** and shows that **x** distributes between 1.5 and 2.5, and that the HH signal penetrates the forest canopy deeper and is returned from the bottom of the forest stronger than the VV–polarized signal. The repeat-pass SAR interferometry relies on the similarity of the signals received at two different times. Signals from the volume scatterer become incoherent as time separates. Only coherent targets could produce signals scattered or double-bounced from tree trunks because these areas are less sensitive to the time separation of the two observations. In this regard, the penetration of the signal through the forest canopy is very important in achieving high-quality repeat-pass SAR interferometry.

Figure 6-20 compares the PALSAR interferometric coherences of HH-HH, VV-VV, and HV-HV observed for the Amazon and Rio Branco on October 20, 2006, and September 6, 2006, respectively. This figure shows that HH and VV coherences are similar and higher than HV coherence. VV has a measured signal penetration of 7% for f_d but could be sufficient for good interferometric coherence even if the time baseline is 46 days. Rio Branco is a dense rainforest region, and low coherence at HV seems to be reasonable. Less dense forest may have more values of f_d and higher

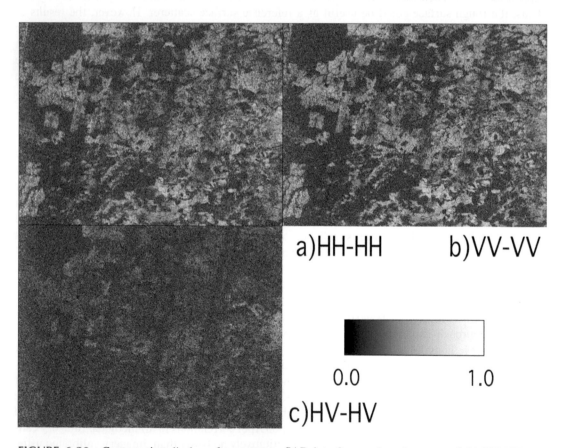

a)HH-HH   b)VV-VV

0.0   1.0

c)HV-HV

FIGURE 6-20 Comparative display of repeat-pass SAR interferometric coherences of (a) HH-HH and (b) VV-VV. The master and slave images were acquired on October 20, 2006, and September 6, 2006, respectively, as 46 days separated in time and 193 m separated in space (e.g., perpendicular baseline). Slightly dark stripes running vertically seem to be interferometric decorrelation due to ionospheric disturbances (Shimada et al. 2008).

coherence. On the other hand, HH polarization contains 23% of the total power as a double-bounce component. This high value could be sufficient enough for good interferometric coherence as seen from Figure 6-20.

6.8 SUMMARY

This chapter proposes a new PolSAR calibration method that applies an incoherent decomposition model to uncalibrated covariance data measured for forest and surface and determines the PDM. The Freeman-Durden model is used to express the polarization-dependent signal reflection from, and penetration through, the forest. Nonlinear equations built for uncalibrated PolSAR data are solved iteratively. This method is applicable to the lower frequency SAR that associates with the polarization-dependent signal penetration through forest canopies. Using the time series PALSAR data acquired for the Amazon Rainforest for about 3 years, we confirm that the proposed method succeeds the PDM estimation and that the calibrated data preserve the polarimetric performance on HH-VV orthogonality, low cross talks, and ideal polarimetric signature for the CR. This chapter also investigates the signal penetration properties of the forest with which the L-band SAR associates.

APPENDIX 6A-1: VOLUME COVARIANCE FOR THE NEUMANN MODEL

The proposed method contains two parameters (e.g., γ and ρ) to express the volume covariance. The Freeman-Durden and Freeman-2 methods easily assign these parameters, that is, $\rho = 1/3$, $\rho = 1/3$, and $\gamma = (1 - \rho)/2$, where ρ is fitted to the measurement, respectively. On the other hand, Neumann's coherency model is represented by orientation randomness, τ, and particle anisotropy, δ. It is necessary to convert these parameters to ρ and γ.

The Neumann coherency matrix from the volume scattering can be approximated as (Neumann et al. 2009; Neumann et al. 2010):

$$T_v = \begin{pmatrix} 1 & (1-\tau)\delta & 0 \\ (1-\tau)\delta^* & \frac{1}{2}|\delta|^2 & 0 \\ 0 & 0 & \frac{1}{2}|\delta|^2 \end{pmatrix} \tag{6A-1.1}$$

Here, δ is the general scattering indicator, and ρ is Freeman's shape parameter (i.e., C_{HHVV}). As an analogy to Freeman and Durden, the volume component can be applied to the estimation of δ:

$$|\delta| = \sqrt{(T_{22} + T_{33})/f_v} \tag{6A-1.2}$$

On the other hand, T_v can be expressed by using the HV basis as follows:

$$T_v = \begin{pmatrix} 1 & \dfrac{I_m(\rho)}{1+R_e(\rho)} & 0 \\ \dfrac{-I_m(\rho)}{1+R_e(\rho)} & \dfrac{1-R_e(\rho)}{1+R_e(\rho)} & 0 \\ 0 & 0 & \dfrac{2\gamma}{1+R_e(\rho)} \end{pmatrix} \tag{6A-1.3}$$

From the measurements, T_{12}, T_{22}, and T_{33} have the range of values shown in Table 6A-1. It should be noted that although the Equations (6A-1.2) and (6A-1.3) are valid for the total sum of the volume,

TABLE 6-A-1
Statistics of the Three Representative Components and
Possible Range of Values

Term	Mean (Standard Deviation)	Range
T_{12}	0.07445(0.0114), −8.03 degrees	0.0630 ~ 0.0858
T_{22}	0.42773(0.0241)	0.4036 ~ 0.4518
T_{33}	0.35228(0.0220)	0.3303 ~ 0.3743

double-bounce, and surface scattering, the parameters were determined by assuming that all the signals were from volume scattering.

Comparing Equations (6A-1.2) and (6A-1.3), we have

$$\rho = \frac{1-|\delta|/2}{1+|\delta|/2} \tag{6A-1.4}$$

and

$$\gamma = \frac{(1-\rho)^2}{1+\rho}. \tag{6A-1.5}$$

Here, we assumed that ρ is real. From these equations and the range of the T parameters, we obtained the range of δ/τ and ρ/γ as shown in Table 6A-2.

APPENDIX 6A-2:   EMPIRICAL DETERMINATION OF THE FOREST COVARIANCE

In addition to the volume covariance determined for three theoretical scattering models, we can propose one more method, which determines an optimum γ and ρ that minimizes the squared sum of the volume covariance defined as

$$E(\gamma,\rho) \equiv \frac{1}{N} \sum \left[\left\{ C_{vvv}(\gamma,\rho) - \tilde{C}_{vvv} \right\}^2 + \left\{ C_{hhvv}(\gamma,\rho) - \tilde{C}_{hhvv} \right\}^2 + \left\{ C_{hvhv}(\gamma,\rho) - \tilde{C}_{hvhv} \right\}^2 \right]. \tag{6A-2.1}$$

Here, $E(\gamma,\rho)$ is the averaged summation of the squared difference between the measured covariance $\tilde{C}$ and the theoretical covariance $C(\gamma,\rho)$ (e.g., Equation [6.11]), and N is the number of samples. For given γ and ρ, the PolCAL determines the PDM, the PDM calibrates the PALSAR data, and then the covariance matrix is calculated. Scanning γ from 0.28 to 0.33 and ρ from 0.28 to 0.38 with an interval of 0.005 for both, the optimum values can be determined as the minimum of Equation (6A-2.1) (See Figure 6A-1). Using the 26 Amazon PALSAR data examples, we have obtained $\rho = 0.325$ and $\gamma = 0.295$.

TABLE 6-A-2
Range of Parameters

f_v	1.00	0.95	0.90	≤ 0.836
δ	0.836 ~ 0.933	0.880 ~ 0.982	0.929 ~ 1.0	1.0
τ	0.893 ~ 0.936	0.940 ~ 0.985	0.992 ~ 1.0	1.0
ρ	0.364 ~ 0.411(0.3875)	0.341 ~ 0.389(0.3650)	0.333 ~ 0.366(0.3495)	0.333
γ	0.246 ~ 0.296(0.2710)	0.269 ~ 0.324(0.2965)	0.294 ~ 0.333(0.3135)	0.333

Note: Averages are in brackets.

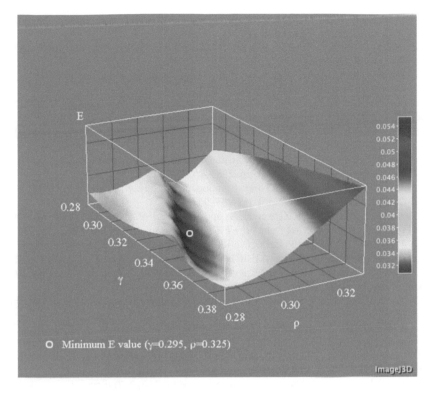

FIGURE 6-A1 Perspective view of the E value on γ and ρ. E is minimized at $\gamma = 0.295$ and $\rho = 0.325$.

APPENDIX 6A-3: SURFACE SCATTERING AND THEORETICAL CONSIDERATIONS

Surface scattering can be theoretically expressed by the following equations. Depending on the dielectric constant, the reflection coefficients from HH polarization and VV polarization are as follows (Freeman and Durden 1998):

$$\alpha_{HH} = \frac{\cos\theta - \sqrt{\varepsilon_r - \sin^2\theta}}{\cos\theta + \sqrt{\varepsilon_r - \sin^2\theta}}$$

$$\alpha_{VV} = \frac{(\varepsilon_r - 1)[\sin^2\theta - \varepsilon_r(1 + \sin^2\theta)]}{(\varepsilon_r \cos\theta + \sqrt{\varepsilon_r - \sin^2\theta})}$$

(6A-3.1)

Here, ε_r is the dielectric constant of the target. This expression shows that the reflection coefficient of a flat surface, even a water surface, has incidence angle dependence and cannot deal the **y** in Equation (6.8), which is the ratio of $\alpha_{HH} / \alpha_{VV}$ as the real value.

REFERENCES

Ainsworth, L., Ferro-Famil, L., and Lee, J. S., 2006, "Orientation Angle Preserving A Posteriori Polarimetric SAR Calibration," *IEEE T. Geosci. Remote*, Vol. 44, No. 4, pp. 994–1003.

Attema, E., Brooker, G., Buck, C., Desnos, Y-L, Emiliani, L., Geldsthorpe, B., Laur, H., Laycock, J., and Sanchez, J., 1997, "ERS-1 and ERS-2 Antenna Pattern Estimates Using the Amazon Rainforest," *Proc. CEOSSAR Workshop on RADARSAT Data Quality*, Saint-Hubert, QC, Canada, February 4–6, 1997.

Bickel, S. H. and Bates, R. H. T., 1965, "Effects of Magneto-Ionic Propagation on the Polarization Scattering Matrix," *Proc. of IEEE*, Vol. 53, No. 8, pp. 1089–1091.

Borgeaud, M., Shin, R. T., and Kong, J. A., 1987, "Theoretical Models for Polarimetric Radar Clutter," *J. Electromagnet Wave*, Vol. 1, No. 1, pp. 73–89.

Cloude, S. R. and Pottier, E., 1996, "A Review of Target Decomposition Theorems in Radar Polarimetry," *IEEE T. Geosci. Remote*, Vol. 34, No. 2, pp. 498–518.

Cote, S., Srivastava, S. K., Le Dantec, P., and Hawkins, R. K., 2005, "Maintaining RADARSAT-1 Image Quality Performance in Extended Mission," *Proc. of 2nd International Conference on Recent Advances in Space Technologies*, Istanbul, Turkey, June 9–11, 2005, pp. 678–681.

Freeman, A., 1992, "SAR Calibration: An Overview," *IEEE T. Geosci. Remote*, Vol. 30, No. 6, pp. 1107–1121.

Freeman, A., 2007, "Fitting a Two-Component Scattering Model to Polarimetric SAR Data from Forests," *IEEE T. Geosci. Remote*, Vol. 45, No. 8, pp. 2583–2592.

Freeman, A. and Durden, S. L., 1998, "A Three-Component Scattering Model for Polarimetric SAR Data," *IEEE T. Geosci. Remote*, Vol. 36, No. 3, pp. 963–973.

Freeman, A., Van Zyl, J. J., Klein, J. D., Zebker, H. A., and Shen, Y., 1992, "Calibration of Stokes and Scattering Matrix Format Polarimetric SAR Data," *IEEE T. Geosci. Remote*, Vol. 30, No. 3, pp. 531–539.

Fujita, F., and Murakami, C., 2005, "Polarimetric Radar Calibration Method Using Polarization-Preserving and Polarization-Selective Reflectors," IEICE Trans. Commun. B., Vol. 88, No. 8, pp. 3428–3435.

Kimura, H., 2009, "Calibration of Polarimetric PALSAR Imagery Affected by Faraday Rotation Using Polarization Orientation," *IEEE T. Geosci. Remote*, Vol. 47, No. 12, pp. 3943–3950.

Klein, J. D., 1992, "Calibration of Complex Polarimetric SAR Imagery Using Backscatter Correlations," *IEEE T. Aero. Elec. Sys.*, Vol. 28, No. 1, pp. 183–194.

Krogager, E., 1990, "A New Decomposition of the Radar Target Scattering Matrix," *Electron. Lett.*, Vol. 26, No. 18, pp. 1525–1526.

Lukowski, T. I., Hawkins, R. K., Cloutier, C., Wolfe, J., Teany, L. D., Srivastava, S. K., Banik, B., Jha, R., and Adamovic, M., 2003, "RADARSAT elevation antenna pattern determination," *Proc. IGARSS 2003, International Geoscience and Remote Sensing Symposium*, Vol. 2, Toulouse, France, July 21–25, 2003, pp. 1382–1384.

Luscomb, A. P., 2001, "POLARIMETRIC PARAMETER ESTIMATION FROM AMAZON IMAGES," *Proc. of the CEOS Calibration Working Group on Calibration and Validation*, Tokyo, April 2–5, 2001, pp. 19–23.

Meyer, F., Bamler, R., Jakowski, N., and Fritz, T., 2006, "The Potential of Low-Frequency SAR Systems for Mapping Ionospheric TEC Distribution," *IEEE Geosci. Remote S.*, Vol. 3, No. 4, pp. 560–565.

Moriyama, T., Shimada, M., and Tadono, T., 2007, "Polarimetric Calibration of ALOS/PALSAR," *Proc. ISAP 2007*, Niigata, Japan, November 5–8, 2007, pp. 776–779.

Neumann, M., Ferro-Famil, L., and Pottier, E., 2009, "A General Model-Based Polarimetric Decomposition Scheme for Vegetated Areas," *Proc. PolinSAR 2009*, Frascati, Italy, January 26–30, 2009, pp. TK–TK.

Neumann, M., Ferro-Famil, L., and Reigber, A., 2010, "Estimation of Forest Structure, Ground, and Canopy Layer Characteristics from Multibaseline Polarimetric Interferometric SAR Data," *IEEE T. Geosci. Remote S.*, Vol. 48, No. 3, pp. 1086–1104.

Nghiem, S. V., Yueh, S. H., Kwok, R., and Li, F. K., 1992, "Symmetry Properties in Polarimetric Remote Sensing," *Radio Sci.*, Vol. 27, No. 5, pp. 693–711.

Quegan, S., 1994, "A Unified Algorithm for Phase and Cross-Talk Calibration of Polarimetric Data-Theory and Observation," *IEEE T. Geosci. Remote*, Vol. 32, No.1, pp. 89–99.

Touzi, R. and Shimada, M., 2009, "Polarimetric PALSAR Calibration," *IEEE T. Geosci. Remote*, Vol. 47, No. 12, pp. 3951–3959.

Shimada, M., 2011, "Model-Based Polarimetric SAR Calibration Method Using Forest and Surface Scattering Targets," *IEEE T. Geosci. Remote*, Vol. 49, No. 5, pp. 1712–1733.

Shimada, M. and Freeman, A., 1995, "A Technique for Measurement of Spaceborne SAR Antenna Patterns Using Distributed Targets," *IEEE T. Geosci. Remote*, Vol. 33, No.1, pp. 100–114.

Shimada, M., Isoguchi, O., Tadono, T., and Isono, K., 2009, "PALSAR Radiometric and Geometric Calibration," *IEEE T. Geosci. Remote*, Vol. 47, No. 12, pp. 3915–3932.

Shimada, M., Muraki, Y., and Otsuka, Y., 2008, "Discovery of Anoumoulous Stripes Over the Amazon by the PALSAR Onboard ALOS Satellite," *Proc. of IGARSS 2008*, Boston, MA, July 7–11, 2008, pp. II 387–390.

Takeshiro, A., Furuya, T., and Fukuchi, H., 2009, "Verification of Polarimetric Calibration Method Inclusing Faraday Rotation Compensation Using PALSAR Data," *IEEE T. Geosci. Remote*, Vol. 47, No. 12, pp. 3960–3968.

Van Zyl, J. J., 1990, "Calibration of Polarimetric Radar Images Using Only Image Parameters and Trihedral Corner Reflectors," *IEEE T. Geosci. Remote*, Vol. 28, No. 3, pp. 337–348.

Van Zyl, J. J., 1993, "Application of Cloude's Target Decomposition Theorem to Polarimetric Imaging Radar Data," *Proc. SPIE 1748, Radar Polarimetry*, San Diego, CA, July 20–25, 1992, pp. 184–191.

Wright, P. A., Quegan, S., Wheadon, N. S., and Hall, C. D., 2003, "Faraday Rotation Effects on L-band Spaceborne SAR Data," *IEEE T. Geosci. Remote*, Vol. 41, No. 12, pp. 2735–2744.

Yamaguchi, Y., Moriyama, T., Ishido, M., and Yamada, H., 2005, "Four-Component Scattering Model for Polarimetric SAR Image Decomposition," *IEEE T. Geosci. Remote*, Vol. 43, No. 8, pp. 1699–1706.

Yueh, S. H, Kwok, R., and Nghiem, S. V., 1994, "Polarimetric Scattering and Emission Properties of Targets with Reflection Symmetry," *Radio Sci.*, Vol. 29, No. 6, pp. 1409–1420.

Zink, M. and Rosich, B., 2002, "Antenna Elevation Pattern Estimation from Rainforest Acquisitions," *Proc. CEOS Working Group on Calibration/Validation*, London, September 24–26, 2002.

7 SAR Elevation Antenna Pattern—Theory and Measured Pattern from the Natural Target Data

7.1 INTRODUCTION

As a planer array, an SAR antenna has a two-dimensional sensitivity in space and directs the radiated power over the image swath. The antenna pattern is designed such that the total signal-to-noise ratio (SNR) is maximized over the swath under the radar parameters (i.e., transmission power). This antenna pattern primarily depends on the antenna size and secondarily on the shape. The directional sensitivity should be calibrated to measure the radar backscatter accurately. A theoretical or on-ground antenna pattern, however, often differs from the inflight antenna pattern, which probably is due to differences in SNR and/or environmental conditions. Thus, the calibration of the inflight antenna pattern is important for obtaining the SAR backscatter accurately. There are two antenna patterns—elevation antenna pattern (EAP) and azimuth antenna pattern (AAP). From the calibration point of view, AAP is important mainly at ScanSAR processing; but EAP is always important at any SAR processing. Thus, in this chapter, we focus on the EAP calibration.

From the previous studies by Moore et al. (Moore and Hemmat 1988; Moore et al. 1986; Shimada and Freeman 1995; Hawkins 1990; and Dobson et al. 1986), it was noted that the distributed target offers the precise measurement of the inflight EAP. This is because a large number of small scatterers contributes to the imaging process, and the natural forest over the flat plain shows the constant radar backscatter over a wide range of the incidence angle.

In this chapter, we will describe the theoretical antenna pattern and measure the inflight EAP using the distributed targets.

7.2 THEORETICAL EXPRESSION

At first, we introduce the theoretical expression for the planer antenna pattern. When the antenna is sized in L_a in azimuth and L_r in range, and has $N \times M$ transmit-receive modules (TRMs) on, the resultant far-field antenna pattern is given by the following equation (Haupt 2010):

$$G(\theta, \psi) = \frac{4\pi L_a L_r}{\lambda^2} \left| \frac{\sin\left((Nkd_e / 2)\cos\theta\right)}{N\sin\left((kd_e / 2)\cos\theta\right)} \right|^2 \left| \frac{\sin\left((Mkd_a / 2)\cos\psi\right)}{M\sin\left((kd_a / 2)\cos\psi\right)} \right|^2 \tag{7.1}$$

where θ is the angle in range direction, ψ the angle in azimuth direction, $k = 2\pi / \lambda$, N the number of the arrays in range, M the number of arrays in azimuth, and d_e (d_a) is the space of the arrays in range and azimuth, respectively. Figure 7-1 shows the coordinate system of the antenna pattern expression.

This equation accurately describes the planer array's antenna pattern and is often used to design the radar system. As shown in Figures 7-2a and 7-2b for the JERS-1 SAR antenna, the two cross sections provide the medium and sharp angular sensitivities, which are called the EAP and the AAP.

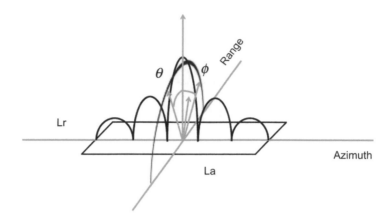

FIGURE 7-1   Antenna pattern coordinate system.

7.3   THE INFLIGHT SAR EAP ESTIMATE

In this analysis, we use the slant range amplitude images that are not antenna pattern corrected. We assume that the raw data are saturation-free and contaminated with the thermal noise. We use JERS-1 SAR and SIR-B data as the L-band SAR because they imaged several distributed targets, such as the Amazon, Illinois, Canada, and so on, from the 1980s to the 1990s. It can be supposed that a distributed target is composed of a huge amount of small point targets (scatterers), and their scattered signals follow the Rayleigh distribution at amplitude and uniform distribution at phase. The received signal is composed of Gaussian signals from the small targets and the internal noise of the SAR receiver. The receiver noise is assumed to be time invariant. The SAR image intensity (correlated image power) is a product of the processor gain and the expected raw signal power. As noted in Chapter 4 and in Freeman and Curlander (1989), each pixel's intensity, P_C, is a sum of the signal, P_{CS}, and the noise, P_{CN} (related coordinates are shown in [A] in Figure 7-3):

$$P_C = P_{CS} + P_{CN} \tag{7.1a}$$

$$P_{CS} = \left(\frac{PRF \cdot \lambda}{2V_g\rho_a}\right)^2 L_W N_L \cdot \frac{P_t\lambda^2\delta_a\delta_r}{(4\pi)^3} \cdot G^2(\phi) \cdot \frac{\gamma^0 \cot\theta}{R^2} \tag{7.2}$$

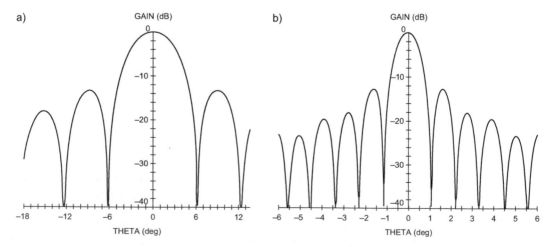

FIGURE 7-2   Elevation antenna pattern (left) and azimuth antenna pattern (right).

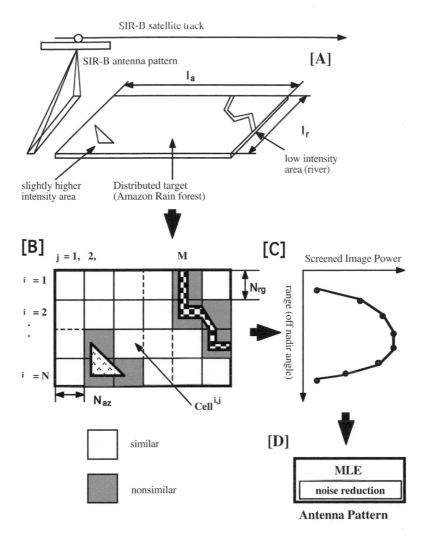

FIGURE 7-3 General processing flow of this study. [A] SAR images the distributed area. [B] The image is divided into many cells, each of which has the size *Nrg* and *Naz*, so that they can be screened using the χ^2 similarity test. [C] Averaged image power versus off-nadir angle is obtained. [D] Finally, the antenna pattern is modeled and fitted to the image powers.

$$P_{CN} = B \cdot R \tag{7.3}$$

$$B = \frac{PRF \cdot \lambda}{2V_g \rho_a} L_W \cdot N_L \cdot \overline{P_N} \tag{7.3a}$$

where V_g is the satellite ground speed; *PRF* is the pulse repetition frequency; L_w is the loss in the peak signal strength due to the azimuth and range reference function weighting; N_L is the number of looks; ρ_a is the azimuth theoretical resolution; δ_a is the azimuth pixel spacing; δ_r is the range pixel spacing; G is the antenna elevation pattern (one way); γ^0 is the gamma-zero (= σ^0 / $\cos\theta$: constant); R is the slant range; θ is the local incidence angle; ϕ is the off-nadir angle; P_t is the transmitted peak power; λ is the wavelength; and $\overline{P_N}$ is the mean noise power. The SAR image has a constant azimuth resolution over a full image swath because of selecting the azimuth correlation time proportional to the slant range. P_{CS} is then proportional to R^{-2} and P_{CN} to R.

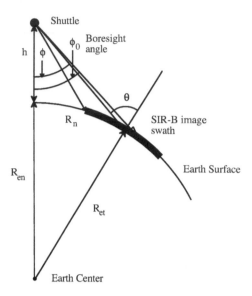

FIGURE 7-4   Coordinate system of SAR imaging, where R_{en} and R_{et} represent the earth radii at sub-satellite point and scene center, respectively.

If we know γ^0, P_{CN} (from system tests or analysis of the range spectrum), and P_C, the antenna pattern can be obtained by a simple arrangement of Equation (7.1a). This formulation requires knowledge of the local incidence angle; to obtain this angle, we may take one of two approaches: selecting a flat and uniform area or utilizing a digital elevation model (DEM) for the scene.

7.3.1   ACCURACY REQUIREMENTS

We first define the accuracy requirement for EAP estimation as 0.1 dB:1 sigma (including random and bias error). Next, the statistical analysis of the distributed target will be expanded as follows.

7.3.2   ERROR CRITERIA FOR INTENSITY AVERAGE

We consider the statistics of the image intensity for an area, A (Figure 7-3), and discuss the feasible error criteria. Each pixel intensity follows the $2N_L$-degree χ^2 distribution function with a mean, P_C, and a variance, P_C^2/N_{EL}, where N_{EL} is the effective number of looks (N_L and N_{EL} are almost similar). If area A is small, the intensity variance only depends on the number of pixels and the distribution function. As area A enlarges, the intensity variance, $\sigma_{C,A}^2$ consists of the second moment of P_C, σ_E^2, and the intensity variance at distributed components, $\sigma_{C,D}^2$:

$$\sigma_{C,A}^2 = \sigma_E^2 + \sigma_{C,D}^2 \tag{7.4}$$

$$\sigma_E^2 = \frac{1}{A} \iint_A (P_C(A) - \overline{P_C(A)})^2 \, dA \tag{7.5}$$

$$\overline{P_C(A)} = \frac{1}{A} \iint_A P_C(A) \, dA \tag{7.6}$$

Since two adjacent pixels are not independent (Shimada and Freeman 1995; Figure 7-5 for JERS-1 SAR), it is necessary to thin out the image at every D pixels so that the auto-correlation ρ_1

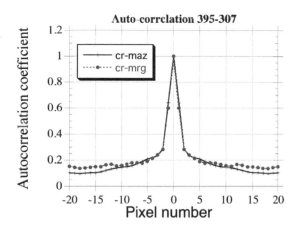

FIGURE 7-5 An example of the autocorrelation coefficient of JERS-1 SAR Amazon data.

is less than 0.1 in the range and azimuth directions. The resultant variance, $\sigma^2_{C,D}$, over N D-spaced samples is given by (see Appendix 7A-1):

$$\sigma^2_{C,D} = \frac{1 + 2\rho_1 + 2\rho_2}{N_L \cdot N} \frac{P^2_C}{N_L} \tag{7.7}$$

where ρ_2 is the auto-correlation in $2D$ distance. Differentiating Equation (7.1) (neglecting the noise) and taking the root sum square of its expansion, we get $\sigma^2_{C,A}$:

$$\sigma^2_{C,A} \leq P^2_C \cdot \left\{ \left(\frac{\Delta \gamma^0}{\overline{\gamma^0}} \right)^2 + \left(\frac{2\Delta G}{\overline{G}} \right)^2 + \ldots \right\} \tag{7.8}$$

where $\overline{G}$ and $\overline{\gamma^0}$ are the estimated antenna pattern and the gamma-zero, respectively. Based on the fact that the radar parameters are constant and γ^0 of the Amazon forest are stable (Shimada et al. 2014), Equation (7.8) can be simplified to:

$$\sigma^2_{C,A} \leq P^2_C \cdot \left(\frac{2\Delta G}{\overline{G}} \right)^2. \tag{7.9}$$

The EAP model error, ΔG, can be the root sum square of the estimated antenna patterns minus the measured antenna patterns:

$$G(\phi) = \overline{G}(\phi) + \Delta G. \tag{7.10}$$

Referring to the error analysis for the SIR-C data (Freeman 1990a; Freeman 1990b; Klein 1990), we apply the following criterion:

$$\left| \frac{\Delta G}{\overline{\overline{G}}} \right| = 0.3 dB (3 \text{ sigma}) \tag{7.11}$$

Arranging Equations (7.4), (7.7), and (7.9), we obtain an inequality, Equation (7.12), that allows the selection of A and the number of the uncorrelated image data points, N, in meeting the antenna pattern measurement requirement, Equation (7.11):

$$\frac{\sigma^2_E}{P^2_C} + \frac{1 + 2\rho_1 + 2\rho_2}{N^2_L \cdot N} \leq \left(\frac{2\Delta G}{\overline{\overline{G}}} \right)^2 \tag{7.12}$$

7.3.3 NUMBER OF DATA TO BE AVERAGED

The left side of Equation (7.12) consists of two types of variances. The first is a continuous function of the location; the second follows some distribution function and becomes smaller with averaging over large data. Many combinations of unknown parameters N and A can exist for Equation (7.12). For simplicity, we assume that the total is shared equally by two terms. The error criteria for each is:

$$\frac{\sigma_E}{P_C} \leq 0.142 dB, \tag{7.13}$$

$$\sqrt{\frac{1 + 2\rho_1 + 2\rho_2}{N_L^2 \cdot N}} \leq 0.142 dB. \tag{7.14}$$

The minimum number of data points N_{min} meeting Equation (7.14) is found by

$$N_{min} = \frac{1 + 2\rho_1 + 2\rho_2}{N_L^2 \cdot \left(10^{0.0142} - 1\right)^2}. \tag{7.15}$$

7.4 SCREENING PROCESS FOR THE DISTRIBUTED TARGETS

To obtain the EAP from the SAR images over distributed targets, the pixels should follow the same distribution function, with the mean as incidence angle-independent or dependent with known functions. Here, we define "uniformity" as a way so that the distribution function does not change with range. The Amazon forest generally satisfies this requirement because the forest backscatters the incoming signal through the volume scattering mechanism (Moore and Hemmat 1988).

Most of the images acquired by SARs show that the Amazon area looks uniform. Looking at the pictures carefully, however, we see some heterogeneous areas filled with rivers and some high-intensity areas. In Figure 7-6, we show pictures of the Canadian forest and images from Colombia.

Removing the nonuniform area in the images increases the credibility of estimation. A similarity test was applied (Figure 7-3 [B]). A whole image is divided into many small rectangular cells sized in N_{az} pixels in azimuth and N_{rg} pixels in range, named $Cell^{i,j}$, where i is the stripe number and j is the column number. By determining the cell size properly within the error criterion, we can eliminate all the nonuniform cells by the similarity test.

7.4.1 SIMILARITY CHECK BY CHI-SQUARE TEST

The similarity test is based on the chi-square test, which asks, "Are these two distributions different?"(Press et al. 1989; Hogg and Craig 1978). Supposing that $C_l^{i,j}$ and $\mathbf{C}_l^i$ are the measured histograms at the lth bin of $Cell^{i,j}$ and ith stripe, the chi-square parameter χ^2 can be used to evaluate this problem:

$$\chi^2 = \sum_{l=1}^{N} \frac{(C_l^{i,j} - C_l^i)^2}{C_l^{i,j} + C_l^i} \tag{7.16}$$

The accumulated chi-square distribution function defined in Equation (7.17) with r degrees of freedom and the gamma function $\Gamma(\cdot)$ gives the criterion X_α, under which statistical value χ^2 exists with the confidence of $\alpha_\%$ (Hogg and Craig 1978):

$$Q\left(\chi^2 \leq X_\alpha \,|\, r\right) = \int_0^{X_\alpha} \frac{1}{\Gamma\left(\frac{r}{2}\right) \cdot 2^{\frac{r}{2}}} \cdot w^{\frac{r}{2}-1} \cdot e^{-\frac{w}{2}} dw \tag{7.17}$$

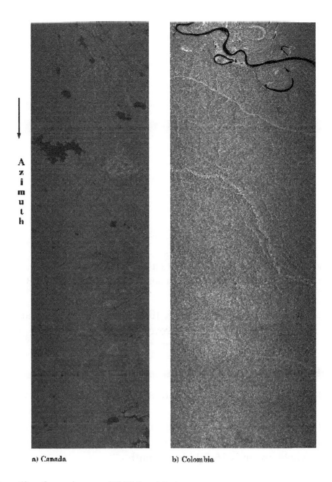

a) Canada   b) Colombia

FIGURE 7-6 (a) Canadian forest image, (b) Colombia image.

If χ^2 is less than or equal to the value X_α with $\alpha\%$ confidence, we can say, "These two distribution functions are not different." Hereafter, four values (95, 97.5, 99, and 100%) are used for the levels of confidence.

7.4.2 Determination of N_{rg}

We consider σ_E^2. In a cell, P_C depends on R, and Equations (7.5) and (7.6) are one-dimensional. From the first-order Taylor expansion of $P_C(R)$ around the center of the stripe, where $R = R_0$, σ_E^2 / P_C^2 is approximated by:

$$\frac{\sigma_E^2}{P_C^2} = \frac{1}{P_C^2 \cdot L_{rg}} \int_{-L_{rg}/2}^{L_{rg}/2} \left(\left. \frac{\partial P_C(R)}{\partial R} \right|_{R=R_0} \right)^2 R^2 \, dR = \frac{1}{P_C^2} \left(\left. \frac{\partial P_C(R)}{\partial R} \right|_{R=R_0} \right)^2 \frac{L_{rg}^2}{12} \tag{7.18}$$

$$L_{rg} = N_{rg} \cdot \frac{C}{f_{sample}} \tag{7.19}$$

where C is the speed of light, f_{sample} the sampling frequency, and L_{rg} the width of the cell in the range direction. A calculation was conducted simulating the Brazil case with a boresight angle of 34.7 degrees and using the published antenna pattern model (Dobson et al. 1986). Figure 7-7 shows

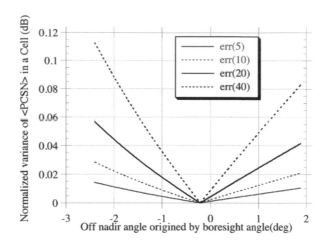

FIGURE 7-7 Simulation of $10*\log_{10}(1+\sigma_E/P_C)$ for Brazil-1 using representative parameters of SIR-B, where the edges of the image swath have higher values, even controlled, less than 0.14 dB.

the σ_E/P_C dependence on the stripe with N_{rg} for 5, 10, 20, and 40. At the near edge, N_{rg} of 40 almost meets the limitation requirement of 0.14 dB. However, we will use 20 for N_{rg} to obtain better results.

7.4.3 Determination of N_{az} and I_a

For the azimuth sizes of the cell (N_{az}) and entire image (I_a), we set the following conditions:

   i. The number of samples calculating the image intensity is greater than N_{min}, Equation (7.15).
   ii. A priori, a ratio of I_a to N_{az}: N_{raz} is set to 16 because the screening process can be simplified. This means the whole image is 16-fold in azimuth.

Conditions (i) and (ii) place the requirement on N_{az} as follows:

$$\frac{N_{raz}}{2} \cdot \left[\frac{N_{az}}{D}\right] \cdot \left[\frac{N_{rg}}{D}\right] \geq N_{min} \tag{7.20}$$

where [] means the Gauss notation. This gives the requirement for N_{az}:

$$N_{az} \geq \left(\frac{2 \cdot N_{min}}{N_{raz}} \cdot \left[\frac{N_{rg}}{D}\right]^{-1} + 1\right) \cdot D \tag{7.21}$$

Once N_{az} is obtained, I_a is simply given by:

$$I_a = N_{raz} \cdot N_{az}. \tag{7.22}$$

7.4.4 Similarity Test Procedure

There are 16 cells in each strip. The procedure (Figure 7-8) is given as follows:

   i. Calculate the intensity histogram at each cell ($C_l^{i,j}$), where l is the histogram bin, and obtain the reference histogram C_l^i from all the associated stripes as:

$$C_l^i = \frac{1}{16}\sum_{j=1}^{16} C_l^{i,j} \tag{7.23}$$

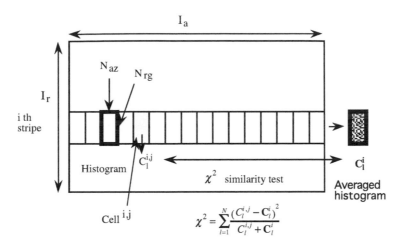

FIGURE 7-8   Screening procedure for each stripe's data in azimuth. The image is segmented in the azimuth direction (horizontal) and range direction (vertical). Ith strip consists of many $N_{az} \times N_{rg}$ sized rectangular segment in parallel to the azimuth direction. Each segment is measured for its histogram and processed in the similarity test to screen the usable data.

ii. Calculate χ^2 between $Cell^{i,j}$ and the ith reference by Equation (7.16).

iii. If χ^2 is less than **Q** at a given confidence, the $Cell^{i,j}$ is assigned as a member of the majority ith stripe.

iv. If more than half of the cells are similar, the stripe is assigned to EAP calculation. If not, the stripe is discarded.

v. This procedure is repeated over the full swath.

Applying this screening process to the image from Colombia, the algorithm successfully picked out the nonuniform points, such as the river running across the swath on the right-hand side; 5% of the data cells in the Colombia image are rejected as nonuniform.

In Figure 7-9, we introduce the JERS-1 SAR image and the screened area pattern.

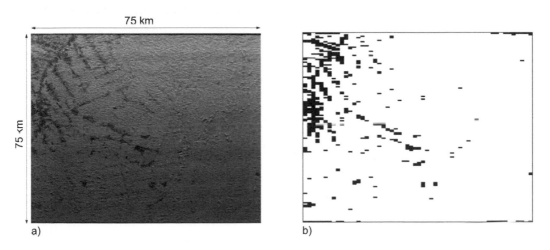

FIGURE 7-9   a) Amazon SAR image acquired by JERS-1 on May 17, 1992, GRS = 395-308, and b) screen process data.

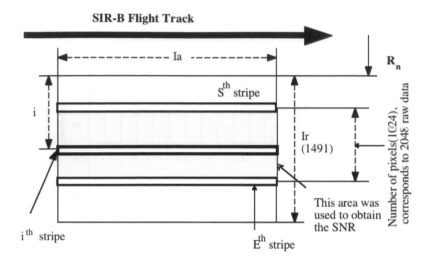

FIGURE 7-10   Image coordinate system including the strip number.

7.5   SAR CORRELATION POWER MODEL FOR EACH STRIPE

We apply the intensity model to the screened stripes (Figures 7-3 and 7-10) as follows:

$$P_{C,i}(m,\mathbf{u}) = \frac{g_m^2\left(\phi_i \big| \mathbf{u}\right)\cot\theta_i}{R_i^2} + B \cdot R_i \tag{7.24}$$

$$\phi_i = \cos^{-1}\left\{\frac{(h+R_{en})^2 + (R_n + N_{rg} \cdot C \cdot (i+0.5)/f_{sample})^2 - R_{et}^2}{2 \cdot (h+R_{en}) \cdot (R_n + N_{rg} \cdot C \cdot (i+0.5)/f_{sample})}\right\} \tag{7.25}$$

$$\theta_i = \sin^{-1}\left\{\frac{(h+R_{en}) \cdot \sin\phi_i}{R_{et}}\right\} \tag{7.26}$$

where i is the stripe number; R_{et} is the Earth radius at the image center; R_{en} is the Earth radius at the nadir of the SAR; h is the orbit height; R_n is the slant range to the near edge; B is the unknown constant; and $g_m(\phi_i \mid \mathbf{u})$ is the mth antenna pattern model with characterization vector, $\mathbf{u}$.

The image intensity model, Equation (7.24), is associated with the antenna pattern models discussed in the next section.

7.5.1   ANTENNA PATTERN MODEL

Here, we prepare three antenna pattern models:

- Model A:

$$g_1\left(\phi \big| \mathbf{u}\right) = 10^{0.1 \cdot \{a \cdot (\phi-\phi_0)^2 + b\}} \tag{7.27}$$

$$\mathbf{u} = \{\phi_0, a, b\} \tag{7.27a}$$

- Model B:

$$g_2\left(\phi \big| \mathbf{u}\right) = 10^{0.1 \cdot \{a \cdot (\phi-\phi_0)^2 + b + c \cdot (\phi-\phi_0)^4\}} \tag{7.28}$$

$$\mathbf{u} = \{\phi_0, a, b, c\} \tag{7.28a}$$

- Model C:

$$g_3(\phi \mid \mathbf{u}) = d \left\{ \mathrm{sin}\, c\left(\frac{\pi L_R \sin(\phi - \phi_0)}{\lambda} \right) + \mathrm{sin}\, c\left(\frac{\pi L_R \sin(\phi + \phi_0)}{\lambda} \right) \right\}^2 \cos(\phi - \phi_0) \qquad (7.29)$$

$$\mathbf{u} = \{\phi_0, d, e, L_R\} \qquad (7.29a)$$

where a, b, c, L_R, and e are the unknowns, and ϕ_0 is the boresight direction.

Model A is similar to the reference model (Moore and Hemmat 1988). Model C is a fit to the ground-based (preflight) antenna pattern for SIR-B, as measured by the antenna manufacturer, Ball Aerospace (Denver, CO). The Model C curve was used in the SIR-B image correlator, with e and L_R fixed at 0.3π and 2.16, for radiometric correction of standard SIR-B products. Model B is a fourth-order polynomial, which is more flexible in the EAP shape than Model A. A simple Taylor expansion of Model C reveals that fourth-order terms are significant, suggesting that the higher order of Model B should provide the needed degree of freedom.

7.5.2 Determination of B

Unless the noise is subtracted, the estimated EAP becomes broader. The SNR, measured from the raw signal spectrum, can be used to estimate the noise or directly implemented as BR. The constant B is determined by Equation (7.30). The total image intensity, $P_{total}(\mathbf{u})$, is

$$P_{total}(\mathbf{u}) = \sum_{i=S}^{E} \int_{R_{n,i}}^{R_{f,i}} \left\{ \frac{g_m^2(\phi \mid \mathbf{u})\cot\theta}{R^2} + B \cdot R \right\} dR, \qquad (7.30)$$

where S (E) is the start (end) stripe in the integral (Figure 7-10); $R_{f,i}$ and $R_{n,i}$ are the slant ranges at the far and near edges of the ith stripe; and $\mathbf{u}$ the unknown parameters. Letting $P_{int}(\mathbf{u})$ be the integral of the first rightside term of Equation (7.30) and the SNR as this integral, $P_{total}(\mathbf{u})$ is given by

$$P_{total}(\mathbf{u}) = P_{int}(\mathbf{u}) + \sum_{i=S}^{E} \frac{B}{2}\left(R_{f,i}^2 - R_{n,i}^2\right)$$

$$= P_{int}(\mathbf{u}) \cdot \left(1 + \frac{1}{SNR}\right) \qquad (7.31)$$

$$= \sum_{i=S}^{E} \overline{P_i} \cdot \Delta R$$

where $\overline{P_i}$ is the ith stripe intensity, and ΔR is the range spacing, C / f_{sample}. Finally, B is given by

$$B = \frac{2 \cdot \displaystyle\sum_{i=S}^{E} \overline{P_i} \cdot \Delta R}{(SNR + 1) \displaystyle\sum_{i=S}^{E} \left(R_{f,i}^2 - R_{n,i}^2\right)}. \qquad (7.32)$$

7.6   MAXIMUM LIKELIHOOD ESTIMATION AND THE SOLUTION

The chi-square $[\chi_m^2(\mathbf{u})]$ is the likelihood of the parameter estimation at the mth model:

$$\chi_m^2(\mathbf{u}) = \sum_{i=S}^{E} \frac{\{\overline{P_i} - P_{C,i}(m,\mathbf{u})\}^2}{\sigma_i^2} \tag{7.33}$$

$$\sigma_i^2 = \sigma_E^2 + \frac{1 + 2\rho_1 + 2\rho_2}{N_L \cdot N_i} \cdot \left(\overline{P_i}\right)^2 \cdot \left(R_{A,i} - 1\right) \tag{7.34}$$

where σ_i^2 is the ith intensity variance, N_i the number of samples, and $R_{A,i}$ the ratio of variance to the square of averaged intensity. The solution of the maximum likelihood estimation (MLE), $\mathbf{u}$, is given by the Lth order simultaneous nonlinear equations, which is derived by differentiating Equation (7.33) with respect to $\mathbf{u}$:

$$\sum_{i=S}^{E} \frac{\overline{P_i} - P_{C,i}(m,\mathbf{u})}{\sigma_i^2} \frac{\partial P_{C,i}(m,\mathbf{u})}{\partial u_k} = 0 \quad k = 1,2,3,...L \tag{7.35}$$

Replacing $\mathbf{u}$ with $\mathbf{u} + \Delta\mathbf{u}$ and expanding around $\mathbf{u}$ simplifies Equation (7.35) as

$$M_k^l \cdot \Delta u_l = N_k \tag{7.36}$$

where

$$M_k^l = \sum_{i=S}^{E} \left[\frac{\{\overline{P_i} - P_{C,i}(m,\mathbf{u})\}}{\sigma_i^2} \cdot \frac{\partial^2 P_{C,i}(m,\mathbf{u})}{\partial u_k \cdot \partial u_l} - \frac{\partial P_{C,i}(m,\mathbf{u})}{\partial u_k} \cdot \frac{\partial P_{C,i}(m,\mathbf{u})}{\partial u_l} \right] \tag{7.37}$$

$$N_k = -\sum_{i=S}^{E} \left[\frac{\{\overline{P_i} - P_{C,i}(m,\mathbf{u})\}}{\sigma_i^2} \cdot \frac{\partial P_{C,i}(m,\mathbf{u})}{\partial u_k} \right]. \tag{7.38}$$

Providing first-order derivatives and applying the Levenberg-Marquardt method (Press et al. 1989), these equations were solved.

7.7   A CASE STUDY FOR THE HIGH DYNAMIC RANGE SAR: SIR-B

We applied the previous method to estimate the inflight SIR-B EAP. The SIR-B images were collected from various areas: the Amazon, Colombia, Sumatra, Canada, and Illinois farmlands (listed in Table 7-1) to investigate the applicability of the proposed method.

Table 7-2 shows the distribution property of ρ_1, ρ_2, and N_{min}. The least uniform image is the Illinois farmland where the correlations extend to ten pixels, which requires a large number of pixels for averaging. The I_a (N_{az}) obtained is less than 928 (58), except for Illinois.

Table 7-3 lists B coefficients for all the locations, at which the noise level measurement is necessary for estimating more accurate antenna 3 dB or 1.5 dB beam width.

TABLE 7-1

Characteristics of Evaluated Images

(a) Summary of Image Data Used

Number of Scenes Evaluated	**8**	
Image data type	Slant range correlated image/not radiometrically converted	
Contents of data amplitude		
Number of bits per pixel	8	(0–255)
Number of looks	4	
Pixel spacing in range/azimuth	9.88/9(m)	

(b) Summary of Image Data Used (cont.)

Area	**Brazil-1**	**Colombia**	**Brazil-2**	**Illinois-1**
Data take[a]	118.30	118.30	118.30	070.10
Scene center	−4,17.4	3,22.0	−4,48.1	38,12.2
(lat., long.)[b]	−64,35.1	−68,54.8	−64,17.4	−88,23.5
Incidence angle[c]	35.6	36.6	35.6	49.0
Altitude (km)	222.41	222.14	222.46	231.92
SNR (dB)	9.22	9.28	9.22	4.48
Record length[d]	4352	4352	4352	4608
Number of records[d]	1491	1491	1491	1441
Area	**Illinois-2**	**Brazil-3**	**Canada**	**Sumatra**
Data take	97.20	118.30	053.20	086.60
Scene center	40,16.8	−3,36.3	55,51.8	−2,49.6
(lat., long.)[b]	−90,56.6	64,58.9	−100,5.0	103,12.2
incidence angle[c]	30.4	34.6	33.7	44.5
Altitude (km)	229.04	222.37	237.79	222.79
SNR (dB)	10.9	8.9	7.8	6.2
Record length[d]	4,608	4,352	4,352	4,096
Number of records[d]	1,492	1,491	1,202	1,697

Note:

[a] All these data were kindly supplied by the SAR Data Catalogue Center in NASA's Jet Propulsion Laboratory (JPL). Data take is a unique number assigned for each SIR-B operation duration.

[b] Lat. and long. are the abbreviations for latitude and longitude in units of deg. and min.

[c] Defined at the center of the image.

[d] Unit is unsigned character.

TABLE 7-2

N_{min}, N_{az}, and I_a

Scene	ρ_1	ρ_2	N_{min}	D	N_{rg}	I_a	N_{az}
Brazil-1	0.08	0.03	316	3	20	368	23
Colombia	0.10	0.05	337	3	20	400	25
Brazil-2	0.06	0.03	306	3	20	368	23
Illinois-1	0.17	0.09	394	10	20	4,112	257
Illinois-2	0.18	0.07	389	10	20	4,064	254
Brazil-3	0.13	0.07	363	3	20	416	26
Canada	0.10	0.05	337	5	20	928	58
Sumatra	0.06	0.02	301	3	20	352	22

Note: N_{min}, D are in pixels. Effective look number N_L is 3.5.

TABLE 7-3

Coefficient B

Scene_ID	Scene_Name	B	SNR(dB)
0	Brazil-1	1.45	9.2
1	Colombia	1.56	9.3
2	Brazil-2	1.47	9.2
3	Illinois	0.42	4.5
4	Illinois-2	0.49	8.9
5	Brazil-3	1.56	10.9
6	Canada	0.38	7.8
7	Sumatra	1.36	6.2

7.7.1 RESULTS

In total, 96 combinations (8 scenes, 3 models, and 4 confidence levels) were examined, and questions about (i) confidence level, (ii) antenna pattern model, and (iii) the best location were evaluated. The measurement for these questions was the residual error (RE):

$$RE = \sqrt{\overline{x^2} - \overline{x}^2}$$

$$\overline{x^2} = \frac{1}{N}\sum_{i=1}^{N}\{10 \cdot \log_{10} \overline{P_i} - 10 \cdot \log_{10} P_{C,i}(m,\mathbf{u})\}^2 \qquad (7.39)$$

$$\overline{x} = \frac{1}{N}\sum_{i=1}^{N}\{10 \cdot \log_{10} \overline{P_i} - 10 \cdot \log_{10} P_{C,i}(m,\mathbf{u})\}$$

7.7.2 CONFIDENCE LEVEL

Figure 7-11 shows the RE dependence on the confidence level for both the Canada and Colombia images. The results indicate that RE increases as the confidence level increases; Colombia has a much smaller RE than Canada. Out of four confidence levels, 99% provide the acceptable RE. Thus, 99% was adopted as the confidence level.

7.7.3 ANTENNA PATTERN MODEL

Model B always provides stable convergence with a smaller RE than Model A or Model C (Figure 7-11). Model A also shows good stability but has a slightly larger RE than Model B. A stable solution is a bit more difficult for Model C, but this is probably due to the larger number of parameters and the initial value dependency on the solution, especially the boresight angle, ϕ_0. Two cases were examined to evaluate MLE accuracy with the given initial boresights: Case A selected the boresight from the telemetry and Case B selected the boresight angle estimated from Model B. Case A provided the RE in the order: Model B, Model A, and Model C (Figure 7-11a). Case B provided the RE in the order: Model B, Model C, and Model A (Figure 7-11b). The effect of the error in boresight angle on SIR-B was previously noted in Dobson et al. (1986). Based on these results, Model B was selected as the best model for the study.

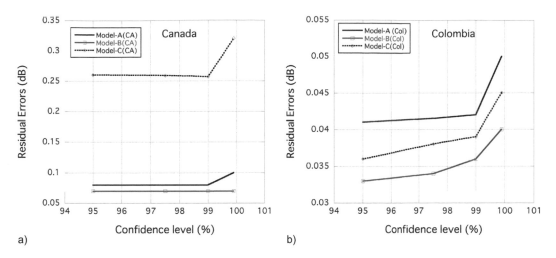

FIGURE 7-11 Residual error dependency on confidence levels in (a) Canadian and (b) Colombian locations.

7.7.4 BEST IMAGE DATA

Harmonization of the screened data and the estimated model (Model B) are shown in Figure 7-12, and their REs are given in Figure 7-13a.

Brazil-1 provides the smallest RE because of data uniformity; Colombia, Sumatra, Canada, and Illinois follow.

Illinois-1 (Figure 7-12d) shows the worst fit, with a wide data spread remaining around the curve. Illinois-2 (Figure 7-12e) contains similar scatterers (mixed agricultural fields), yet it has a smaller RE. There are two explanations. First, Illinois-1 covers only 2 degrees of the off-nadir, whereas Illinois-2 has 5 degrees. Second, Illinois-1 has an SNR of 4.5 dB, and Illinois-2 has an SNR of 10.9 dB. This gives Illinois-2 a greater chance of being a good fit.

Brazil-3 includes a large river in the image. This is successfully excluded by the screening process, as evidenced by a gap in Figure 7-12f. Canadian screened data fit well with the estimated antenna pattern, except for a few points near the edge and peak. The final image is from the Sumatran forest (Figure 7-12h). Although the data cover only 3 degrees off-nadir at the larger boresight, there is a fairly small residual error of 0.04 dB. Based upon this information, we select Brazil-1 as our best image data for the rest of the study.

7.7.5 EFFECT OF NOISE REDUCTION

A simulation study on the SNR versus antenna beam width was conducted using Model A and Model B with the screened Brazil-1 having 99% confidence. Here, SNR was set between 5 and 50 dB, and the related noise was subtracted. Figure 7-13b tells us that the beam width at SNR = 9 dB should be correct if the SNR is indeed 9 dB. The beam width at 50 dB of SNR is the same as the measurement that would result if no noise reduction was performed.

i. Model-B without noise reduction provides 6.5 degrees at 3 dB beamwidth; at 0.5 degrees wider estimates than those with noise reduction.
ii. Model-A without noise reduction gives similar results to Moore and Hemmat (1988).

7.7.6 RELATION BETWEEN THE ANTENNA BEAM WIDTH AND OFF-NADIR RANGE

The MLE accuracy depends on the unknown function property, the number of data, and the off-nadir angle range with which the function is associated. Figure 7-13c indicates the relationship

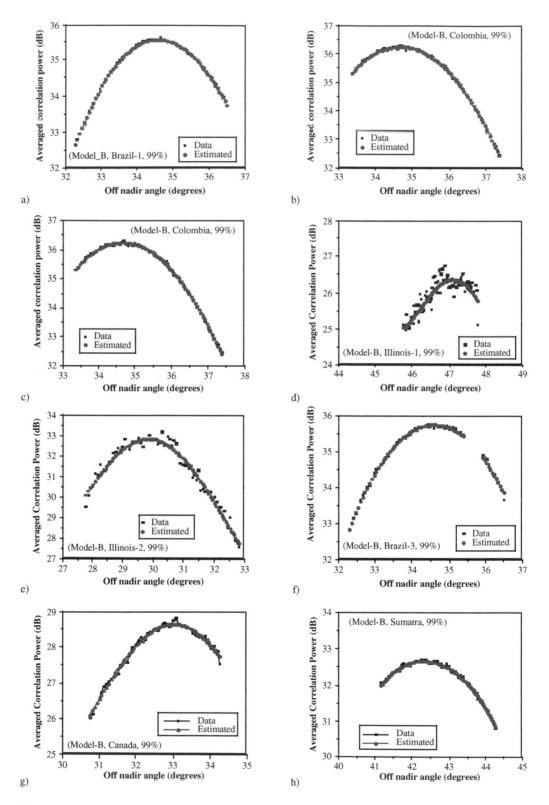

FIGURE 7-12 Data fitting for Brazil-1, Colombia-1, Brazil-2, Illinois-1, Illinois-2, Brazil-3, Canada, and Sumatra locations by Model B with 99% confidence.

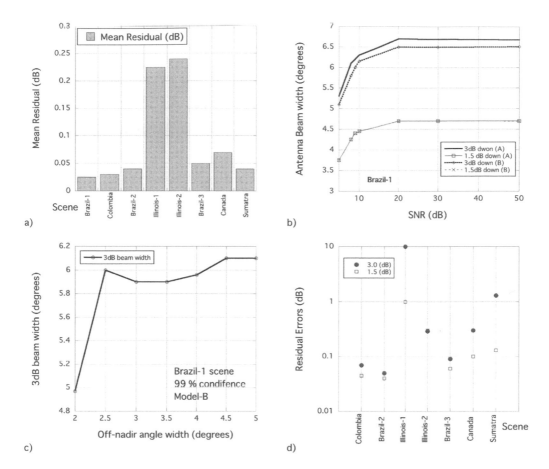

FIGURE 7-13 (a) Mean residual distribution over eight evaluation locations. (b) Relationship between antenna beam and SNR of the image data, where two lines are very close for 1.5 dB width. Four lines depend on the Methods A and B, and 1.5 dB/3.0 dB for antenna gain width. (c) Estimated beam width and data window width used for Brazil-1. (d) Repeatability of evaluation with regard to residual error.

between the beam width and the off-nadir angle range with Model B and Brazil-1 screened at the 99% confidence. This indicates that the minimum requirement for the off-nadir range is at least 2.5 degrees.

7.7.7 Repeatability of the Antenna Pattern Fit

The recommended inflight elevation antenna pattern for SIR-B is as follows:

$$10 \cdot \log_{10} g_2(\psi \mid \psi_0, a, b, c) = a \cdot (\phi - \phi_0)^2 + b + c \cdot (\phi - \phi_0)^4 \tag{7.40}$$

where:
$a, b, c = -0.286, 41.203, -0.00487$
Residual error = 0.024 dB
ϕ_0, ϕ = boresight angle, off-nadir angle
Beam width (3 dB down) = 6.0 degrees
Beam width (1.5 dB down) = 4.4 degrees

We calculate the root mean square (RMS) error between two antenna patterns, one of which is the target and the other of which is the reference defined in Equation (7.40), and we call this

TABLE 7-4

Summary of the Antenna Pattern Width

Item	1.5 dB Beam Width (deg)	3.0 dB Beam Width (deg)
This study	4.4 degrees	6.0 degrees
Ball Aerospace Data	4.8 degrees	6.7 degrees
JPL	4.2 degrees	5.9 degrees
Moore (1988)	5.0 degrees	6.9 degrees

repeatability. This error is defined as the standard deviation between two antenna patterns, including residual errors, as follows:

$$RMSerror = \sqrt{\Delta G_{ref}^2 + \overline{\Delta G_{target}^2} + (\overline{G_{ref}} - \overline{G_{target}})^2} \tag{7.41}$$

where ΔG_{ref} is the residual EAP error of reference antenna pattern, ΔG_{target} the residual error of the target scene, $\overline{G_{ref}}$ the antenna pattern model for the reference, and $\overline{G_{target}}$ the antenna pattern model derived from the target; $\overline{G_{ref}}$ and $\overline{G_{target}}$ are calculated at the 3-dB and 1.5-dB down points and correspond to the RMS errors shown in Figure 7-13d. For the Amazon, the RMS error is less than 0.08 dB at the 3-dB down points. This error increases to 0.3 dB for the Canadian forest and is worse than 1-dB for Sumatra and Illinois-1, with both having poor SNRs.

7.7.8 COMPARISON

Table 7-4 compares results of this study with previous SIR-B antenna pattern estimates. It shows that this result is 0.7 degrees smaller than that from Ball Aerospace and 0.9 degrees smaller than Moore's (1988) result. The possible reasons for the difference of 0.9 degrees are:

   i. *The model for the antenna pattern*—The antenna model with the best fit is confirmed as the fourth-order power model. If a second-order model is used, the lack of freedom gives a 0.2 degrees wider estimation in the 3-dB width result.

   ii. *Noise reduction*—Noise increases the total power in the image. The image with noise also gives a 0.5 degrees wider estimation.

Figure 7-14 shows the angular dependence of the three antenna patterns in comparison with Model B. This shows that Moore's pattern (1988) and that of Ball Aerospace differ from the Model B more than 0.3 to 0.5 dB and may cause stripes at the border of two neighboring images.

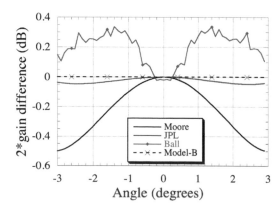

FIGURE 7-14 Angular dependencies of the three antenna patterns in comparison with Model B. Twice the difference is shown on the vertical axis because of the direct relationship with γ^0.

7.8   SUMMARY

This chapter showed a method for obtaining the inflight EAP of SAR based on slant range SAR images obtained mainly over the Amazon Rainforest. Using this result, the SIR-B antenna pattern has been estimated to have a 3-dB beam width of 6.0 degrees. The conclusions about this methodology and about the parameters to be selected for its use are as follows:

  i. **Area of interest:** The Amazon Rainforest provides a good enough distributed target because of uniformity.
 ii. **Signal-to-noise ratio:** Select the image that has a good SNR.
iii. **Noise reduction:** Noise reduction is necessary to properly estimate the EAP.
 iv. **Antenna pattern model:** The recommended antenna pattern model is Model B (fourth-order polynomial) for SIR-B type antennas.
  v. **Boresight angle:** It is necessary to select the boresight angle so that the widest range of off-nadir angles can be included in the image power pattern to obtain a better estimate by MLE.
 vi. **Error criterion:** The final results showed that the antenna error (residual) normally is less than 0.08 dB for the Amazon. The error criteria for the antenna pattern (0.3 dB as 3 sigma) is confirmed to be appropriate.
vii. **Repeatability:** Repeatability of EAP from the Amazon data is less than 0.1 dB RMS.

APPENDIX 7A-1:   AVERAGING THE PARTIALLY CORRELATED SAMPLES AND THEIR STANDARD DEVIATIONS

Suppose that there are N data sets (X_i), each of which is governed by the same distribution function, $P(X_i)$, where $i = 1, 2...., N$.

1. Each data set (X_i) is composed of M elements,

$$x_k, k = 1, \ 2, ..., M \tag{7A-1.1}$$

2. The N data sets have the same variance, σ^2, where x_i is its mean, and suffix i stands for the ith data set.

$$\sigma_i^2 = \overline{(x_i - \overline{x_i})^2} = \sigma^2, i = 1, 2, ..., N \tag{7A-1.2}$$

3. The correlation coefficient is symmetrical (like the autocorrelation function).

$$\rho_{i,j} = \left\{ \begin{array}{ll} \rho_l & l = |i - j| \\ 1 & i = j \end{array} \right\} \tag{7A-1.3}$$

$$\rho_{i,j} = \frac{\sigma_{i,j}^2}{\sigma_i \cdot \sigma_j} \tag{7A-1.4}$$

$$\sigma_{i,j}^2 = \overline{(x_i - \overline{x_i}) \cdot (x_j - \overline{x_j})} = \overline{x_i \cdot x_j} - \overline{x_i} \cdot \overline{x_j} \tag{7A-1.5}$$

The average over M data, z, and its variance, σ_z^2, is calculated as follows:

$$z = \overline{x_i} \tag{7A-1.6}$$

$$\sigma_z^2 = \overline{z^2} - \overline{z}^2$$

$$= \frac{1}{N^2} \left\{ \left(\sum_{i=1}^{N} \overline{x_i^2} + \sum_{i>j,i=2}^{N} 2\overline{x_i \cdot x_j} \right) - \left(\sum_{i=1}^{N} \overline{x_i}^2 + \sum_{i>j,i=2}^{N} 2\overline{x_i} \cdot \overline{x_j} \right) \right\}$$

$$= \frac{1}{N^2} \left\{ \sum_{i=1}^{N} \left(\overline{x_i^?} - \overline{x_i}^? \right) + \sum_{i>j,i=2}^{N} 2(\overline{x_i \cdot x_j} - \overline{x_i} \cdot \overline{x_j}) \right\} \qquad (7A\text{-}1.7)$$

$$= \frac{1}{N^2} \left(\sum_{i=1}^{N} \sigma_i^2 + \sum_{i>j,i=2}^{N} 2\rho_{i,j} \cdot \sigma_i \cdot \sigma_j \right)$$

Using this relation, we have:

$$\sigma_z^2 = \frac{1}{N^2} (N\sigma^2 + 2N\overline{\rho}\sigma^2) = \frac{\sigma^2}{N}(1 + 2\overline{\rho}) \qquad (7A\text{-}1.8)$$

where

$$\overline{\rho} = \frac{1}{N} \sum_{i>j,i=2}^{N} \rho_{|i-j|} = \frac{1}{N} \sum_{l=1}^{N} (N-1)\rho_l \qquad (7A\text{-}1.9)$$

The variance, σ_z^2, is then given by:

$$\sigma_z^2 = \frac{1 + 2\overline{\rho}}{N} \sigma^2 \qquad (7A\text{-}1.10)$$

REFERENCES

Dobson, M. C., Ulaby, F. T., Brunfeldt, D. R., and Held, D. N., 1986, "External Calibration of SIR-B Imagery with Area Extended and Point Targets," *IEEE T. Geosci. Remote,* Vol. 24, No. 4, pp. 453–461.

Freeman, A., 1990a, "SIR-C Calibration Plan: An Overview," JPL Report, JPL-D-6997, NASA, Jet Propulsion Laboratory, California Institute of Technology, Pasadena, CA.

Freeman, A., 1990b, "Spaceborne Imaging RADAR-C/SIR-C Ground Calibration Plan," JPL Report, JPL-D-6999, NASA, Jet Propulsion Laboratory, California Institute of Technology, Pasadena, CA.

Freeman, A. and Curlander, J. C., 1989, "Radiometric Correction and Calibration of SAR Images," *Photogramm. Eng. Rem. S.,* Vol. 55, No. 9, pp. 1295–1301.

Haupt, L. R., 2010, *Antenna Arrays: A Computational Approach,* Wiley, Hoboken, NJ.

Hawkins, R. K., 1990, "Determination of Antenna Elevation Pattern for Airborne SAR Using the Rough Target Approach," *IEEE T. Geosci. Remote,* Vol. 28, No. 5, pp. 896–905.

Hogg, R. V. and Craig, A. T., 1978, *Introduction to Mathematical Statistics,* Macmillan, New York, 1978.

Klein, J. D., 1990, "Spaceborne Imaging RADAR-C/Engineering Calibration Plan," JPL Report JPL-D-6998, NASA, Jet Propulsion Laboratory, California Institute of Technology, Pasadena, CA.

Moore, R. K. and Hemmat, M., 1988, "Determination of the Vertical Pattern of the SIR-B Antenna," *Int. J. Remote Sens.,* Vol. 9, No. 5, pp. 839–847.

Moore, R. K., Westmoreland, V .S., Frank, D., and Hemmat, M., 1986, "Determining the Vertical Antenna Pattern of a Spaceborne SAR by Observation of Uniform Targets," in *Proc. IGARSS '86 Symposium,* Vol. 1, Zurich, Switzerland, September 8–11, 1986, pp. 469–472.

Press, W. H., Flannery, B. P., Teukolsky, S. A., and Vetterling, W. T., 1989, *Numerical Recipes in C: The Art of Scientific Computing,* Cambridge University Press, Cambridge, UK, pp. 487–490 and 517–547.

Shimada, M. and Freeman, A., 1995, "A Technique for Measurement of Spaceborne SAR Antenna Patterns Using Distributed Targets," *IEEE T. Geosci. Remote,* Vol. 33, No. 1, pp. 100–114.

Shimada, M., Itoh, T., Motooka, T., Watanabe, M., Shiraishi, T., Thapa, R., and Lucas, R., 2014, "New Global Forest/Non-Forest Maps from ALOS PALSAR Data (2007-2010)," *Remote Sens. Environ.,* Vol. 155, pp. 13–31, http://dx.doi.org/10.1016/j.rse.2014.04.014

8 Geometry/Ortho-Rectification and Slope-Corrections

8.1 INTRODUCTION

SAR observes the Earth's surface under all weather and day–night conditions and provides both amplitude and phase for application purposes. Foreshortening, range and azimuth shift, layover, radiometric variation due to the slope, and shadowing lead the geometric and radiometric distortions.

When an imaging sensor (radar or optical) observes the land surface at squint directions, the target area is distorted by two modulations—geometry and radiometry.

The first issue involves geometry. If the targets were viewed from the center of the Earth, the image should look differently from the original sensor data. Robust conversion from a sensor-coordinate to a map-coordinate enhances the utilization of the sensor data because merging with a variety of data improves the data interpretation. This process is called ortho-rectification and can be achieved when the pixel height and the sensor position are well known. While ortho-rectification depends on the imaging methods (i.e., radar or optical), the principle is the same. To fully utilize the SAR data, accurate ortho-rectification is crucial.

The second issue is that SAR image intensity is modulated by terrain. From the radar equation, four backscattering coefficients are derived as the radar measure: the normalized radar cross section (NRCS), sigma-naught (σ^0), gamma-naught (γ^0), and beta-naught (β^0). Because the coefficients are normalized by the radar illumination area, they would be accurate only when the terrain height and resultant slope are known. Otherwise, the Earth ellipsoid is used for calculating the illumination area and the radar measure becomes incorrect. Thus, calculation of the correct radar illumination area is essential.

The accuracy of geometry and radiometry for SAR ortho-rectification and slope-correction relies on pixel height. While the SAR provides accurate ranging and azimuth location in the SAR coordinate, the height affects the location accuracy of the ortho-plane or map coordinate. This is because the Doppler frequency of the target varies with the height and causes an azimuth shift, and also because the radar ranging causes a range shift depending on height (only when the SAR imaging uses the nonzero Doppler center). The slant-range SAR intensity is radiometrically and geometrically modified in three ways: foreshortening, layover, and shadowing. The DEM provides the pixel height in equi-angular or equi-area map coordinates. Ortho-rectification requires the SAR pixel to be linked to the DEM (height).

In this chapter, we describe the theoretical background of geometric and radiometric distortions associated with SAR imaging. First, we describe the relationship among the SAR image shift, the pixel height, and the Doppler shift as well as generation of the DEM-Based Simulated SAR Image (DSSI). Second, we describe correction of the terrain-induced radiometric variation (i.e., slope correction) and the ortho-rectification of the SAR image (Shimada 2010).

8.2 GEOMETRIC CONVERSION BY THE FORWARD PROJECTION

As shown in Figure 8-1, the SAR imaging process projects (forward projection) all the targets or the scatterers on the topographic Earth's surface onto the slant range image plane (SRIP). This forward conversion is successful all the time. But, the backward projection is successful most of the areas except the layover or the shadowing areas. We consider the relationship of these forward and backward projections and also the geometric conversion of pixels to latitude and longitude. In SAR

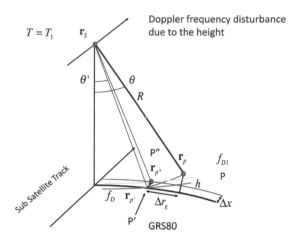

FIGURE 8-1   Coordinate system of the SAR image shows the projection of the nonzero height target (P) to P" over the GRS80 through P'. Here, three points are listed in Table 8-1; $\mathbf{r}_p$ is the vector expression of target point P.

processing, a pixel is uniquely identified by a Doppler frequency and a range distance in the range-Doppler plane. There are two important issues: range time compression and dual azimuth-and-range shifts of the nonzero height target.

8.2.1   RANGE AND AZIMUTH SHIFTS

Figure 8-1 describes: (1) A point P locates at height h over the Earth ellipsoid with a given latitude and longitude (the off-nadir angle to P is θ), and to which (2) the SAR imaging process, associated with the Doppler centroid (f_D) and Doppler chirp rate (f_{DD}), applies.

P is characterized by a Doppler frequency f_{D1}, differing from f_D by Δf_D (only when both h and f_d are nonzero). SAR imaging projects P onto the GRS80 plane in two steps: (1) moves (foreshortens) P to P' as intercepted over GRS80 under the same slant range distance (Rs) and f_D and (2) shifts P' to P'' in Δx, meeting the Doppler frequency shift (Δf_D). Another expression is P is range-shifted to P', which is the intersection point between the GRS80 and a (curved) line defined by a given slant range and f_D plus a shift in azimuth Δx due to Δf_D. This line normally has a complex form, and a part of a plane formed by a triangle, P-SAR-SSP, only if $f_D = 0$ with circular orbit around the Earth as a spherical surface. Here, the SSP is the sub-satellite point, which intersects the GRS80 in the line connecting the satellite and the center of the Earth. (Table 8-1 summarizes these three positions.)

The three methods previously discussed may be used to determine the location of the pixel on the SRIP as the back-projection process.

TABLE 8-1

Summary of the Point Movement in the Range Azimuth Domain

Notation	Position	Height	Doppler	Slant Range
P	$\mathbf{r}_p$	h	f_{D1}	R_S
P'	$\mathbf{r}_{p'}$	0	f_D	R_S
P"	$\mathbf{r}_{p''}$	0	f_{D1}	R_S

Note: Position with respect to Earth's center.

8.2.2 POSITION DETERMINATION

8.2.2.1 Rigorous Method

Accurate position determination of the point requires a full description of the surface on which the pixel is located, which is the digital elevation model. There are three equations that the points should satisfy:

$$f_{D1} = \frac{2f_0}{c}\left(\mathbf{u}_s - \omega \times \mathbf{r}_p\right)\frac{\left(\mathbf{r}_p - \mathbf{r}_s\right)}{\left|\mathbf{r}_p - \mathbf{r}_s\right|} \tag{8.1}$$

$$R_S = \left|\mathbf{r}_p - \mathbf{r}_s\right| = \frac{c}{2}\left(n \cdot f_{prf} + \Delta t_{off} + \frac{i}{f_{sample}}\right) \tag{8.2}$$

$$\mathbf{r}_p = R_E\left(\varphi, \lambda\right) + \left\{h\left(\varphi, \lambda\right) + h_{geoid}\left(\varphi, \lambda\right)\right\} \tag{8.3}$$

Here, f_{D1} is a measured Doppler frequency of the target P; c is the light speed; f_0 is the carrier frequency; $\mathbf{u}_s$ is the satellite velocity in the Earth-centered inertial (ECI) reference system; ω is the Earth's rotation vector; $\times$ is the vector product; $\mathbf{r}_p$ is the position vector of the target P; x_p, y_p, z_p are their three-dimensional components; $\mathbf{r}_s$ is the position vector of the satellite; and R_S is the slant range between the satellite and the target, P. The second formula in Equation (8.2) is from the radar parameters where n is the integer offset, f_{prf} the pulse repetition frequency, Δt_{delay} the internal time delay in the SAR, i the data address in the sampling window, and f_{sample} the sampling frequency.

Equation (8.3) represents the surface topography where R_E is the radius of the GRS80, $h(\varphi, \lambda)$ the height from the geoid, $h_{geoid}(\varphi, \lambda)$ the geoid height, φ the geodetic latitude, λ the longitude, and e the ellipticity.

The aforementioned equations can be solved for φ and λ, except for layover or shadowing regions, where the handling of the singular convergence is the issue.

8.2.2.2 Simplified (Robust) but Accurate Method

Replacing $\mathbf{r}_p$ n the numerator of Equation (8.1) by $\mathbf{r}_p - \mathbf{r}_{p'} + \mathbf{r}_{p'}$, and assuming that Earth's rotation speeds at $\mathbf{r}_p$ and $\mathbf{r}_s$ are almost the same, Equation (8.1) can be expanded to

$$f_{D1} \cong \frac{2f_0}{c}\left(\mathbf{u}_s - \omega \times \mathbf{r}_{p'}\right)\frac{\left(\mathbf{r}_{p'} - \mathbf{r}_s\right)}{\left|\mathbf{r}_p - \mathbf{r}_s\right|} + \frac{2f_0}{c}\left(\mathbf{u}_s - \omega \times \mathbf{r}_p\right)\frac{\left(\mathbf{r}_p - \mathbf{r}_{p'}\right)}{\left|\mathbf{r}_p - \mathbf{r}_s\right|}. \tag{8.4}$$

The first and the second terms have f_D and Δf_D as follows:

$$f_{D1} = f_D + \Delta f_D \tag{8.5}$$

The target with a Doppler frequency shift, Δf_D, is imaged in the azimuth direction by

$$\Delta x = -\frac{\Delta f_D}{f_{DD}}v_g \tag{8.6}$$

The target with a height is also range shifted by

$$\Delta r_g \cong \frac{h}{\tan\theta_{inci}} \tag{8.7}$$

Thus, the target P' is shifted to P'' with Δx in the azimuth direction caused by a Doppler shift; θ_{inci} is the normal incidence angle of the target and v_g the satellite speed on the ground in azimuth. It should be noted that Δr_g is an approximated range shift; a more detailed range shift is given in Appendix 8A-1, which should be used for precise determination.

A reference target point with zero height can be given by solving the following equations:

$$f_D = \frac{2}{\lambda}\left(\mathbf{u}_s - \omega \times \mathbf{r}_{p'}\right)\frac{\left(\mathbf{r}_{p'} - \mathbf{r}_s\right)}{\left|\mathbf{r}_{p'} - \mathbf{r}_s\right|} = a_0 + a_1 R_S \tag{8.8}$$

$$\frac{x_{p'}^2}{R_a^2} + \frac{y_{p'}^2}{R_a^2} + \frac{z_{p'}^2}{R_b^2} = 1 \tag{8.9}$$

$$R_b = R_a\sqrt{1 - e^2 \sin^2 \varphi} \tag{8.10}$$

$$e^2 = \frac{R_a^2 - R_b^2}{R_a^2} \tag{8.11}$$

Here, Equation (8.8) is given for P' with the vector form of $\mathbf{r}_{p'}$ and $x_{p'}$, $y_{p'}$, $z_{p'}$ as their three components; a_0 and a_1 are the Doppler models in the SAR imaging; R_a is the equatorial radius of the Earth ellipsoid; R_b is the polar radius; and e is the ellipticity. For a given R_S from Equation (8.2), $\mathbf{r}_{p'}$ can be accurately determined iteratively.

Thus, the real scattering target is projected onto the SAR image plane via two steps: (1) shifting Δx in the azimuth direction and (2) shifting Δr_g in the range direction. The former is caused by the Doppler frequency shift due to the squint angle observation and the nonzero Doppler center of the SAR antenna, and the latter is caused by nonzero pixel height. Both shifts present difficulties in determining the pixel locations without knowing the pixel height. In other words, knowledge of pixel height can allow these distortions to be corrected.

Both azimuth and range shifts rely on DEM accuracy. The associated location errors in azimuth and range due to the height error Δh can be expressed as follows:

$$\Delta x_h = -\frac{2f_0}{f_{DD}c}\left(\mathbf{u}_s - \omega \times \mathbf{r}_p\right)\frac{\left(\mathbf{r}_{p''} - \mathbf{r}_p\right)}{\left|\mathbf{r}_p - \mathbf{r}_s\right|}$$

$$= -\frac{2f_0}{f_{DD}c}\left(\mathbf{u}_s - \omega \times \mathbf{r}_p\right)\frac{\left(\mathbf{r}_{p'} - \mathbf{r}_p\right)\left(\mathbf{r}_{p''} - \mathbf{r}_p\right)}{\left|\mathbf{r}_p - \mathbf{r}_s\right|\left(\mathbf{r}_{p'} - \mathbf{r}_p\right)} \tag{8.12}$$

$$\cong \Delta x \frac{\Delta h}{h / \tan \theta_{inci}}$$

$$\Delta r_{gh} = \frac{R_E}{R_E + h}\frac{1}{\tan \theta_{inci}}\Delta h \tag{8.13}$$

Here, Δx_h is the additional azimuth error and Δr_{gh} is the same for the range. The first term, Δx, equals 0.01 and is negligible where $\Delta h \sim 10$ m, $h \sim 4{,}000$ m, and $\theta_{inci} \sim 35°$. The second term becomes slightly larger than the height accuracy depending on the normal incidence angle.

8.2.2.3 Simple (but Less Accurate) Position Determination

Rewriting Equations (8.2), (8.8), and (8.9) gives:

$$f_{D1} = \frac{2f_0}{c}\left(\mathbf{u}_s - \omega \times \mathbf{r}_p\right)\frac{\left(\mathbf{r}_p - \mathbf{r}_s\right)}{\left|\mathbf{r}_p - \mathbf{r}_s\right|}$$

$$R_S = \left|\mathbf{r}_p - \mathbf{r}_s\right| = \frac{c}{2}\left(n \cdot f_{prf} + \Delta t_{off} + \frac{i}{f_{sample}}\right)$$

$$\frac{x_p^2}{R_a^2} + \frac{y_p^2}{R_a^2} + \frac{z_p^2}{R_b^2} = 1, \text{ respectively.}$$

Equation (8.9) requires any pixel to be on the Earth ellipsoid. This is not realistic but offers the simplest way for position determination with non-negligible geometric errors. These equations are solved by two to three iterations.

8.2.3 COMPRESSION OF THE RANGE SCALING

In a case where the Doppler center frequency is nonzero but rather small, a hybrid processing convolved with the secondary range compression is conducted for focus-preserved imaging. This implies a change of light speed or time scale, and the scaling factor should be implemented for slant range calculation (Curlander and McDonough 1991; Jin and Wu 1984):

$$t_O = t_N \cdot \frac{f_{DD}}{f_{DD} + k_0\left(\dfrac{f_D}{f_0}\right)^2} \tag{8.14}$$

where t_O is the original timing and t_N the new time coordinate, f_{DD} the Doppler chirp rate, f_D the Doppler frequency, k_0 the chirp rate of the transmitted signal, and f_0 the transmission carrier frequency.

8.3 SLOPE-CORRECTION AND ORTHO-RECTIFICATION

In this section, we investigate the incidence angle dependence of the sigma-naught, re-correction of the antenna elevation gain, radiometric normalization of the layover region, comparison of the one-pass and three-pass interpolations for generation of the ortho-rectification image, and, finally, the procedure for ortho-rectification processing.

8.3.1 SLOPE-CORRECTED SIGMA-NAUGHT (σ^0) AND GAMMA-NAUGHT (γ^0)

The terrain surface, as the target of the SAR observation, inclines to the radar line of sight and is contaminated with the surface roughness (Figure 8-2). This area, A, is projected onto the area A' on the Earth's surface. Normally, A and A' are different. Thus, the NRCS measurement using A becomes incorrect. On the other hand, A is modulated by the terrain slope, and it is essential to characterize the target feature in an image. There are three expressions: the normalized radar cross section (NRCS, sigma-naught, or σ^0); the gamma-naught (γ^0): σ^0 divided by the cosine of the local incidence angle; and the beta-naught (β^0): σ^0 divided by the sine of the local incidence angle. Investigations (Ulaby et al. 1982) mention that the σ^0 decreases with the incidence angle except for the distributed target, the Amazon Rainforest.

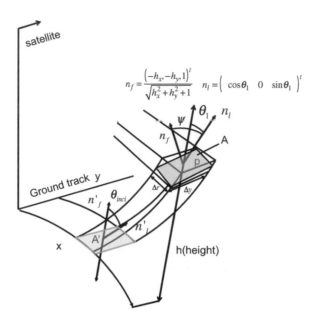

FIGURE 8-2 Scattering geometry. "A" is a part of the target surface that is intercepted by one SAR pixel ($\Delta r \Delta y$). The angle (ψ) between the surface normal $\mathbf{n}_f$ and slant range normal ($\mathbf{n}_l$) is a key for slope correction and differs from θ between $\mathbf{n}_r$ and ellipsoidal surface normal $\mathbf{n}_{f'}$.

An area of each pixel on the terrain mostly differs from the one on the GRS80 because it changes with the orientation of the tangential surface. Correction of the pixel area ratio, defined as the illumination area-correction factor (IACF), reduces the local incidence angle dependence of σ^0 but not perfectly. This is because the σ^0 of a natural target has its own local incidence angle dependence. By defining this dependence as the local incidence angle-correction factor (LICF) or a sigma-naught model (σ^0_{inci}), further correction of LICF could reduce or eliminate the incidence angle dependence of σ^0. Thus, the standard SAR image can be converted to the slope-corrected sigma-naught ($\tilde{\sigma}^0$):

$$\tilde{\sigma}^0 = \sigma^0 \cdot SCF \tag{8.15}$$

$$SCF = IACF \,/\, LICF \tag{8.16}$$

$$IACF \equiv \frac{\cos \psi}{\sin \theta_{inci}} \tag{8.17}$$

$$\mathbf{n}_f = \frac{1}{\sqrt{h_x^2 + h_y^2 + 1}} \begin{pmatrix} -h_x & -h_y & 1 \end{pmatrix}^t \tag{8.18}$$

$$\mathbf{n}_l = \begin{pmatrix} \cos \theta_1 & 0 & \sin \theta_1 \end{pmatrix}^t \tag{8.19}$$

$$\cos \psi = \mathbf{n}_f \cdot \mathbf{n}_l = \frac{\sin \theta_1 - \cos \theta_1 \cdot h_x}{\sqrt{h_x^2 + h_y^2 + 1}} \tag{8.20}$$

Here, σ^0 is the sigma-naught from the SAR image; ψ is an angle between the local normal vector ($\mathbf{n}_f$) and the normal vector ($\mathbf{n}_l$) perpendicular to the radar line of sight (LOS) within a plane in LOS and the center of the Earth (COE) (Shimada and Hirosawa 2000); θ_1 is the angle between the $\mathbf{n}_f$ and the position vector at the center of the pixel; θ_{inci} is the normal incidence angle at the GRS80; h is the height, h_x is its x derivative, and h_y is its y derivative. Note that, for the pixel on GRS80, ψ and θ_{inci} are complementary to 90 degrees and $\sin\theta_{inci}$ equals $\cos\psi$. It should be noted that, until recently, the standard SAR products mainly have been expressed by σ^0 or β^0 (e.g., JAXA has been adopting the σ^0 since the 1990s as well as the Sigma-SAR, while RADARSAT systems have been adopting the β^0), and they can be exchanged with trigonometric relations.

The slope corrected γ^0 is often used for forest data analysis because the diffuse reflection somewhat reduces the incidence angle dependence (Shimada and Ohtaki 2010). Thus, the slope-corrected γ^0 and σ^0 are convertible, but there is a slightly different expression for the slope correction, as seen in Equation (8.15)

$$\gamma^0 \equiv \frac{\sigma^0}{\cos\theta_{local}}\frac{\cos\psi}{\sin\theta_{inci}} = \frac{\sigma^0}{\cos\theta_{local}} IACF. \tag{8.21}$$

$$\theta_{local} = \cos^{-1}\left(\frac{(\mathbf{r}_s - \mathbf{r}_p)}{|\mathbf{r}_s - \mathbf{r}_p|}\cdot n_l\right) \tag{8.22}$$

Here, $\mathbf{r}_s$ and $\mathbf{r}_p$ are satellite position and target position, respectively. There are several studies modeling the σ^0 from the natural target; Goering et al. (1995) described the target dependence of forest as the diffuse scatterer and bare soil as the specular scatterer; Bayer et al. (1991) and Hinse et al. (1988) proposed a polynomial of the cosine of the local incidence angle; and Ulaby et al. (1982) showed the incidence angle dependence for several varieties of vegetation. Here, we tried to obtain a simple expression in a way that the slope corrected σ^0 could be incidence angle-independent. Using the PALSAR HH and JERS-1 SAR data, we obtained the LICF as a function of local incidence angle:

$$LICF \text{ or } \sigma^0_{inci} = 10^{d\cdot\theta_{local}} \tag{8.23}$$

We derived $d = -0.008$ for PALSAR and -0.006 for JERS-1 SAR when θ_{local} is expressed in degrees. Although we examined cosine dependence, Equation (8.23) showed the smallest variance and the best model. Between two coefficients, γ^0 provides more uniform backscattering coefficients than $\tilde{\sigma}^0$ and mainly is used as the unit for the slope-corrected SAR image.

8.3.2 ANTENNA ELEVATION PATTERN RE-CORRECTION

The lower level of SAR products, such as SLC or 1.5, are corrected for the antenna elevation pattern over the GRS80. Ortho-rectification can correct the exact antenna pattern at the target pixel height. The following is the correction parameter:

$$\gamma^0_{ORT} = \gamma^0_{SLT}\frac{G^2_{ele}(\theta')}{G^2_{ele}(\theta)} \tag{8.24}$$

Here, the suffix *ORT* represents the ortho-rectified value and SLT the slant range, θ' is the off-nadir angle calculated for the GRS80, and θ the real off-nadir angle (Figure 8-1); G^2_{ele} is the antenna elevation pattern.

8.3.3 RADIOMETRIC NORMALIZATION AT THE LAYOVER REGION

The layover or near layover area shows extremely high backscatter because most of the signals from the natural relief converge to one pixel or a small number of pixels. Because ortho-rectification expands the images with the converged pixel values as if seen from the center of the Earth to outer space, this brightness depends on the total area associated with the backscattering. Thus, one correction is as follows:

$$CFL = \frac{c}{2 f_{sample} \alpha R \sin \theta_0} \tag{8.25}$$

where CFL is a normalization factor, θ_0 the minimum local incidence angle at the layover region, and α the range of off-nadir angle that associates with the layover region. It should be noted that this normalization does not assure radiometric accuracy and works only for the suppression of brightness.

8.3.4 DEM-BASED SAR SIMULATION IMAGE (DSSI)

In this method, a simulated SAR image from the DEM (DSSI) plays an important role in determining the time and azimuth offsets. We show how to create a DSSI in Figure 8-3.

For a given azimuth time and a given slant range (R_s'), the parameters—satellite position, local incidence angle, illuminated area, and the slant range between the satellite and the target point (R_s) are calculated as follows in (a) of the first step.

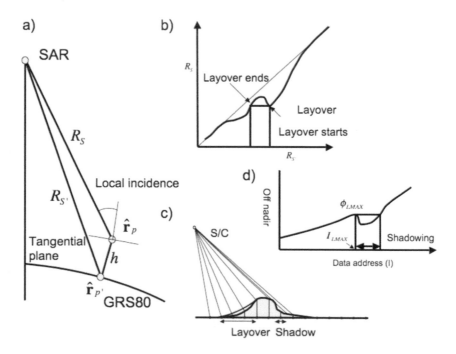

FIGURE 8-3 Coordinate system for the local incidence angle map and the mask information. (a) Relationship between the target point and the zero-height target point where the local incidence angle is shown as the one-dimensional angle; (b) the relationship between two slant ranges, and their singularity gives the layover; (c) shows the location of the layover and the shadowing; and (d) the shadowing condition.

1st Step: When decreasing R_s' from the far to the near (changing I from I_{max} to I_{min}), three areas are classified: the normal, layover, and shadowing areas (Figure 8-3).

a. Normal area: $\{I \mid R_{S,I} < R_{S,I+1} \cap \phi_I < \phi_{I+1}\}$ (8.26)

b. Layover area: $\{I \mid I < I_{LMIN} \cap R_{S,I} > R_{S,LMIN}\}$ (8.27)

c. Shadowing area: $\{I \mid I > I_{LMAX} \cap \phi_I < \phi_{LMAX}\}$ (8.28)

Here, I is the data address in range; ϕ_I is the off-nadir angle; $\cap$ is the conditional operation of *AND*; | means such that I_{LMIN} is the data address that gives the local minimum of slant range, $R_{S,LMIN}$; and I_{LMAX} is the data address that gives the local maximum of the off-nadir angle, ϕ_{LMAX}.

2nd Step: Scanning the azimuth time for a whole scene, we obtain two-dimensional local incidence angles and an illumination mask of layover, shadowing, normal, and ocean. Using the sigma-naught model in Equation (8.29), we can generate a draft DSSI:

$$\sigma_m^0 = 10^{d \cdot \theta_{local}} \tag{8.29}$$

3rd Step: In principle, $\mathbf{r}_s$ and DEM are correct, then $\mathbf{r}_{p'}$ is also correct, and the two images (SAR image and DSSI) coordinate well. In case either the DEM and/or $\mathbf{r}_s$ is incorrect, they do not coordinate well, and this leads to incorrect ortho-rectification. To prevent this, we may adjust the two time offsets—the range time offset (Δt_1), which is a delay in the receiver timing, and the azimuth time offset (ΔT_1), which is the orbit shift in the azimuth direction in such a way that the regenerated DSSI is co-registered with an SAR image. Two parameters are determined iteratively. Finally, the SAR image pixel is linked precisely with the DEM (and height). It should be noted that the ellipsoid height is the height from the geoid (e.g., SRTM, GTOPO30) plus geoid height.

8.3.5 PROCESS DESCRIPTION

Since a nonzero height target is shifted in the range and azimuth directions, ortho-rectification and geocoding are a nonlinear transformation of a slant range image to a map image. There are two methods: (1) step-by-step generation of an azimuth-shifted image, a range-shifted image, and projection onto the map, and (2) direct conversion of the slant range to the final map using the address manipulation formula. Calculation is simpler in the first method, but image quality suffers from multiple interpolations. The second method, with an interpolation, preserves image quality, although the address manipulation is complex. We adopt the second method because of the image quality being maintained.

8.3.6 PROCEDURE FOR ORTHO-RECTIFICATION AND DSSI GENERATION

Ortho-rectification consists of three steps (Figure 8-4):

Step-1 obtains the pixel height at the given slant range (R_s) relating R_s' and R_s (Figure 8-3a). The solution is found by using a lookup table iteratively at a given azimuth time. This step also provides the information on layover, shadowing, normal, and also the ocean, which is identified using the water mask information from SRTM-DEM. By scanning the azimuth time over an image, the maps for height, range, and azimuth shifts as well as DSSI are generated. The SAR image is rearranged based on the height-dependent azimuth shift referring to the aforementioned map. If the orbit and the DEM are accurate enough, the DSSI

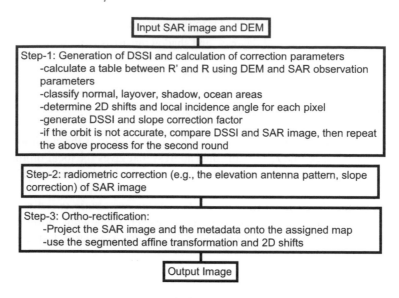

FIGURE 8-4   Flowchart showing the processes of generating the DSSI and ortho-rectifying the SAR data.

moves to Step-2. Otherwise, the DSSI and SAR image are compared, two time offsets are calculated, and the DSSI is updated for Step-2.

Step-2 prepares two factors, IACF and LICF, for all the pixels in the slant range image and also prepares the additional correction factor for the elevation antenna pattern.

Step-3 converts the slant range image to the final map: slant range, geocode, and geo-reference, correcting the range and azimuth shifts for each pixel. The SAR images and metadata are also associated with the final map.

LICF modeling and preparing the higher or resolution comparable DEM are important to express clearly the image edges and sharp lines at the layover regions, which works to co-register with the SAR image. We prepare two types for the SAR images and the DSSI (Table 8-2). Type 1 is an image pair of an SAR slant range image (including complex data) without any correction and the DSSI corrected both for azimuth and range shift. Type 2 is an image pair of the SAR image corrected for the azimuth shift and the DSSI corrected for the range shift. The azimuth shift correction for the complex data may cause the phase processing, and hence the DInSAR processing, to be degraded. Therefore, this correction for the complex data is not included in Table 8-2.

TABLE 8-2
DSSI and SAR Images

Type Number	SAR Image and Associated Corrections	DSSI and Associated Corrections
1	Slant range image (including complex data) without any corrections	Slant range DSSI, which is corrected for range and azimuth shifts based on Eqs. (8.12) and (8.13)
2	Slant range (amplitude) image that is corrected for azimuth shift based on Eq. (8.6)	Slant range DSSI that is corrected for the range shift based on Eq. (8.13)

Note: These corrections can be made using cubic convolution interpolation.

8.4 EXPERIMENTS AND EVALUATIONS

To evaluate the previously mentioned methods, we use two SARs (Table 2-3): the JERS-1 SAR observing Mt. Fuji and the surrounding area from the off-nadir of 35.1 degrees acquired on July 28, 1998, which is not the yaw-steering attitude control, and the ALOS PALSAR observing Mt. Fuji from 41.5 degrees and acquired on August 9, 2006, which adopts the yaw-steering attitude control. These two satellites are good examples for showing the height dependence of the Doppler shift and azimuth shift; the Doppler frequency for the JERS-1 SAR at the equator is roughly ± 2,000 Hz (for the descending and ascending orbits) and the Doppler frequency for the ALOS PALSAR is less than 100 Hz globally. We then evaluated the Doppler shift dependence on the height and resultant azimuth image shift. We selected Mt. Fuji, Japan, and the foothill region for this evaluation as elevation ranges from 0 m to 3776 m. There is a small highland area that allows the deployment of the point target for geometric evaluation.

In addition, we used the ALOS semi-global test sites, which were defined within the framework of the ALOS/PALSAR CAL/VAL (Shimada et al. 2009), for deployment of the calibration instruments. They were selected within ± 60 degrees of latitudes because of the DEM availability. Geometric accuracy of the ortho-rectification was measured using their corner reflectors (CRS) and their true geo-locations.

8.4.1 EVALUATION OF AZIMUTH SHIFT

8.4.1.1 JERS-1 SAR

Azimuth shift and the resultant correction were evaluated using the JERS-1 SAR data, which were acquired on July 28, 1998, with the off-nadir angle of 35.1 degrees in the descending orbit. Figure 8-5 shows (a) the azimuth shift-corrected SAR image and (b) a DSSI using the Type 2 method (refer to Table 8-2), both of which have 24 m of azimuth pixel size and 8.78 m of slant range pixel size. Both images look similar at a glance. Figure 8-6 shows the (a) Doppler frequency shift and (b) resultant azimuth shift. These images are similar and only differ in terms of their Doppler chirp rate and range dependence, as seen from Equation (8.6). To closely observe the range and azimuth variation of these two parameters, Doppler frequency shift and azimuth shift, two lines (A and B) were selected (Figure 8-5a), with these aligned in the range and azimuth directions and passing the summit of Mt. Fuji. Figure 8-7 shows (a) the azimuth profile of Doppler frequency shift and azimuth

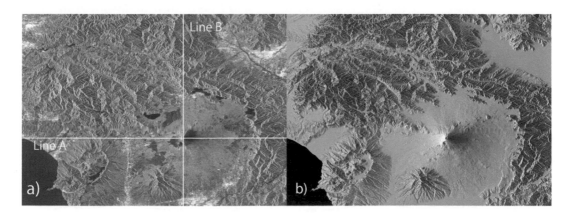

FIGURE 8-5 Comparison of (a) an SAR image corrected for the azimuth pixel shift due to height variation and (b) a DEM-based simulated SAR image, which is only corrected for the range shift. Here, the vertical axis is in the azimuth direction and the horizontal axis is in the range direction (left is near range and right is far range). The image is a four azimuth-look slant range image with an azimuth pixel size of 24 m and a range pixel size of 8.78 m.

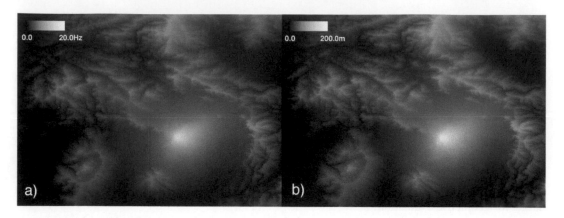

FIGURE 8-6 (a) Doppler shift of the target area at Mt. Fuji and (b) the corresponding azimuth shift image at Mt. Fuji.

pixel shift and (b) the range profile of these two parameters along these two lines. The summit of Mt. Fuji at the height of 3,776 m was shifted by 179.6 m in azimuth direction. This was calculated from Equation (8.6) using the Doppler shift of 16.022 Hz at the summit of Mt. Fuji, the Doppler chirp rate of −626.5 Hz/s, and the ground satellite speed of 7.02 km/s. Here, the Doppler center frequency for the SAR imaging approximated 1,760 Hz; thus, the Doppler shift was positive. This means that the SAR image and DSSI cannot be co-registered unless the azimuth shift is corrected. It also means that the geometric accuracy cannot be assured even when the orbital position accuracy is improved.

8.4.1.2 PALSAR

A similar evaluation was conducted for the PALSAR data, which were acquired on August 9, 2006, at an off-nadir angle of 41.5 degrees in descending node. The PALSAR data are shown in Figure 8-8 where (a) is the SAR image corrected for the azimuth shift, (b) is the DSSI, (c) is the azimuth cross section of Doppler and azimuth image shift, and (d) is the range cross section of Doppler and azimuth image shift. The azimuth and the range pixel sizes are 18 m and 4.68 m, respectively. The

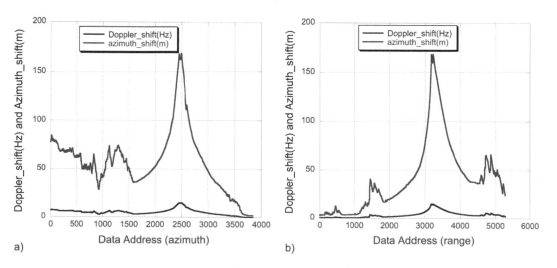

FIGURE 8-7 (a) Range cross section of the Doppler frequency and the azimuth shift and (b) azimuth cross section of the Doppler frequency and the corresponding azimuth shift.

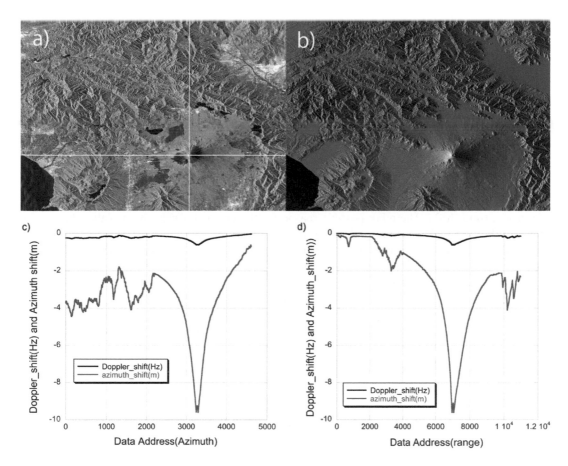

FIGURE 8-8 (a) PALSAR image acquired over Mt. Fuji corrected for the azimuth shift, (b) the DSSI, (c) Doppler shift and azimuth shift on azimuth address across the summit, and (d) Doppler and azimuth shifts on range address across the summit. The azimuth and range pixel sizes of the PALSAR image are 18 m and 4.68 m, respectively.

images for the Doppler and the azimuth shifts are not shown because values are small. The azimuth shift at the summit of Mt. Fuji was only −9.5 m. From this, we can estimate that the world's highest mountain (Mt. Everest at 8,776 m) may be azimuth shifted by approximately 30 m. The yaw steering can also suppress the deviation of the pixel location within some tens of meters in the azimuth direction. This means that yaw-steering attitude control is the only way to reduce the geometric error and generate high-quality SAR images. It should be noted that the Doppler center frequency for the processing was set at −97 Hz and, as a result, the Doppler frequency shift is negative.

8.4.2 GEOLOCATION ACCURACY EVALUATION USING THE CALIBRATION SITES

We evaluated the geometric accuracy of data ortho-rectified using the proposed method based on three types of data: (1) CRs, (2) the Geospatial Information Authority of Japan (GSI) DEM, and (3) the SRTM DEM. CRs are often used for evaluation of the geometric and radiometric accuracy of the slant range image. While the SRTM DEM is prepared semi-globally within ± 60 degrees of latitude with 90-m spacing, the GSI of Japan provides a DEM for Japan with 50-m spacing. In Shimada et al. (2009), the geometric accuracy of standard PALSAR images has been reported with an RMSE equal to 9.7 m of easting and northing components by using 572 Ground Control Points

(GCPs) deployed worldwide without any parameter tunings. These CRs are prepared with highly accurate positions (within sub-meters for latitude, longitude, and height from the GRS80 or geoid plane).

The geometric accuracy of the ortho-rectified SAR image was evaluated under the following conditions: (a) the range and azimuth time shifts were calculated by co-registration of the DSSI and the SAR image regardless of the orbit data accuracy associated, and (b) co-registration was not conducted and the time shifts were not updated. The evaluation results using PALSAR data and based on the GCPs showed that the geometric accuracy under condition (b) was always better (smaller) than that under condition (a). A reason for this may be because the area-matching method for the DSSI and SAR image performs well with a linear featured area at the mountain ridge and bright part or layover area of the SAR image, which is also the mountain peak. The DSSI generated with a backscattering model on the incidence angle does not express that part well, and it generates slightly erroneous time shifts. The other possible reason for the erroneous generation is that the speckle noise in the four-look SAR image does not coincide with the speckle-free DSSI. Here, four-look means that one SAR pixel is generated by averaging four contiguous pixels in the azimuth direction. While PALSAR data do not require co-registration processing, this is mandatory for JERS-1 SAR data because of the low orbital position accuracy.

Geometric accuracies and their dependence on three off-nadir angles (i.e., 21.5, 34.3, and 41.5 degrees) were evaluated using the CRs deployed globally (Figure 8-9). These figures were created only from the results obtained under condition (b). Each off-nadir case has the geometric accuracy measurement for ortho-rectification (left) and the slant range image (right). The geolocation error (RMSE) is defined by the following equation:

$$RMSE \equiv \sqrt{\frac{1}{N}\sum_{i=1}^{N}\left\{(\Delta x_i)^2 + (\Delta y_i)^2\right\}}$$

$$\Delta x_i = R \cdot \cos\varphi_{i,0} \cdot (\lambda_i - \lambda_{i,0}) \qquad (8.30)$$

$$\Delta y_i = R \cdot (\varphi_i - \varphi_{i,0})$$

Here, N is the number of measurements, suffix i the sample number, R the equatorial radius at the target point, $\varphi_{i,0}$ the true latitude of the ith measurement and φ_i is that measured from the SAR data, λ_i the true longitude, and $\lambda_{i,0}$ the SAR measurement; Δx_i and Δy_i are the easting error and northing error, respectively.

Table 8-3 summarizes the geometric accuracies of the ortho-rectified image and the slant range image as a reference. Each configuration has three values: the RMSE of the geolocation accuracy,

TABLE 8-3
Geolocation Accuracy Summary for Ortho-Rectification and Slant Range Images

Off-Nadir Angle (°)	Geolocation Accuracy (m): Average (Standard Deviation, Numbers)	
	Ortho-Rectified Image	Slant Range Image
21.5	17.383 (7.211, 21)	13.19 (5.267, 28)
34.3	11.925 (7.266, 104)	8.244 (4.716, 124)
41.5	9.488 (5.127, 50)	7.286 (4.017, 56)
Total value in RMSE	12.103 (6.718, 175)	8.885 (4.619, 208)

Note: Values in each element are RMSE defined by Equation (8.30) (standard deviation, number of samples).

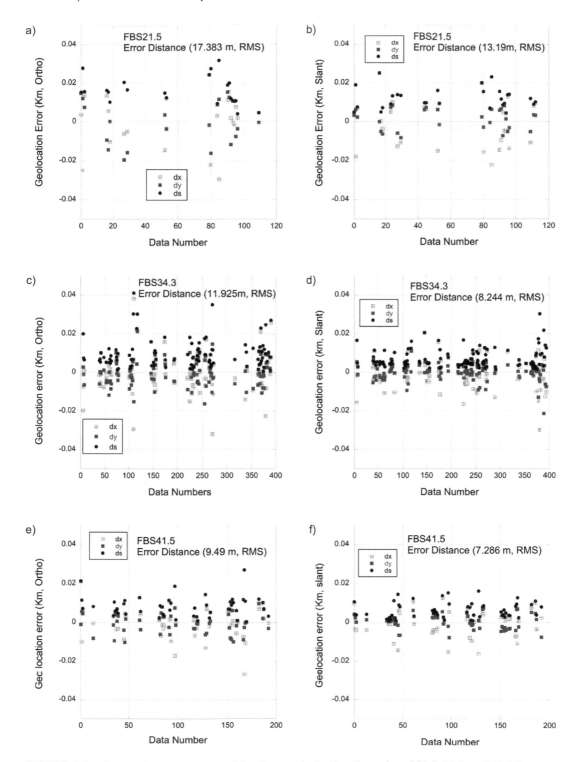

FIGURE 8-9 Geometric errors measured for three typical off-nadir angles of 21.5, 34.3, and 41.5 degrees. Each off-nadir case has figures for the ortho-rectification image (left) and the slant range case (right); Δx is the easting error, Δy the northing error, and Δs is the root mean square error. Here, the data number in the x-axis is the identification of the GCP and is almost aligned to the time series.

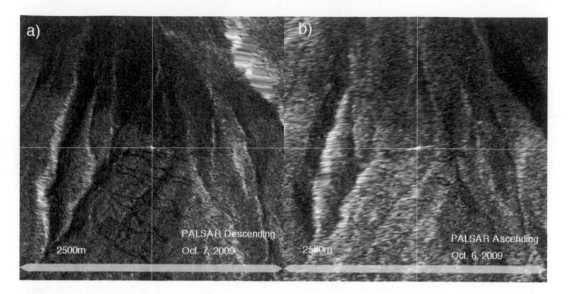

FIGURE 8-10   Two examples of geolocation evaluations: (a) Mt. Fuji in a descending path and (b) Mt. Fuji in an ascending path. The two cross points of the vertical and horizontal lines show the point calculated from the ground truth data, while the bright point shows the real ARC.

the standard deviation of the geolocation, and the number of samples. In general, the ortho-rectified image has a larger geometric error than the slant range image. There are two reasons for this:

1. The ortho-rectification accuracy depends on DEM accuracy, and there is a height difference of one to several meters between the ground truth value and the corresponding DEM.
2. While the ortho-rectified image is generated from the amplitude slant range image by bilinearly interpolating the surrounding pixel values to the corresponding output plane, the CR point is not perfectly preserved, and the pixel location that provides the largest response may become slightly inaccurate.

SLC data can, however, be used for sub-pixel interpolation to estimate the CR location. All the experiments show that the RMSE is larger than the standard deviation. Since the geometric error of RMSE is given by *sqrt* (standard deviation^2 +average^2), the geometric accuracy can be improved when we reduce the average component.

To evaluate the geometric accuracy, two ortho-rectified images were selected from the Mt. Fuji calibration site, where ARC was deployed at a high altitude area in a descending orbit on October 7, 2009, and in an ascending orbit on October 6, 2009 (Figure 8-10). The cross point of the two horizontal and vertical lines shows the calculated location of the ARC using ancillary data generated from the ortho-rectification process and true latitude and longitude of the ARC corrected for the internal time delay of 66 nanoseconds (ns). The cross point and the ARC position meet quite well. The true geolocation of the ARC is N35°20'12.97" and E138°43'57.02" on the ascending path and N35°20'12.97" and E138°43'55.78" on the descending path. The altitude is 2412.449 m de, and from the ortho-rectified image, the brightest point is located at N35°20'13.09" and E138°43'55.90" and at an altitude of 2,410.625 m (for the descending path) and N35°20'13.00", E138°43'57.03" and 2,410.000 m (for the ascending path; Table 8-4).

The ARC and the cross point of the two orthogonal lines agree well for both images. Therefore, the proposed ortho-rectification method properly corrects the height-induced azimuth and range shifts. Here, the 90-m spaced SRTM DEM was used for this ortho-rectification. As listed in Table 8-4, the height difference measured between the ground truth and the SRTM DEM is around 2 m. The ARC

TABLE 8-4

Comparison of the Ortho-Rectified Geometric Accuracy

No.	Latitude	Longitude	Height (m)	Δx (m)	Δy (m)	Δs (m)
Ascending	N35°20'13.00"	E138°43'57.03"	2410.625	0.253	3.092	3.102
Truth (asc)	N35°20'12.97"	E138°43'57.02"	2412.449			
Descending	N35°20'13.09"	E138°43'55.90"	2410.000	3.040	3.711	4.797
Truth (desc)	N35°20'12.97"	E138°43'55.78"	2412.449			

Note: Ascending data was acquired on October 6, 2009; descending data was acquired on October 7, 2009; "asc" and "desc" refer to ascending and descending orbits, respectively.

was deployed within a highland and flat area, and the 90-m resolution DEM is sufficiently accurate for the ortho-rectification. However, ortho-rectification of a small area with high relief requires a high-resolution DEM.

8.4.3 RADIOMETRIC NORMALIZATION AND ORTHO-RECTIFICATION

The radiometric normalization effect was evaluated using two PALSAR images. The first image is the Mt. Fuji region, which contains a variety of targets (i.e., steep mountain slopes, more gently sloping hilly areas, and lakes with the land cover largely forest but also with some urban areas), and it was acquired on August 9, 2006 (Figure 8-11). The second image is a mountainous region of Papua New Guinea, a mostly forested area on hilly terrain, and it was acquired on July 16, 2008 (Figure 8-12). Both target areas are associated with a large variation of terrain heights ranging from 0 to 4,000 m. For Mt. Fuji, a 50-m GSI-DEM was available, although for Papua New Guinea, only a 90-m SRTM DEM was available.

To evaluate range dependence of the normalization effect, we show two slant range images with and without slope corrections. In addition, two horizontal lines and two vertical lines were selected for quantitative evaluation of the slope correction effect. They are drawn in Figure 8-13a and b for Mt. Fuji and Figure 8-13c and d for Papua New Guinea. As seen in Figures 8-11 and 8-12, intensity variations due to topography that appeared in the standard SAR image were largely removed. Small rectangular areas surrounded by white lines in Figures 8-11 and 8-12 will be introduced later as the slope-corrected and ortho-rectified image examples.

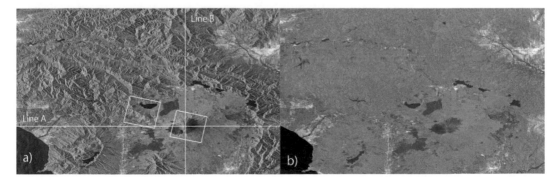

FIGURE 8-11 ALOS PALSAR images of Mt. Fuji (a) before and (b) after slope correction (for layover and cross-section area).

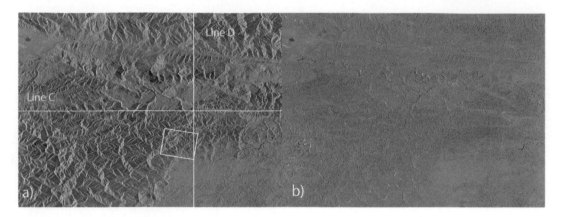

FIGURE 8-12 ALOS PALSAR images of a mountainous area in Papua New Guinea (a) before and (b) after correction for layover and cross-sectional area. Center coordinates are S0°44"17.75" in latitude and E133°0'23.14" in longitude.

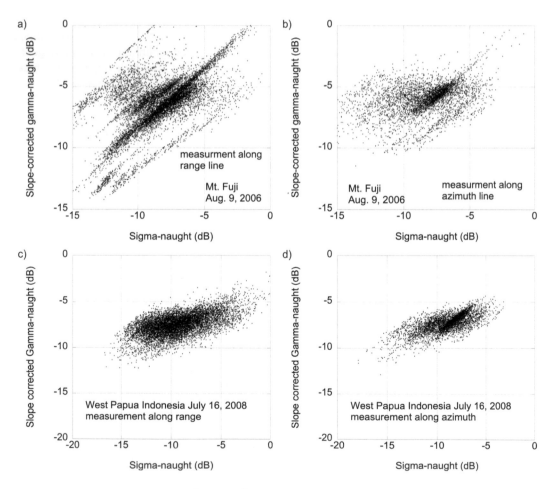

FIGURE 8-13 (a) Slope-corrected γ^0 versus σ^0 along (a) Line A for Mt. Fuji, (b) Line B in Mt. Fuji, (c) Line C in Papua New Guinea, and (d) Line D in Papua New Guinea.

Next, those images were ortho-rectified, slope-corrected, and map-projected using the proposed method and were compared with the images produced without slope correction. To closely view the images, an image section of 10 km east to west by 10 km north to south from the PALSAR standard image of 70 km × 70 km was extracted.

Two areas were selected for Mt. Fuji—the summit of Mt. Fuji (Figure 8-14a and b) and Lake Yamanaka and its southern hilly area (Figure 8-14c and d). Most of the terrain-induced σ^0 variation was removed, but the textures were preserved. The proposed slope correction method accounted for both the illumination area variation associated with the terrain slope within the pixel area and the local incidence angle dependence of the backscattering coefficients. Although the σ^0 model empirically derived for PALSAR and JERS-1 in Equation (8.23) could be used for the slope correction routine, the slope-corrected γ^0 of Equation (8.21) is used here because of better intensity

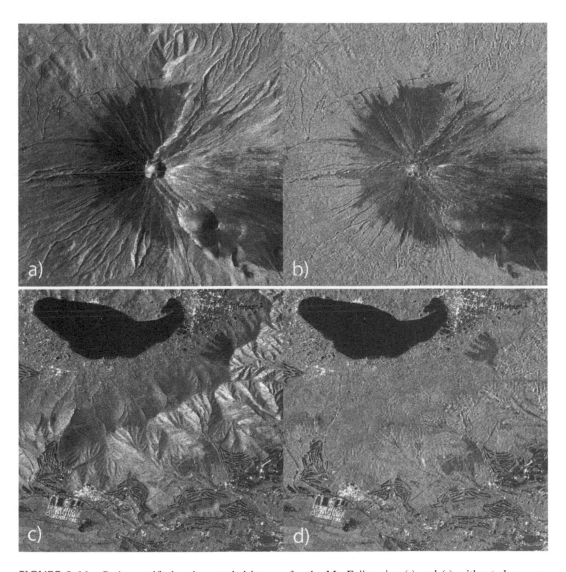

FIGURE 8-14 Ortho-rectified and geocoded images for the Mt. Fuji region (a) and (c) without slope correction in σ^0 and (b) and (d) with slope correction in γ^0. The image dimensions are 10 km and 10 km in the horizontal and vertical directions, respectively. These two areas correspond to two rectangular areas in Figure 8-10a.

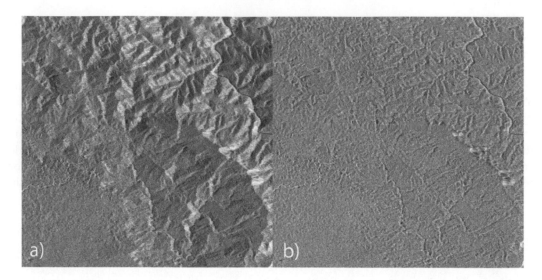

FIGURE 8-15 Ortho-rectified and geocoded images for mountainous area of Papua New Guinea (a) without slope correction in σ^0 and (b) with slope correction in γ^0. The image dimensions are 10 km and 10 km in the horizontal and vertical directions, respectively. This area corresponds to a rectangular area in Figure 11a.

suppression. For Papua New Guinea (Figure 8-15a and b), the proposed method also suppressed the variation in sigma-naught within the mountainous areas, although some brighter anomalies are still evident. These are associated with layover that cannot be expressed by the current accuracy and resolution of the DEM. A more accurate DEM, which could be generated by TanDEM-X, for example, is required.

8.5   DISCUSSION

This section discusses (a) the azimuth shift for co-registering the SAR image and the DSSI, (b) the validity of the offset tunings, (c) radiometric normalization using local slope correction, and (d) the availability of the proposed method for ortho-rectification and slope correction of the SAR (slant range) images.

8.5.1   AZIMUTH SHIFT AND CORRECTIONS

As discussed earlier, the Doppler shift induced by terrain height and the SAR processor with nonzero Doppler center frequency (NZD-SP) causes an azimuthal image shift. Two cases were demonstrated. JERS-1 SAR, representing the yaw-steering satellite, indicated an azimuthal pixel shift of 200 m at the Mt. Fuji summit, resembling 500 m or more for Mt. Everest. The shift is proportional to a ratio of Doppler shift to Doppler chirp rate and is independent of the frequency. However, PALSAR, representing the yaw-steering satellite, is experienced with Doppler frequency of less than 100 Hz and, in adopting NZD-SP, causes the azimuthal shift of 10 m at the Mt. Fuji summit. The azimuth beam width of 1.5 degrees requires 2,000 Hz of Doppler bandwidth and a bit higher PRF. The zero-Doppler SAR imaging (ZD-SP) is simply implemented, but it could cause the azimuth ambiguity. Thus, the NZD-SP is preferable even though the geometric conversion becomes rather complex.

8.5.2   VALIDATION OF THE OFFSET TUNINGS

PALSAR geometry is more accurate at no timing offset update than at timing offset update. It is recommended that the timing offset update is only necessary when the orbit is less accurate or the SAR receiving window delay is not determined precisely.

Because the JERS-1 orbit was so erroneous (180 m in azimuth and 50 m in cross track as three-sigmas), two timing offsets were necessarily determined. A current spaceborne SAR is linked with GPS, so the orbit and timing are accurate. ALOS, as one example, has four orbits: (a) the estimated orbit for the data reception has an accuracy of several tens of meters, (b) S-band range and range rate data as the determination orbits with accuracies of several tens of meters, (c) real-time single-frequency GPS-based orbital data with an accuracy of 30 m, and (d) high-precision positions with an accuracy of 40 cm as the three sigmas (Shimada et al. 2009). The timing offset determination is not necessary.

8.5.3 RADIOMETRIC NORMALIZATION BY THE SLOPE CORRECTION

Balancing the illumination area ratio expands the area compression by foreshortening at the fore-slope area and compression of the area expansion within the backslope area. Quantitative evaluation of this method showed the effectiveness of correcting the intensity modulation by the terrain slope. However, it remained difficult to correct the intensity jump at the layover region because that requires accurate calculation of the number of intersections between the given slant range and the terrain, and the exact accumulation of each backscattering contribution. It also requires a precise DEM with high spatial resolution. When the radiometric correction for the layover is not complete, the reliability of this area is reduced and an identification mask could be implemented.

8.5.4 EVALUATION OF THE PROPOSED METHOD

We evaluated the proposed method in terms of geometry and radiometry, using γ^0 instead of σ^0, as well as the robustness of the correction procedure.

1. From 208 PALSAR images and the global CRs or ARCs, a geometric accuracy of 12.103 m (RMSE) was measured. One ARC at an altitude of 2410 m depicted 5-m accuracy. From this, we determine that the ortho-rectification method can be fully usable in the GIS.
2. The slope correction compensates the radiometric variation due to terrain height variation, but the performance is limited by the DEM resolution. The 90-m DEM successfully corrects the hilly terrain areas but does not do this for the mountainous terrain areas with steep slopes. Higher resolution DEM may greatly assist the slope correction.
3. The robustness of the procedure depends on the automation and stability of the processing. A fundamental component of ortho-rectification is determining a pixel (or height) dependent range and azimuth shifts, and the proposed method non-iteratively calculates these in a deterministic way for all of the pixels. However, the approach is compromised when inaccurate orbital data are used, and determination of a mean range and azimuth time offset then becomes necessary. Within flat terrain, the DSSI and SAR image cannot be registered effectively without manual assistance, although the use of optical data mosaics (e.g., Landsat panchromatic data) may assist this process in the future. As the slope correction is not an iterative process, a more robust processing method has been provided.

8.6 SUMMARY

This chapter introduced the methods for pixel location determination and an ortho-rectification and slope correction. The method prepares:

a. Azimuthal shift determination due to the height-induced Doppler shift
b. Range shift and metadata using the lookup table relating to layover, shadowing, and normal clarification
c. Range and azimuth shift determination by co-registering the DSSI and the SAR image
d. The use of γ^0 and empirical determination of the incidence angle dependence of the σ^0 for DSSI

In each case, processing reliability depends on DEM accuracy. The method was applied to JERS-1 SAR and ALOS/PALSAR, with these representing non-yaw (with less accurate orbits) and yaw-steered (with highly accurate orbits) satellites, respectively. For the JERS-1 SAR, the two-dimensional shift showed satisfactory co-registration between the SAR image and the DSSI. For PALSAR, the geolocation accuracy was less at the larger off-nadir angle of 41.5 degrees, but the RMSE for 21.5, 34.3, and 41.5 degrees using the CRs deployed globally was 12.103 m. The geolocation accuracy of the proposed method was well maintained even for an altitude of more than 2,400 m, and the slope correction significantly suppressed the intensity modulation caused by terrain variation. Implementation of ortho-rectification and slope correction allows the generation of image products that can be used to support a wide range of applications.

APPENDIX 8A-1:   ACCURATE FORESHORTENING ESTIMATION

From Figure 8A-1, we have

$$\phi_2 = \cos^{-1}\left[\frac{\left\{R_E\left(\phi_2\right)+h\right\}^2 + R_X^2 - R_S^2}{2R_X \cdot \left\{R_E\left(\phi_2\right)+h\right\}}\right], \tag{8A-1.1}$$

where ϕ_2 is the angle between the SAR-CEO and CEO, R_X the distance between the satellite and the center of the Earth, R_E the slant range given by the radar system, R_S the distance between the center of the Earth and the target on the GRS80, and h the height of the target from the GRS80.

When R_E is a continuous function of the angle ϕ_2, Equation (8A-1.1) can be solved. Here, we prepare R_E as the polynomial function of ϕ for solving Equation (8A-1.2):

$$\phi = \cos^{-1}\left(\frac{R_E^2 + R_X^2 - R_S^2}{2R_X R_E}\right). \tag{8A-1.2}$$

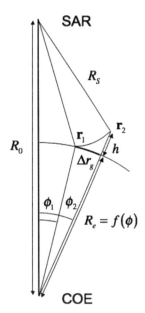

FIGURE 8A-1   Coordinate system applied for calculation of foreshortening.

Because a distance, Δr_g, along the GRS80 is ϕ_2 dependent as Equation (8A-1.3) and can be a polynomial,

$$\Delta r_g = \int_{\phi_1}^{\phi_2} R_E(\phi)d\phi. \qquad (8A\text{-}1.3)$$

REFERENCES

Bayer, T., Winter, R., and Schreier, G., 1991, "Terrain Influences in SAR Backscatter and Attempts to their Correction," *IEEE T. Geosci. Remote*, Vol. 29, No. 3, pp. 415–462.

Curlander, J. C. and McDonough, R., 1991, *Synthetic Aperture Radar, Systems and Signal Processing*, Wiley, Hoboken, NJ.

Goering, D. J., Chen, H., Hinzman, L. D., and Kane, D. L., 1995, "Removal of Terrain Effects from SAR Satellite Imagery of Arctic Tundra," *IEEE T. Geosci. Remote*, Vol. 33, No., 1, pp. 185–194.

Hinse, M., Gwyn, Q. H. J., and Bonn, F., 1988, "Radiometric Correction of C-Band Imagery for Topographic Effects in Regions of Moderate Relief," *IEEE T. Geosci. Remote*, Vol. 26, No. 2, pp. 122–132.

Jin, M. Y. and Wu, C., 1984, "A SAR Correlation Algorithm which Accommodates Large-Range Migration," *IEEE T. Geosci. Remote*, Vol. GE-22, No. 6, pp. 592–597.

Shimada, M., 2010, "Ortho-rectification and Slope Correction of SAR Data Using DEM and Its Accuracy Evaluation," *IEEE JSTARS special issue on Kyoto and Carbon Initiative*, Vol. 3, Issue 4, 2010, pp. 657–671.

Shimada, M. and Hirosawa, H., 2000, "Slope Corrections to Normalized RCS Using SAR Interferometry," *IEEE T. Geosci. Remote*, Vol. 38, No. 3, pp. 1479–1484.

Shimada, M., Isoguchi, O., Tadono, T., and Isono, K., 2009, "PALSAR Radiometric and Geometric Calibration," *IEEE T. Geosci. Remote*, Vol. 47, No. 12, pp. 3915–3932.

Shimada, M. and Ohtaki, T., 2010, "Generating Continent-Scale High-Quality SAR Mosaic Datasets: Application to PALSAR Data for Global Monitoring," *IEEE J-STARS Special Issue on Kyoto and Carbon Initiative*, Vol. 3, No. 4, pp. 637–656.

Ulaby, F. T., Moore, R. K., and Fung, A. K., 1982, *Microwave Remote Sensing: Active and Passive, Volume II: Radar Remote Sensing and Surface Scattering and Emission Theory*, Addison-Wesley, Boston.

9 Calibration—Radiometry and Geometry

9.1 INTRODUCTION AND CALIBRATION SCHEME

The term calibration has two meanings: an action undertaken to derive a correct relationship between the input and output in terms of the geophysical or engineering unit—the normalized radar cross section (NRCS), or an action undertaken to convert the output to the correct input or NRCS by using the calibration model. Thus, "an object to be calibrated" is composed of the SAR (hardware) and the image (or raw data) as shown in Figure 9-1. The calibration model contains the SAR (hardware) characteristics and the processing model, radiometry, geometry and the polarimetry. Figure 9-2 shows the various calibration processes and their components, calibration sources (reference sources), and their mutual correlations. Antenna pattern is involved in radiometry and polarimetry. The first two components are closely connected; the antenna elevation pattern is commonly applied to the SAR image.

9.2 RADIOMETRY AND POLARIMETRY

In this section, we will summarize the radiometry, the polarimetry, and the antenna pattern previously discussed in Chapters 3 through 7 and derive a simple equation to relate the NRCS and the image intensity as follows.

9.2.1 SAR IMAGE EXPRESSION

In Chapter 4, we derived

$$P_C = \left(\tau f_s \frac{PRF}{2V_g \rho_a} \lambda \right)^2 \frac{P_t (G\lambda)^2}{(4\pi)^3} N_l \frac{\sigma^0}{\sin\theta \cdot R^2} \delta_a \delta_r (SCR \cdot D) + N_{oise}$$

$$= A \cdot \frac{G^2(\theta)}{R^2 \sin\theta} \sigma^0 + N_{oise}$$

(9.1)

$$A = \left(\tau f_s \frac{PRF}{2V_g \rho_a} \lambda \right)^2 \frac{P_t \lambda^2}{(4\pi)^3} N_l \delta_a \delta_r (SCR \cdot D).$$

(9.2)

Here, P_C is the image intensity, N_{oise} the noise level (intensity), $G(\theta)$ the antenna elevation pattern, θ the incidence angle, R the slant range distance, and A the constant that contains all the related parameters.

Correcting $G(\theta)$, $\sin\theta$, and R^{-2} at the SAR imaging process and converting the intensity to the amplitude as denoted by "DN" (digital number), Equation (9.1) has two equalities

$$DN = \sqrt{B \frac{R^2 \sin\theta}{G^2(\theta)} P_C} = \sqrt{B(A \cdot \sigma^0 + N_{oise})}.$$

(9.3)

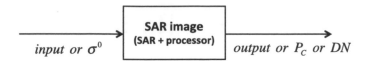

FIGURE 9-1   Input-output of an SAR image.

The SAR image has a large dynamic range, and the amplitude image needs at least 16-bit digits. To enable expression of the image from dark to bright, scaling is necessary. "*B*" is a constant to fit the image (amplitude rather than intensity) within a 16-bit dynamic range. Equation (9.3) is converted to

$$\sigma^0 = \frac{1}{A}\left(\frac{1}{B}DN^2 - N_{oise}\right). \tag{9.4}$$

The recent SARs have noise levels negligibly smaller than the observation signal. Introducing the ensemble average $\langle\cdot\rangle$ and obtaining the mean value of the radar backscatter by suppressing the speckle, we have,

$$\sigma^0 \cong \frac{1}{AB}\cdot\langle DN^2\rangle = CF\cdot\langle DN^2\rangle \tag{9.5}$$

where A and B are the constants, and CF the calibration factor replacing AB. In general, DN distributes in log-normal; B could be selected by a given constant value; and A can be adjusted (or calibrated) to the known scattering reference. This could be also expanded to a logarithmic expression as follows:

$$\sigma^0[dB] = 10\log_{10}\langle DN^2\rangle + CF \tag{9.6}$$

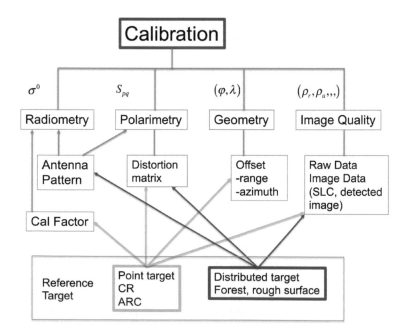

FIGURE 9-2   Calibration scheme and correlation diagram, where σ^0 is the backscattering coefficients; S_{pq} is the scattering matrix; (φ,λ) are the latitude and longitude of the position; and ρ_r,ρ_a are the range and azimuth resolutions.

This can be expanded to the single look complex (SLC) products as

$$\sigma^0_{1.1\ product} = 10 \cdot \log_{10} \left\langle I^2 + Q^2 \right\rangle + CF - A_0 \tag{9.6'}$$

Here, *DN* is the digital number of the amplitude image, which is a level 1.5 product (16-bit unsigned short integer) and *I* and *Q* are the real and imaginary parts of the SLC product (level 1.1). The conversion factor A_0 is set to 32.0 and *CF* is negative in decibel.

9.2.2 Full Polarimetric Expression

In Chapter 6, we derived the polarimetric calibration formula:

$$\mathbf{Z} = Ae^{\frac{-4\pi r}{\lambda}j} \mathbf{D}_r \cdot \mathbf{F} \cdot \mathbf{S} \cdot \mathbf{F} \cdot \mathbf{D}_t + \mathbf{N}$$

$$\mathbf{Z} = \begin{pmatrix} Z_{hh} & Z_{hv} \\ Z_{vh} & Z_{vv} \end{pmatrix}, \mathbf{D}_r = \begin{pmatrix} 1 & \delta_2 \\ \delta_1 & f_1 \end{pmatrix}, \mathbf{F} = \begin{pmatrix} \cos\Omega & \sin\Omega \\ -\sin\Omega & \cos\Omega \end{pmatrix} \tag{9.7}$$

$$\mathbf{S} = \begin{pmatrix} S_{hh} & S_{hv} \\ S_{vh} & S_{vv} \end{pmatrix}, \mathbf{D}_t = \begin{pmatrix} 1 & \delta_3 \\ \delta_4 & f_2 \end{pmatrix}, \mathbf{N} = \begin{pmatrix} N_{hh} & N_{hv} \\ N_{vh} & N_{vv} \end{pmatrix},$$

where **Z** is the uncalibrated distortion matrix containing four scattering components but corrected for the antenna elevation pattern and the free space decay, **S** the calibrated scattering matrix, $\mathbf{D}_r$ the receive distortion matrix, $\mathbf{D}_t$ the transmit distortion matrix, **F** the Faraday rotation matrix with Ω as the Faraday rotation angle, and **N** the noise. *A* is the calibration parameter convolved with the antenna elevation pattern and the other radar parameters.

After the polarimetric calibration, Equation (9.7) is converted to

$$\mathbf{S} = \mathbf{F}^{-1} \cdot \mathbf{D}_r^{-1} \frac{1}{A} e^{\frac{4\pi r}{\lambda}j} (\mathbf{Z} - \mathbf{N}) \cdot \mathbf{D}_t^{-1} \cdot \mathbf{F}^{-1}$$

$$\cong \frac{1}{A} e^{\frac{4\pi r}{\lambda}j} \mathbf{F}^{-1} \cdot \mathbf{D}_r^{-1} \cdot \mathbf{Z} \cdot \mathbf{D}_t^{-1} \cdot \mathbf{F}^{-1}. \tag{9.8}$$

"S" contains four components and is the "polarimetrically calibrated" (SLC) data. $\mathbf{F}^{-1} \mathbf{D}_r^{-1} \mathbf{D}_t^{-1}$ in Equation (9.8) is the unitary transformation. Equation (9.8) is the same as Equation (9.7) except for the unitary matrix. Thus, polarimetric SLC can be converted to the sigma-zero by

$$\begin{pmatrix} DN_{hh}^2 & DN_{hv}^2 \\ DN_{vh}^2 & DN_{vv}^2 \end{pmatrix} = \frac{1}{AB} \cdot \begin{pmatrix} \sigma_{hh}^0 & \sigma_{hv}^0 \\ \sigma_{vh}^0 & \sigma_{vv}^0 \end{pmatrix} + N_{oise}. \tag{9.9}$$

Then, sigma-zero is given by:

$$\begin{pmatrix} \sigma_{hh}^0 & \sigma_{hv}^0 \\ \sigma_{vh}^0 & \sigma_{vv}^0 \end{pmatrix} = CF \cdot \begin{pmatrix} \left\langle DN_{hh}^2 \right\rangle & \left\langle DN_{hv}^2 \right\rangle \\ \left\langle DN_{vh}^2 \right\rangle & \left\langle DN_{vv}^2 \right\rangle \end{pmatrix} \tag{9.10}$$

$$\begin{pmatrix} \sigma_{hh}^0 & \sigma_{hv}^0 \\ \sigma_{vh}^0 & \sigma_{vv}^0 \end{pmatrix} [dB] = 10 \cdot \log_{10} \begin{pmatrix} \left\langle DN_{hh}^2 \right\rangle & \left\langle DN_{hv}^2 \right\rangle \\ \left\langle DN_{vh}^2 \right\rangle & \left\langle DN_{vv}^2 \right\rangle \end{pmatrix} + CF[dB]. \tag{9.11}$$

Real values of $\mathbf{D}_r$, $\mathbf{D}_t$ over the Phased-Array L-Band Synthetic Aperture Radar (PALSAR) were shown in Chapter 6. ALOS-2 (Advanced Land-Observing Satellite) and Pi-SAR-L2 can be examined in a similar manner.

Dual polarization data cannot be polarimetrically calibrated straightway due to polarization insufficiency. When the cross talks are negligible and Faraday rotation is quite low, the scattering matrix becomes

$$\mathbf{S} \cong \frac{1}{A} e^{\frac{4\pi r}{\lambda} j} \mathbf{D}_r^{-1} \cdot \mathbf{Z} \cdot \mathbf{D}_t^{-1}$$

$$\left(\begin{array}{cc} S_{hh} & S_{hv} \\ S_{vh} & S_{vv} \end{array} \right) = \left(\begin{array}{cc} Z_{hh} & Z_{hv}/f_1 \\ Z_{vh}/f_2 & Z_{vv}/f_1 f_2 \end{array} \right).$$

(9.12)

Recent SARs have nearly zero cross talks but Faraday rotation depends on the frequency and the solar activity. At the L-band, Faraday rotation is quite large (0 to approximately 20 degrees depending on the time of the day and the geographic location), so the previous assumption might not be valid.

9.2.3 Inflight Antenna Calibration

All the SAR antenna patterns are measured on the ground before launch. They often change considerably, and this probably is due to huge vibrations during the launch and the conditional difference between the ground and space (Chapter 7). Recalibration of in-orbit antenna pattern is mandatory as the Committee on Earth Observation Satellites (CEOS) SAR CAL/VAL subgroup recommends analyzing the Amazon forest images that provide the incidence-angle-independent gamma-naught (Moore and Hemmat 1988; Dobson et al. 1986; Hawkins 1990). There are two antenna patterns: the range antenna pattern (RAP) and the azimuth antenna pattern (AAP).

9.2.3.1 Range Antenna Pattern (RAP)

To estimate the reliable antenna patterns using the SAR images observed for the natural forest, the nonuniform areas—deforestation, rivers, and different brightness—are excluded by using an F-distribution test with a confidence greater than 99.5%, and the least square minimization determines the coefficients:

$$G_R(\phi) = \alpha + \beta(\phi - \phi_0)^2 + \gamma(\phi - \phi_0)^4 \, [dB]$$

(9.13)

Here, ϕ is the off-nadir angle, ϕ_0 the bore-site off-nadir, and G_R is the antenna gain in decibels (see Chapter 7; Shimada and Freeman 1995).

9.2.3.2 Azimuth Antenna Pattern (AAP)

Accurate AAP is essential for suppressing ScanSAR scalloping in look summation (Shimada 2009). An azimuthal average of the SAR images over the uniform area and the polynomial model on the azimuth angle is the best approximation for the azimuth antenna pattern:

$$G_A(\varphi) = \sum_{i=0}^{n} a_i \cdot (\varphi - \varphi_0)^i \, [dB]$$

(9.13')

where a_i is the coefficient and φ the azimuth angle (see Chapter 5); n is selected around 10.

9.2.4 DETERMINATION OF THE CALIBRATION FACTOR (CF)

There are two methods for determining the calibration factor (CF): equalizing the two-dimensional integral of the impulse response from the corner reflector (CR), from which the ground clutter is subtracted, and the radar cross section of the CR (Gray et al. 1990) or using the Amazon forest for which gamma-naught is constant at −6.5 dB for HH polarization (Shimada 2005). The second method is simpler because the Amazon is large, flat, and a test site can be easily selected. We combine two methods to cover a wide range of incidence angles that SAR provides. Using the first approach, two methods are available: the integral method shown in Equation (9.14) and the peak method shown in Equation (9.15). In general, the integral method is more reliable than the peak method because the latter needs the exact resolution.

$$CF_{int\,eg} = \frac{RCS \cdot \sin\theta}{\iint\limits_{Area} \left(DN^2 - DN_N^2\right) dA} \tag{9.14}$$

$$CF_{peak} = \frac{RCS \cdot \sin\theta}{\rho_a \cdot \rho_r \left(DN^2 - DN_N^2\right)} \tag{9.15}$$

Here, RCS is the radar cross-section of the CR, θ the incidence angle, DN the digital number of the SAR amplitude image, ρ_{the} azimuth resolution, ρ_r the range resolution, suffix N the background value, and A_{rea} the area for the integral. If the calibration sites are selected on a dark site, the side lobe does not reach further than 200 m, even using a radar cross section of the target at 37 dBm^2. The integral method was proposed by Gray et al. (1990) such that the integral of the response preserves the scattering energy regardless of the image focus. Figure 9-3 compares the calibration integrals and the peaks.

Figure 15-6 shows the temporal variation of the gamma-zero of the L-band SAR (PALSAR) for the natural forest at 15 different global regions (Shimada et al. 2014a). It shows that gamma-zero of any forest area is region dependent but constant for multiple years, and that the second method is acceptable, requiring the correction of inter-regional differences.

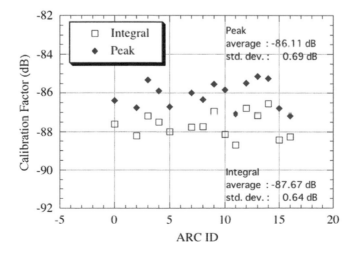

FIGURE 9-3   Stability of the calibration coefficients at the various CRs for two methods, integral and peak.

9.2.5 EXPRESSION OF THE SAR PROCESSING TO SAR CALIBRATION

The raw data are subjected to range and azimuth compression, whereas the strip data are processed by the correlation algorithm—range Doppler; the browse or ScanSAR data are processed by spectral analysis (SPECAN), and the complex data output follows (Figure 9-4)

$$
Z_{pq} = \left(
\begin{array}{cc}
\left\{ \dfrac{R\sqrt{\sin\theta_{inci}}}{G_R^{p,q}\left(\phi_{off}, beam\right)}\left(V_{pq} \oplus f_{INF} \oplus f_{rg}^{+}\right)\right\}_{RC} \oplus f_{az}^{*} & \text{correlation} \\[4ex]
\displaystyle\sum_{Nlook} F^{-1}\left[\left\{\dfrac{R\sqrt{\sin\theta_{inci}}}{G_R^{p,q}\left(\phi_{off}, beam\right)}\left(V_{pq} \oplus f_{INF} \oplus f_{rg}^{*}\right)\right\}_{RC} \cdot f_{az}^{*}\right] & \text{SPECAN}
\end{array}
\right)
\tag{9.16}
$$

$$
f_{rg} = \exp\left(\pi k_{t,model}\,t^2 j\right)
\tag{9.17}
$$

$$
f_{az} = \exp\left(2\pi j\left(\frac{f_{dd}}{2}T^2 + f_d \cdot T\right) + m\pi j\right) : \left(\begin{array}{l} m=1: if\ p=V \\ m=0: if\ p=H \end{array}\right)
\tag{9.18}
$$

$$
f_{INF}(\omega) = \frac{1}{1 - B_0/B_1} \cdot \left\{\begin{array}{l} 1: if\ spectrum\ at\ \omega\ is\ similar\ to\ SAR\ transmitted\ one \\ 0: if\ spectrum\ at\ \omega\ is\ not\ similar\ to\ SAR\ transmitted\ one \end{array}\right\}
\tag{9.19}
$$

$$
\mathbf{V}_{pq} = \frac{1}{\sqrt{G_{MGC} \cdot \left(P_t/\overline{P_t}\right)}\left\{1 - S_a(t,T)\right\}}\left[(\mathbf{v}-\overline{\mathbf{v}})\frac{\sigma_I}{\sigma_Q}\right]_{pq} \frac{\tau_0 f_{s0}}{\tau f_s}\sqrt{\frac{B_w}{B_{w0}}}G_1(beam, \bmod e)
\tag{9.20}
$$

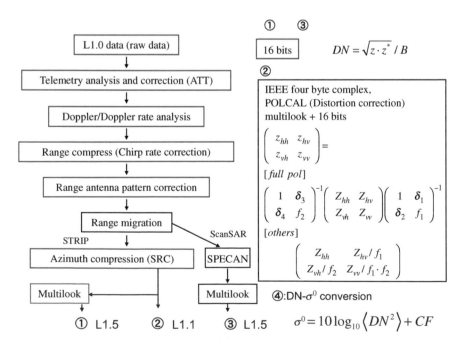

FIGURE 9-4   Process flow chart of PALSAR image generation and image calibration, where SRC is the secondary range compression, DN is the digital number, σ^0 is the sigma-naught, and CF is the calibration factor.

$$P_{pq} \cong \left(\begin{array}{ll} a \dfrac{G_R^{2pq}}{R^2 \sin \theta_{inci}} \sigma_{pq}^0 + N_{pq} & correlation \\[3mm] b \dfrac{G_R^{2pq}}{R^4 \sin \theta_{inci}} \sigma_{pq}^0 + N_{pq} & SPECAN \end{array} \right) \tag{9.21}$$

Here, Z_{pq} in Equation (9.16) expresses the slant range complex image of all the modes at p transmit polarization and q receive polarization, where p and q can be selected as either horizontal (H) or vertical (V); $\oplus$ indicates the correlation in the time or space domain; f_{rg} is the range reference function in Equation (9.17); f_{az} is the azimuth reference function in Equation (9.18), where transmission polarization of V for polarimetry requires a phase shift of π to co-register the H and V images; f_{INF} is the notch filter applied in the frequency domain in Equation (9.18), where $\oplus f_{INF}$ should be interpreted as multiplication in the frequency domain; and $\{\cdot\}_{RC}$ is the range curvature correction. Here, F is Fourier transformation, B_0 is the total number of the frequency bins zero padded, and B_1 is the total number of the frequency bins (e.g., sampling frequency); $k_{t,mode}$ is the mode-dependent chirp rate of the transmitted signal, f_D the Doppler center frequency, and f_{DD} the Doppler chirp rate; G_R is the range antenna pattern, R the slant range, θ_{inci} the incidence angle, ϕ_{off} the off-nadir angle, *beam* a beam number that varies from 0 to 131, and N_{look} the number of looks that is associated with the azimuth antenna pattern that is not shown in Equation (9.16); t is the fast time and T is the slow time.

In Equation (9.20), V_{pq} is the raw data corrected for the receiver gain (G_{MGC}), saturation rate (S_a), and analog-to-digital converter (ADC) imbalance between I and Q; σ_I and σ_Q are the gains (standard deviation); v and $\bar{v}$ are the raw and mean values, respectively; and P_t is the transmission peak power. The correlation gain reduction due to saturation is recovered by segmenting the whole image into small pieces (Shimada 1999).

Here, G_1 (beam, mode) is a newly introduced gain offset for unifying the calibration and adjusting the processor gain variation across the different beams by referring to FBS343HH, which is the representative PALSAR observation mode:Fine Beam single at HH polarization; τ is the pulse width; f_s is the sampling frequency; and Bw is the bandwidth. Suffix 0 shows the reference value, with $B_{w0} = 28$ MHz, $f_{s0} = 16.0$ MHz, and $\tau_0 = 27.0$ μs; G_1 will be determined using the response from the corner reflector for strip mode and using the distributed target for SPECAN.

Image intensity (power), P_{pq} in Equation (9.21), is inversely proportional to R^2 since the azimuth integration time is proportional to the slant range (to maintain the same azimuth resolution across the track), while the SPECAN is inversely proportional to R^4 since the deramping in azimuth uses a constant integration time across the track. Noise is associated with the high-bandwidth radar and with the non-zero thermal temperature in the receiver instrument. However, the lower value may be ignored when measuring σ^0. Processor gains "a" and "b" differ from the image-formation algorithm and antenna beams (refer to Chapters 3 and 4).

9.3   GEOMETRIC CALIBRATION

As discussed in Chapter 8, the pixel location is obtained by solving three equations on Doppler frequency, a distance between the SAR and the pixel, and the constraints that the pixel locates at known height over the Earth's ellipsoid. The resultant pixel position is a continuous function of the pixel (p) and line (l) as

$$(\varphi, \lambda) = \big(g(l, p), f(l, p) \big), \tag{9.22}$$

where φ and λ are the latitude and the longitude, and $g()$, and $f()$ are the connecting functions.

The previous distance depends on the atmospheric density and the ionosphere. Exact distance can be calibrated using the CR on the ground: geometric calibration determines the correct pixel location in a SAR image.

SAR imaging projects the non-zero height target onto the Earth's surface: GRS80. Let $\mathbf{r}_{pg}$ be the position with height z, and $\mathbf{r}'_{pg}$ the corresponding position on GRS80, $\mathbf{r}_s$ the satellite position, and f_D the Doppler frequency model; $\mathbf{r}'_{pg}$ can be obtained iteratively meeting the following equations:

$$\left| \mathbf{r}_{pg}(z) - r_s \right| = \left| \mathbf{r}'_{pg} - r_s \right| \tag{9.23}$$

$$f_D \left(r_{pg}(z) \right) = f_D \left(r'_{pg} \right) \tag{9.24}$$

The non-zero height is shifted in two ways—foreshortening in the range direction and azimuth walk due to the height-induced Doppler frequency shift. The resultant geometric error (Δ) is given by

$$\Delta = \left| r'_{pg} - r_p \right|. \tag{9.25}$$

Two parameters, the range time offset (Δt) and the azimuth time offset (ΔT), are highly correlated with the geometric errors in the east-west and north-south directions and are determined in a way to minimize the geometric error between the ascending and descending data. These tunings are called the "offset tunings in range and azimuth directions."

After the parameter determination, the aforementioned $g()$ and $f()$ can be given empirically as the polynomial equations:

$$\varphi = \sum_{i=0}^{4} \sum_{j=0}^{4} a_{i,j} \cdot \left(l - l_c\right)^{4-j} \cdot \left(p - p_c\right)^{4-i} + \varphi_c$$

$$\lambda = \sum_{i=0}^{4} \sum_{j=0}^{4} b_{i,j} \cdot \left(l - l_c\right)^{4-j} \cdot \left(p - p_c\right)^{4-i} + \lambda_c$$

$$\tag{9.26}$$

$$p = \sum_{i=0}^{4} \sum_{j=0}^{4} c_{i,j} \cdot \left(\lambda - \lambda_c\right)^{4-j} \cdot \left(\varphi - \varphi_c\right)^{4-i} + p_c$$

$$l = \sum_{i=0}^{4} \sum_{j=0}^{4} d_{i,j} \cdot \left(\lambda - \lambda_c\right)^{4-j} \cdot \left(\varphi - \varphi_c\right)^{4-i} + l_c$$

where a, b, c, and d are coefficients determined for each scene and the values suffixed c are defined at the scene center.

9.4   IMAGE QUALITIES

Two types of data can be evaluated: raw data and SLC.

9.4.1   RAW DATA

By analyzing the raw data, the hardware characteristics and the potential performance can be estimated. Key information is frequency spectrum, signal-to-noise ratio (SNR), I-Q average, orthogonality of I-Q channels, I-Q gain ratio, saturation rate, radio frequency interference (RFI), and chirp rate from the replica (recent SARs have a stable chirp signal generator and the range compression based on this does not cause further error).

9.4.1.1   Signal-to-Noise Ratio (SNR)

SNR is calculated as a ratio of the average power spectrum at a plateau area, $S + N$, and the noise area, N, which is measured between the bandwidth and the sampling frequency, minus one. All of the modes are measured around 8 dB and exceed 6 dB, compared to approximately 3 to 5 dB of JERS-1 SAR (Shimada et al. 1993) because PALSAR has more transmission power

$$SNR = \frac{\overline{P}_{S+N}}{\overline{P}_N} - 1 \tag{9.27}$$

In general, PALSAR-2 has more than 12 dB of SNR because the transmission power has been increased.

9.4.1.2 Average of I and Q

The null number for the ADC can be calculated by the following equations:

$$\overline{I} = \frac{1}{N}\sum_{I=0}^{N-1} I, \ \overline{Q} = \frac{1}{N}\sum_{I=0}^{N-1} Q \tag{9.28}$$

There are some trends in these stabilities: JERS-1 is less stable; Pi-SAR-L1, which is composed of eight individual ADCs, has different null levels at each ADC; and ALOS-2 has very stable values—all of which are evidence of the progress in digital hardware technology.

9.4.1.3 Orthogonality of the ADC

$$\Delta\phi = \cos^{-1}\left(\frac{\overline{(I-\overline{I})\cdot(Q-\overline{Q})}}{\sqrt{\overline{(I-\overline{I})^2}}\sqrt{\overline{(Q-\overline{Q})^2}}}\right) \tag{9.29}$$

Here, I and Q are the averaged I and Q.

9.4.1.4 I-Q Gain Ratio

This value can be calculated by

$$G_{I/Q} = \frac{\left\langle(Q-\overline{Q})^2\right\rangle}{\left\langle(I-\overline{I})^2\right\rangle} \tag{9.30}$$

9.4.1.5 Saturation

While the saturation reduces the correlation power, the measured saturation ratio implies the reduced power component:

$$S_a = \frac{h[0]+h[N-1]}{\sum\limits_{i=0}^{N-1} h[i]} \times 100[\%] \tag{9.31}$$

When the initial PALSAR operation started in early 2006 with the automatic gain control (AGC) mode, the data were severely saturated by 20% or more due to the functional limitations of AGC. Selecting the manual gain control (MGC) with the appropriate level, the average saturation rates dropped to between 0.4% and 2.4%.

9.4.1.6 Frequency Spectrum

The FFT of raw data in range direction shows the frequency spectrum of all the associated components of the signal reception: the SAR transmission property and the RFI. RFI in the SAR received signal causes the white noise in the image and reduces the SAR image quality. Recently, the increase of the L-band communication: broadcasting, communication system on the ground, air-traffic monitoring radar and more, tends to contaminate with the SAR data. Only the method to reduce the interference is to develop the appropriate bandpass filter.

9.4.2 Single Look Complex (SLC)

9.4.2.1 Impulse Response Function (IRF) and Resolution

The two-dimensional and two one-dimensional IRF is used for measuring range and azimuth resolution as the 3 dB down beam width. IRF can be calculated by using the FFT oversampling of factor 8 or 16, depending on the resolution requirement, which applies the zero padding in the frequency domain. The point is that the FFT needs to be performed for the SLC product not for the amplitude product.

9.4.2.2 Side Lobes

Two side lobe ratios are defined to specify the resolution measure: peak-to-side-lobe ratio (PSLR) and integrated side-lobe ratio (ISLR). The former is the ratio of the first side-lobe to the main beam peak, and the latter is 1.0 minus the relative main-lobe energy divided by all the energy, where the border between the main lobe and the others is the null point between the main beam and the first side lobe.

9.4.2.3 Noise-Equivalent Sigma-Zero (NESZ)

NESZ represents the maximum radar sensitivity and is defined as the minimum sigma-naught observed at each incidence angle. Ideally, the NESZ is measured from the shadowed area in a steep mountain. In reality, the minimum search along a strip provides the incidence angle dependence of NESZ.

9.4.2.4 Ambiguity

Azimuth ambiguity appears along a track by

$$\frac{f_{prf}}{f_{DD}} v_g.$$

(9.32)

PALSAR's shorter antenna requires a higher PRF and reduces the azimuth ambiguity (AA) relative to JERS-1 SAR. Thus, AA is not often seen. However, range ambiguity (RA) sometimes appears at the image edge because the neighboring pulse return received through the antenna side lobe causes line-like noise due to improper range curvature. The measured RA of this line noise is −23 dB, but the specification is −16 dB. Future SARs should improve on these values (refer to Chapter 3).

9.4.2.5 Cross Talk between HH and HV or VV and VH

For the full polarimetry case, cross talk measurement is very sensitive to the accuracy of the distortion matrix and the Faraday rotation. Here, we prepare two measures, the normalized cross-correlation between HH and HV or VH and VV for the distributed target, and the power ratio of HV to HH or VH to VV for the IRF, where each power value is corrected for background noise:

$$cross_1 = 10 \cdot \log_{10} \left(\frac{\left| \left\langle s_{hv} s_{hh}^* \right\rangle \right|}{\sqrt{\left\langle s_{hv} s_{hv}^* \right\rangle} \cdot \sqrt{\left\langle s_{hh} s_{hh}^* \right\rangle}} \right)$$

(9.33)

$$cross_2 = 10 \cdot \log_{10} \left(\frac{P_{hv} - P_{sur,hv}}{P_{hh} - P_{sur,hh}} \right)$$

9.4.2.6 Irregularities

Radio frequency interference (RFI), ionospheric irregularity, Faraday rotation, and tropospheric irregularities are caused by human and natural activities (will be discussed in Chapter 14).

9.4.2.7 Radiometric Stability

The calibration factor (CF) is calculated by using the method described earlier. Temporal stability is often used for the calibration stability.

9.4.2.8 Geometry

A CR is used for providing the accurate geolocation reference—latitude and longitude, and their statistical evaluation can qualify the SAR products.

9.5 CALIBRATION SOURCES

9.5.1 Artificial Calibration Sources

CRs, ARCs, and ground-based receivers represent the calibration instruments (Figure 9-5).

9.5.1.1 Corner Reflectors (CR)

As listed in Table 9-1, CRs are made of circular, planar, dihedral, trihedral, or other shapes of metallic plates (Ulaby et al. 1982; Freeman et al. 1988; Freeman 1990). A trihedral CR is often used for calibrating like-polarization and full polarimetry because the larger RCS and wider beam width allow ease of deployment. A dihedral CR calibrates the cross-polarization. A pentagonal CR was proposed to improve the reflected wave that may be degraded by the possible multipath between the ground and the leaves of the CR (Sarabandi and Chiu 1994). The disadvantages of CRs are

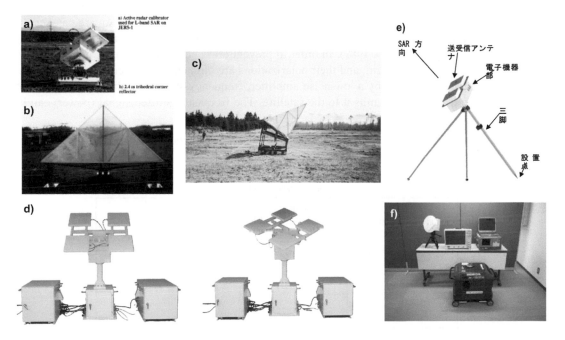

FIGURE 9-5 All the calibration instruments used for the L-band SAR: (a) ARC for JERS-1 SAR, (b) 2.4-m CR for JERS-1, (c) 3.0-m CR for PALSAR, (d) polarimetric ARC for PALSAR, (e) compact ARC for PALSAR-2, and (f) receiver for PALSAR-2 and Pi-SAR-L2.

TABLE 9-1

Summary of the Corner Reflectors

No.	Type of Target	Peak RCS	Half-Power Beam Width	Comments
1	Rectangular plate	$\dfrac{4\pi A^2}{\lambda^2}$	$0.44\lambda/a$	Large σ_{max}, very narrow beam width
2	Circular plate	$\dfrac{4\pi A^2}{\lambda^2}$	$0.44\lambda/b$	Large σ_{max}, very narrow beam width
4	Trihedral triangular corner reflector	$\dfrac{4\pi a^4}{3\lambda^2}$	30–40 degrees	σ_{max}, about 3 dB lower than that of a flat plate with the same aperture
5	Trihedral square corner reflector	$\dfrac{12\pi a^4}{\lambda^2}$	30–40 degrees	σ_{max}, about 3 dB lower than that of a flat plate with the same aperture
6	Dihedral corner reflector	$\dfrac{16\pi a^2 b^2}{\lambda^2}$	~40 degrees in elevation $\lambda/2\,b$ in azimuth	Narrow beam width in azimuth

Note: Refer to Ulaby et al. (1982) and Sarabandi and Chiu (1993); A is the area; λ is the wavelength; a, b is the length of leaves.

uncertainty due to undulation of the leaf by the wind (Bird et al. 1993) and a relatively small radar cross section, especially for spaceborne SARs. Figure 9-5b and c shows the 2.4-m deployable CR for the JERS-1 SAR and the 3.0-m permanently deployed CR used for ALOS/PALSAR and ALOS-2/PALSAR-2, respectively.

9.5.1.2 Frequency Tunable Active Radar Calibrator (ARC)

The ARC consists of an antenna, a receiver, and a transmitter. Here, we introduce the ARC developed for JERS-1 SAR: The deployment is shown in Figure 9-5a, a block diagram appears in Figure 9-6, and the characteristics are outlined in Table 9-2. The antenna consists of two square patches; its azimuth and elevation beam widths are 30 degrees (3 dB down), so there is a wide allowance for the antenna setting angle (NASDA 1989). In order to prevent cross coupling, the receive and transmit antennas are separated by 50 cm, and their polarizations are aligned perpendicularly. The receiver amplifies the received signal by a low-noise amplifier, frequency-shifts it in the phase controller/shifter, re-amplifies, and transmits it to the satellite. The frequency-shifted signal (power) can be

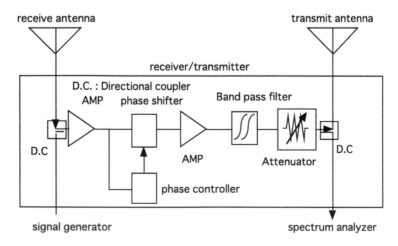

FIGURE 9-6 Block diagram of the ARC for JERS-1 SAR.

TABLE 9-2

Active Radar Calibrators and Receiver

Satellite	JERS-1	PALSAR	PALSAR-2	PALSAR-2	PALSAR-2	PALSAR-2
Type	ARC	ARC (PARC)	ARC	ARC	GC	REC
Radar cross section (dBm²)	15–60 (5 dB step)	15 ~ 60	25: spotlight 30: 3-m strip 35: 6-m strip 40: 10-m strip 52.7: strip		25: spotlight 30: 3-m strip 35: 6-m strip 40: 10-m strip	
Max. transmit power (dBm)		22.5 (19.5)	18.9		–	
Max. receive power (dBm)		−44.5 (−41.5)	−38.1		–	none
Frequency shift	20, 40, 80, 180 (Hz)	none	none		none	–
Antenna beam width	(30, 32.4) deg.	–	–		–	–
Frequency (MHz)	1,215–1,300	1,256–1,284	1,256–1,284		1,215–1,300	1,215–1,300
Receiver monitor	yes (+ spectrum analyzer)	yes (+ speana + 100 MHz ADC)	–		–	100 MHz ADC + spectrum analyzer
Stability	<± 0.5 dB	<± 0.1	<± 0.2		<± 0.2	
Battery	2 h'	external engine	2 h'		2 h'	
Units	2	1	1		2	1
Satellite tracking	no	yes (program)	no		no	no
Off nadir angle			9.9 ~ 50.8		9.9 ~ 50.8	9.9 ~ 50.8
Temperature (°C)	–	−10 ~+50	−10 ~+50		−10 ~+50	−10 ~+50
Humidity (%RH)	–	35 ~ 100	35 ~ 100		35 ~ 100	35 ~ 100
All weather	no	yes	no		no	no

Note: GC = geometric calibrator.

monitored by a spectrum analyzer through the directional coupler. Gain can be measured by using a signal generator and spectrum analyzer. A step-wise selective attenuator can set the radar cross section from 15 to 60 dBm^2 in 5 dBm^2 step and frequency shifts of 0, 20, 40, 80, and 180 Hz. While the SAR pulse is being detected, the phase controller mixes the signal (S_{ARC}) whose phase changes linearly with time.

9.5.1.3 Polarimetric Active Radar Calibrator (PARC)

PARC was developed for calibrating the PALSAR polarimetry and the SAR images (Table 9-2 and Figure 9-5d). To be a stable signal source, it contains a temperature-resistant gain controller (Partier unit) and allows the stable signal emission of 60.0 dB with the variation of ±0.1 dB for the outside temperature range of −10 to 50 Celsius, and it can track the satellite in directing the antenna peak during the satellite passage. The other characteristics can be seen in the specification. Disadvantages are the heavy weight (500 kg in total) and the internal delay of the PARC reaches 20 m. One more disadvantage is that the polarization orthogonality between HH and HV is not exactly 90 degrees. This issue affects the polarimetric calibration; a trihedral CR satisfies the orthogonality and becomes the robust calibration instrument.

9.5.1.4 Portable ARC

ALOS-2 ARC improved the ALOS/PARC at two points—it has a lighter weight (20 kg) and an almost zero internal delay of less than 1 m within 1 cm accuracy. The RCS is stabilized using the thermally insulated units as before. A pair of ARCs measure the cross talk in combining the H and V polarizations. These improvements have increased the functionality of the ARC (Figure 9-5f).

9.5.1.5 Ground-Based Receivers

A receiver in Figure 9-5f, consisting of a 100-MHz ADC, amplifier, digital data recorder, and antenna, measures the property (intensity, phase, and chirp rate) of each pulse transmitted from the SAR and estimates the azimuth antenna pattern of the SAR; Figure 9-7 shows the up-down chirp property and azimuth antenna pattern of PALSAR-2.

9.5.2 NATURAL FOREST IN THE AMAZON

Since the volume scattering works dominantly in the dense forest, the SAR images for the rainforest are another calibration source where the gamma-zero shows a constancy over a wide range of incidence angles (Figure 9-8, Figure 9-9), or even a slight decrease with incidence angle, and can be fully expressed by a mathematical distribution function (Figure 3-32).

9.5.3 INTERNAL CALIBRATION

PALSAR prepares three internal calibration sources—chirp replica, noise data, and rotating-element electric-field vector (REV) measurements (Mano and Katagi 1982). PALSAR-2 has similar functions.

9.5.3.1 Chirp Replica

Chirp signals generated (D/A converted) by digital chirp seeds for a pulse duration are injected to the receiver and signal converter to measure the re-digitized chirp signal for range compression and on-orbit stability.

9.5.3.2 Noise Measurements

There are three no-transmission modes to measure noise received from the outside and/or generated inside the PALSAR. Noise 1 measures the noise level with a specified single

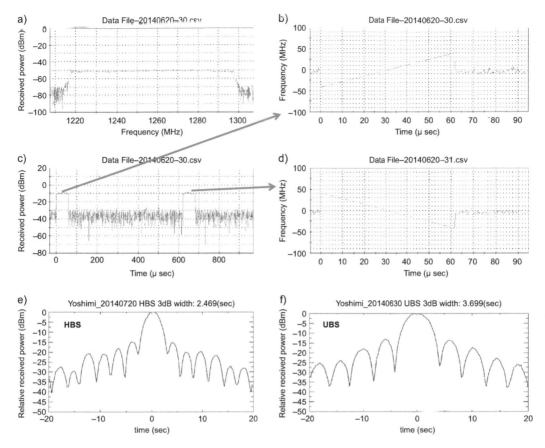

FIGURE 9-7a–f   Direct pulse receptions and their pulse-to-pulse analysis (PALSAR-2) and azimuth pattern as the slow time receiving intensity.

transmit-receive module (TRM) disconnected from the antenna. Noise 2 is to measure it with all of the TRMs disconnected from the antenna. Noise 3 is to measure the total noise with no signal transmission.

9.5.3.3   REV

An REV is used to check the condition of cach of the 80 TRMs. First, the TRMs are activated one-by-one at different phase shifts (11.25 degrees, plus 32 steps to 360 degrees). The emitted signal is received by the extra 12 antennas attached on the PALSAR antenna plane, and the received signal pattern is compared with the theoretical one. If a discrepancy is found, a problem has occurred.

In operation, chirp measurements taken over 30 s are defined as calibration slots. Calibration slots are added at the beginning and end of the observation slot, whose duration follows the user requirements.

9.6   CALIBRATION SUMMARY FOR THE EXISTING SARs

We have three L-band SAR satellites and one airborne SAR calibrated during their mission lives. Here, we introduce their representative results. Most have been published in various journals and discussed in workshops.

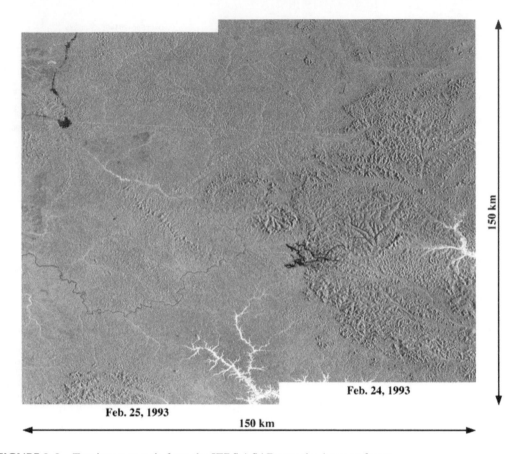

FIGURE 9-8   Two image mosaic from the JERS-1 SAR over the Amazon forest.

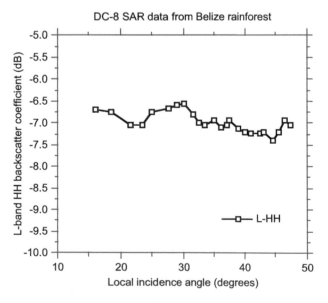

FIGURE 9-9   Incidence angle dependency of Airborne Synthetic Aperture Radar (AIRSAR) data over Belize (Shimada and Freeman 1995).

TABLE 9-3

JERS-1 SAR Calibration Summary

Item	Azimuth	Spec. or Remarks	Range	Spec. or Remarks
Resolution	6.5 m (0.3)	6.1	9.6 m (0.1)	8.9
PSLR	−15.56 dB (5.9)		−15.53 dB (5.2)	
ISLR	−8.70 dB (4.9)			
Ambiguity	22 dB		20 dB	
Geometric accuracy	40.0 m		104	111 (RSS)
Stability of calibration factor (dB)	0.6 (integral)	0.9	0.7 (peak)	0.9
Radiometric accuracy (dB)	1.1 (integral)	1.2	1.4 (peak)	1.2
Uniformity (dB)	0.2	1 sigma	0.10	0.1 dB
Saturation (%)	< 5%			
RFI (%)	27			
I-Q gain	< 0.3 dB			
SNR	5–7 dB			
I-Q orthogonality	1.5 degrees			

Note: SAR images were acquired in the single transmitter mode (325 W). Resolution, PSLR, and ISLR were calculated by evaluating the impulse responses from ARCs.

The following tables summarize the calibrated performances: JERS-1 SAR in Table 9-3, ALOS/PALSAR in Table 9-4, ALOS-2/PALSAR-2 in Table 9-5, and Pi-SAR-L2 in Table 9-6.

9.6.1 JERS-1 SAR

The first Japanese spaceborne SAR onboard JERS-1, the JERS-1 SAR, highlights land and sea-ice monitoring from geological and ecological standpoints due to the L band wavelength and the bigger off-nadir angle, which suppresses the foreshortening effect. The global monitoring mission of JERS-1 SAR's all-weather observation and data recording capability requires the absolute calibration of JERS-1 SAR, which forms a chain of the target, the JERS-1 SAR instrument, the ground processor, and its product, so that a JERS-1 SAR image can express the scattering intensity of the target with better accuracy, and the application study can progress.

The evaluation of JERS-1 SAR's characteristics in the commissioning phase was conducted precisely and summarized that its performance was confirmed as almost the same as the specification, except for the small SNR of 5 to 6 dB over land and the certain saturation of the raw data. This unexpected degradation might be caused by the low transmission power. However, the correlated image generally seems to be of good quality. Calibration and validation of JERS-1 SAR is necessary to meet the requirements of the user community.

Since April 1992, the calibration and validation of JERS-1 SAR has been conducted using the SAR images, calibration instruments, and evaluation tools. The calibration factor (CF) is −68.51 dB with an accuracy of 1.86 dB (1 sigma) for the full swath of the SAR images. The image quality was also recognized as well designed (Shimada 1996; Shimada 1994).

9.6.2 ALOS/PALSAR

We summarize the results obtained from the calibration of PALSAR on ALOS. All the imaging modes: single, dual, and full polarimetric strip mode as well as ScanSAR, were calibrated and validated using a total of 572 calibration points collected worldwide and distributed targets selected primarily from the Amazon forest (Figure 9-10). Through raw-data characterization, antenna-pattern

TABLE 9-4

ALOS/PALSAR Calibration Summary

Items	Measured Values		No. of Data	Specification
Geometric accuracy	9.7 m (RMS): STRIP mode		572	100 m
	70 m (RMS): SCANSAR			
Radiometric accuracy	0.219 dB (1 sigma) from Amazon forest			1.5 dB
	0.76 dB (1 sigma) from CRs		572	1.5 dB
	0.17 dB (1 sigma: Sweden CRs)		16	1.5 dB
	−34 dB (Noise equivalent sigma-zero for HV)			−23 dB
	−32 dB (as a minimum of FBD-HH)			
	−29 dB (as a minimum of FBS-HH)			
Polarimetric calibration	VV/HH ratio	1.013 (0.062)*	81	0.2 dB
	VV/HH phase diff	0.612 deg (2.66)*		5 deg.
	Cross talk	−31.7 (4.3)		−30 dB
Resolution	azimuth	4.49 m (0.1 m) *	572	4.5 m
	range (14 MHz)	9.6 m (0.1 m) *		10.7 m
	range (28 MHz)	4.7 m (0.1 m)*		5.4 m
Side lobe	PSLR in azimuth	−16.6 dB	572	−10 dB
	PSLR in range	−12.6 dB		−10 dB
	ISLR	−8.6 dB		−8 dB
Ambiguity	azimuth	not appeared		16 dB
	range	23 dB		16 dB
Transmission power	Sum of 80 TRM	2,220 W		2,000 W
Raw data	Saturation	0.4 ~ 2.4%		−
	I/Q orthogonality	1.6 degrees		−
	SNR	7.0 ~ 9.5 dB		−
	I-Q gain ratio	1.00		−
Calibration Factor	−83.0			−83.0

Note: A (B)* represents an average value of A and a standard deviation of (B). PSLR is peak-to-sidelobe ratio, and ISLR is integrated side-lobe ratio.

estimation using the distributed target data, and polarimetric calibration using the Faraday rotation-free area in the Amazon, we performed the PALSAR radiometric and geometric calibration and confirmed that the geometric accuracy of the strip mode is 9.7 m root mean square (RMS), the geometric accuracy of ScanSAR is 70 m, and the radiometric accuracy is 0.76 dB from a CR analysis and 0.22 dB from the Amazon data analysis (standard deviation). Figure 9-11 shows the inflight antenna pattern derivations. Polarimetric calibration was successful, resulting in a VV/HH amplitude balance of 1.013 (0.0561 dB) with a standard deviation of 0.062 and a phase balance of 0.612 degree with a standard deviation of 2.66 degrees in azimuth and range.

Figure 9-12 shows the IRF of the selected CR and temporal variation of the CF. Through a calibration-update analysis, the calibration became more robust and simplified, depending only on a single CF for all of the modes. The calibration factor *CF* is determined to be −83.0, with a standard deviation of 0.76 dB (Figure 9-13). The gamma-zero incidence angle dependence is shown in Figure 9-14.

NESZ is found in the data over Greenland (Figure 9-15) with −25 dB for FBS343HH. Wind-slick regions in Hawaii provided an NESZ of −29 dB for FBS343HH, −32 dB for FBD343HH, and −34 dB for FBD343HV. The former is obtained at the Greenland ice sheet and is 2 dB lower than the specification. The latter is 11 dB better than the specification. Most of the current spaceborne SARs

TABLE 9-5

ALOS-2/PALSAR-2 Calibration Summary

Item		Results	No. of Data	Specification
Geometric accuracy	High-resolution spotlight	5.34 m (L 1.1)/6.73 m (L 2.1)	127/129	20 m
(RMSE)	ScanSAR	60.77 m (L1.1)/29.93 m (L2.1)	7/8	100 m
Radiometric	CR	1.31 (CF: −81.60)	120	1.0 dB
accuracy	Amazon forest	0.406(CF: −82.34)	30	1.0 dB: −6.84
	NESZ(F/H/U)	−41.1 (F)/−36.0 (H)/−36.6 (U)		dB@Amazon−26.0 (F)/
	HH	−49.2 (F)/−46.0 (H)		−28.0 (H)/
	HV			−24.0 (U)
Polarization	VV/HH-gain	1.0143(σ:0.06)	6	1.047
	VV-HH-phase (deg.)	0.350(σ:0.286)		5 deg.
	cross talk (dB)	−43.7(σ:6.65) hv/hh		−30 dB
		−44.0(σ:7.10) vh/vv		−30 dB
		−48.2(σ:6.05) corr		−30 dB
Resolution	Spotlight	0.79(σ:0.028)/1.66(σ:0.04)	3	1.00 × 1.1/1.78
azimuth/range	Ufine [3 m]	2.81(σ:0.034)/1.70(σ:0.022)	35	2.75 × 1.1/1.78
	High Sens. [6 m]	4.06(σ:0.108)/3.53(σ:0.317)	28	3.75 × 1.1/3.57
	High Resol. [10 m]	5.05(σ:0.110)/5.36(σ:0.126)	61	5.00 × 1.1/5.36
Side lobe	PSLR(AZ)	−16.20 dB(σ:2.53)	124	−13.26 dB + 2 dB
	PSLR(RG)	−12.59 dB(σ:1.84)		−13.26 dB + 2 dB
	ISLR	−8.80 dB(σ:3.23)		−10.16 dB + 2 dB
Ambiguity	AZ	23 ~ 14 dB (average: 20)	7	> 20 ~ 25 dB
	RG	No confirmation		> 25 dB
Calibration factor	−83.0			−83.0
Raw data	Saturation	< 0.5%		
	I-Q orthogonality	1.5 degrees		
	SNR	12–13 dB		
	I-Q gain ratio	1.0022		

have an NESZ of −23 dB, and it has been confirmed that PALSAR has the minimum value among them. The reason for HV exceeding HH is that HH uses a larger attenuator than HV.

The average PSLRs are −16.6 dB in azimuth and −12.6 dB in range, the latter of which is similar to the rectangular-window case. The azimuth value exceeds the range value because the azimuth antenna pattern is not compensated for in the image-generation phase. The resolutions in both directions are equivalent to those of the theoretical rectangular window case (Shimada et al. 2009; Shimada 2011; Shimada 2010).

9.6.3 ALOS-2/PALSAR-2

Here, we summarize the performance of the ALOS-2/PALSAR-2 confirmed during the initial calibration and validation phase of August 4, 2014 through November 20, 2014. During this phase, all the PALSAR-2 modes were evaluated for raw data and the quality of the SAR images, and the SAR images were calibrated geometrically and radiometrically using the natural forest in the Amazon with the CRs deployed globally. In total, 58 antenna beams from the six modes—spotlight (84 MHz), ultrafine (84 MHz), high sensitive-full polarimetry (42 MHz), high resolution, ScanSAR narrow (350 Km), and ScanSAR wide (490 Km) were calibrated using the Amazon forest. Geometric accuracy of the standard product is 5.34 m root mean square error (RMSE), and radiometric stability

TABLE 9-6
Pi-SAR-L2 Calibration Summary (as of February 28, 2013)

Item	Measured Data	No. of Data	Specification
Geometric accuracy (m)	~ 10 m (RMS) under evaluation	22	10 m
Radiometric Accuracy	1.16 dB (1 sigma) from CRs	22	1.0 dB
	−36 ~ −43 dB (NESZ for HH VV)	11 scenes	< 35dB
	−45 ~ −53 dB (NESZ for HV, VH) (20°–60°)	11 scenes	
Polarimetric accuracy (m)	VV/HH 1.0213 (0.0228)	22	< 0.2 dB
	VV/HH Phase 1.638° (2.142)		< 5°
	Cross talk −32.463 (CHV/HH)		< −0 dB
	−36.767 (CVH/VV)		
	−38.616 (natural target)		
Resolution	Azimuth 1.01 m (0.25)	22	< 0.8 m
	Range 1.80 m (0.06)		< 1.76 m
Sidelobe	PSLR in azimuth −9.05 dB (3.42)	22	
	PSLR in range −12.5 dB (1.13)		
	ISLR −7.04 dB (1.26)		
Ambiguity	Azimuth not confirmed		
	Range not confirmed		
Calibration factor	CF_1: −79.882 (1.16)	53	
	A: 81		

is 0.4 dB using the Amazon data. The other parameters of the SAR image qualities (i.e., resolution, NESZ, PSLR, etc.) meet the requirement as to SAR image quality. In this evaluation phase, other SAR qualities were evaluated (i.e., Interferometric Synthetic Aperture Radar [InSAR], polarimetry, forest observation, and so on). This paper briefly summarizes the PALSAR-2 initial calibration and validation results (Shimada et al. 2014b).

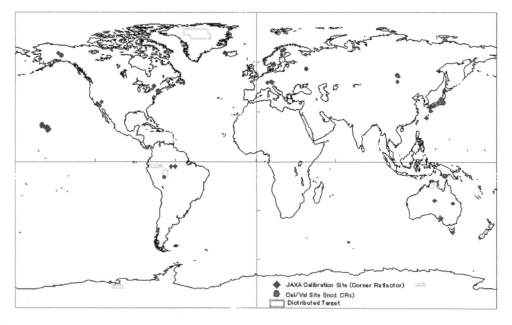

FIGURE 9-10 Worldwide PALSAR calibration sites. Blue and red points indicate corner reflectors, and yellow rectangular areas indicate natural targets (i.e., areas of the Amazon, Antarctica, and Greenland).

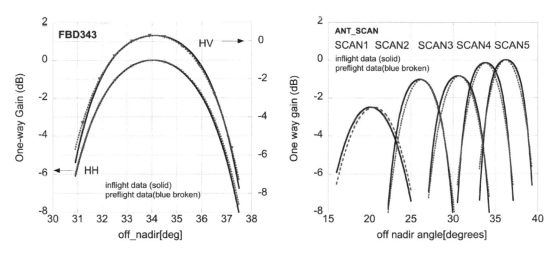

FIGURE 9-11 Comparison of the antenna patterns before and after launch. Bold lines plot inflight measurements; thin blue lines plot preflight ground measurements.

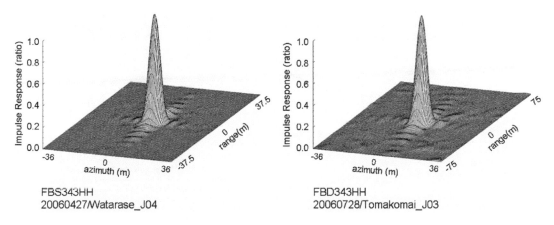

FBS343HH
20060427/Watarase_J04

FBD343HH
20060728/Tomakomai_J03

FIGURE 9-12 Three-dimensional view of the impulse response from the CRs. Left: FBS343HH at Watarase test site, acquired on April 27, 2006. Right: FBD343HH at Tomakomai test site, acquired on July 28, 2006.

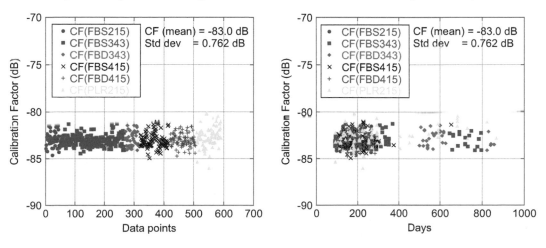

FIGURE 9-13 Distribution of the calibration factors: (left) all modes and (right) long-term variation of CF. Numbers on the *x*-axis identify data sets containing a CR response observed by PALSAR.

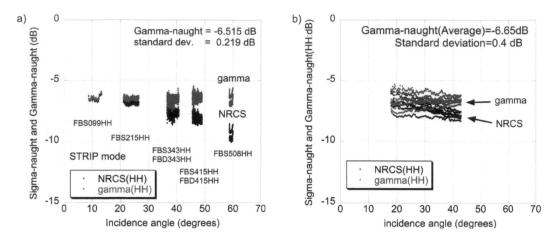

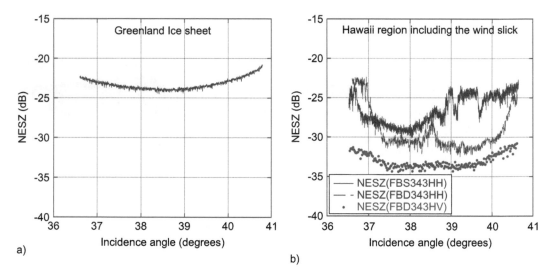

FIGURE 9-14 Gamma-naught and sigma-naught versus incidence angle plotted for strip mode (left) and ScanSAR data (right). Both data sets were collected from the Amazon Rainforest.

FIGURE 9-15 Noise equivalent sigma-naught versus incidence angle. The left figure is for Greenland, observed by FBS343HH, and the right is for Hawaii, observed by FBS343HH and FBD343HV.

9.6.4   Pi-SAR-L/Pi-SAR-L2

In 2012, Pi-SAR-L2 was upgraded for Pi-SAR-L, the JAXA's first L-band airborne fully polarimetric SAR, originally developed in 1996 and operated from 1997 to 2010 for the polarimetric studies. Upgraded by bandwidth expansion to 85 MHz for all transmit-antenna-receive, high-quality in-phase 8-bit AD convertors, and perfect synchronization of the inertial measurement unit (IMU) and SAR data, Pi-SAR-L2 has a higher performance level than the Pi-SAR-L for most of the radiometric and the geometric parameters. Pi-SAR-L2 was calibrated using CRs and Tomakomai forest data during 2012 to 2013 and also later on. The fact that NESZ is as low as −54 dB for HV and −44 dB for HH and VV polarizations has been confirmed. These performances assist the Pi-SAR-L2 application in various areas—especially with disasters. Table 9-6 shows the calibration summary (Shimada et al. 2013a; Shimada et al. 2013b). Because they are more advantageous in higher radiometric performance, airborne SARs can be the pathfinder to the spaceborne SARs.

9.7 SUMMARY

In this chapter, we summarized the calibration procedures and the representative results for the Japanese SARs. Key components were: (1) radar equation, (2) polarimetric equation, (3) antenna pattern model, (4) calibration method, (5) calibration instruments, and (6) application to the 4 L-band SAR data.

REFERENCES

Bird, P. J., Keyte, G. E., and Kenward, D. R. D., 1993, "Calibration of ERS-1 SAR," *Proc. 1993 SAR Calibration Workshop* (CEOS SAR CAL/VAL), Noordwijk, The Netherlands, September 20–24, 1993, pp. 257–281.

Dobson, M. C., Ulaby, F. T., Brunfeldt, D. R., and Held, D. N., 1986, "External Calibration of SIR-B Imagery with Area Extended and Point Targets," *IEEE T. Geosci. Remote*, Vol. 24, No. 4, pp. 453–461.

Freeman, A., 1990, "SIR-C Calibration Plan: An Overview," JPL Report, JPL-D-6997, NASA, Jet Propulsion Laboratory, California Institute of Technology, Pasadena, CA.

Freeman, A., Curlander, J. C., Dubois, P. D., and Klein, J., 1988, "SIR-C Calibration Workshop Report," JPL Center for Radar Studies Publication No. 88-003, Jet Propulsion Laboratory, California Institute of Technology, Pasadena, CA.

Gray, A. L., Vachon, P. W., Livingstone, E., and Lukowski, T. I., 1990, "Synthetic Aperture Radar Calibration Using Reference Reflectors," *IEEE Trans. Geosci. Rem. Sens.*, Vol. 28, No. 3, pp. 374–383.

Hawkins, R. K., 1990, "Determination of Antenna Elevation Pattern for Airborne SAR Using the Rough Target Approach," *IEEE T. Geosci. Remote*, Vol. 28, No. 5, pp. 896–905.

Mano, S., and Katagi, T., 1982, "A Method for Measuring Amplitude and Phase of Each Radiating Element of a Phased Array Antenna," *Trans. IEICE B*, Vol. J65-B, No. 5, pp. 555–560.

Moore, R. K. and Hemmat, M., 1988, "Determination of the Vertical Pattern of the SIR-B Antenna," *Int. J. Remote Sens.*, Vol. 9, No. 5, pp. 839–847.

NASDA Contract Report CDA-3-727, "Development of the Active Radar Calibrator for JERS-1/ERS-1 SAR," Mitsubishi Electric Corp., Tokyo, 1989.

Sarabandi, K. and Chiu, T. C., 1994, "An Optimum Corner Reflector for Calibration of Imaging Radars," *Proc. CEOS SAR Calibration Workshop*, Ann Arbor, Michigan, September 28–30, 1994, pp. 52–79.

Shimada, M., 1994, "Absolute Calibration of JERS-1 SAR Image and Evaluation of its Image Quality (in Japanese)," *J. Remote Sens. Soc. Japan*, Vol. 14, No. 2, pp. 143–154.

Shimada, M., 1996, "Radiometric and Geometric Calibration of JERS-1 SAR," *Adv. Space Res.*, Vol. 17, No. 1, pp. 79–88.

Shimada, M., 1999, "Radiometric Correction of Saturated SAR Data," *IEEE T. Geosci. Remote*, Vol. 37, No. 1, pp. 467–478.

Shimada, M., 2005, "Long-Term Stability of L-band Normalized Radar Cross Section of Amazon Rainforest Using the JERS-1 SAR," *Can. J. Remote Sens.*, Vol. 31, No. 1, pp. 132–137.

Shimada, M., 2009, "A New Method for Correcting ScanSAR Scalloping Using Forest and Inter-SCAN Banding Employing Dynamic Filtering," *IEEE T. Geosci. Remote*, Vol. 47, No. 12, pp. 3933–3942.

Shimada, M., 2010, "On the ALOS/PALSAR Operational and Interferometric Aspects (in Japanese)," *J. Geod. Soc. Japan*, Vol. 56, No. 1, pp. 13–39.

Shimada, M., 2011, "Model-Based Polarimetric SAR Calibration Method Using Forest and Surface Scattering Targets," *IEEE T. Geosci. Remote*, Vol. 49, No. 5, pp. 1712–1733.

Shimada, M., and Freeman, A., 1995, "A Technique for Measurement of Spaceborne SAR Antenna Patterns Using Distributed Targets," *IEEE T. Geosci. Remote*, Vol. 33, No. 1, pp. 100–114.

Shimada, M., Isoguchi, O., Tadono, T., and Isono, K., 2009, "PALSAR Radiometric and Geometric Calibration," *IEEE T. Geosci. Remote*, Vol. 47, No. 12, pp. 3915–3932.

Shimada, M., Itoh, T., Motooka, T., Watanabe, M., Tomohiro, S., Thapa, R., and Lucas, R., 2014a, "New Global Forest/Non-Forest Maps from ALOS PALSAR Data (2007-2010)," *Remote Sens. Environ.*, Vol. 155, pp. 13–31, http://dx.doi.org/10.1016/j.rse.2014.04.014

Shimada, M., Kawano, N., Watanabe, M., Motooka, T., and Ohki, M., 2013a, "Calibration and Validation of the Pi-SAR-L2," *Proc. APSAR 2013*, Tsukuba, Japan, September 23–27, 2013, pp. 194–197.

Shimada, M., Nakai, M., and Kawase, S., 1993, "Inflight Evaluation of L Band SAR of JERS-1," *Can. J. Remote Sens.*, Vol. 19, No. 3, pp. 247–258.

Shimada, M., Watanabe, M., and Motooka, T., 2014b, "Initial Calibration and Validation of the ALOS-2/ PALSAR-2," *Proc. 58th Space Science and Technology Conference*, Nagasaki, Japan, November 12–14, 2014.

Shimada, M., Watanabe, M., Motooka, T., Shiraishi, T., Thapa, R., Kawano, N., Ohki, M., Uttank, A., Sadly, M., and Rahman, A., 2013b, "Japan - Indonesia PI-SAR-L2 Campaign 2012," *Proc. 34th Asian Conference on Remote Sensing*, Bali, Indonesia, Oct. 20–24, 2013,

Ulaby, F., Moore, R., and Fung, A., 1982, *Microwave Remote Sensing, Active and Passive, Volume II: Radar Remote Sensing and Surface Scattering and Emission Theory*, Addison-Wesley, Boston, pp. 767–779.

10 Defocusing and Image Shift due to the Moving Target

10.1 INTRODUCTION

SAR imaging of a moving target suffers from defocusing and blurring of the image due to the Doppler and f_{DD} shift. A frequency tunable active radar calibrator (ARC) is an appropriate instrument for SAR calibration that can yield a larger radar cross section (RCS, or σ) and move the point target in a desirable area isolated from brighter targets. However, the larger frequency shift worsens the image resolution as well as causing dislocation and could reduce the calibration accuracy. Using an ARC and JERS-1 SAR, we evaluate the impact of such a shifted condition on the characterization of impulse response function (IRF) as well as calibration issues on the comparison of the peak method and integral method and, finally, we propose a range of frequency shift. In this chapter, we simulate a moving object using a frequency tunable ARC and evaluate the position accuracy and radiometric sensitivity (Shimada et al., 1999).

10.2 THEORY

10.2.1 COORDINATE SYSTEM

A spaceborne SAR flies on a circular orbit and transmits cascaded pulses at pulse repetition frequency (PRF) to targets fixed to the Earth. In this study, these targets are the ARC and the background (Figure 10-1), and they move with the Earth in the inertial coordinate system (Curlander and McDonough 1991). The essence of an SAR is the possibility of carrying out compression processing of a signal's phase history in the azimuth coordinate. Therefore, the distance between the SAR and the target must be expressed as accurately as possible. Let $\mathbf{r}_s$ and $\mathbf{r}_p$ represent the satellite position and the target position, respectively, and then their distance, R, is given by

$$R \equiv \left| \mathbf{r}_s - \mathbf{r}_p \right|. \tag{10.1}$$

The round-trip time from the ARC and nearby scatterers changes nonlinearly as the satellite moves, and this is called the range curvature migration, for which the first-order time dependency is the range walk and the second or higher order dependency is the range curvature (Raney 1971). Range walk correction is one of the key points for forming well-focused SAR images (Van de Lindt 1977; Wu 1976; Jin and Wu 1984), although the attitude control in yaw reduces the range walk. Thus, we consider both the range curvature and range walk for model derivation and use non-yaw-steered SAR data that employ a relatively large range walk for the evaluation.

10.2.2 RECEIVED SIGNAL

Figure 10-2 shows the range-azimuth coordinate system for SAR observation. The (slow) azimuth time coordinate, T, corresponds to motion in the azimuth direction, and the (fast) range time coordinate, t, corresponds to the pulse propagation in a perpendicular direction. Azimuth time origin, $T = 0$, is defined as when the ARC is observed in the antenna azimuth peak gain

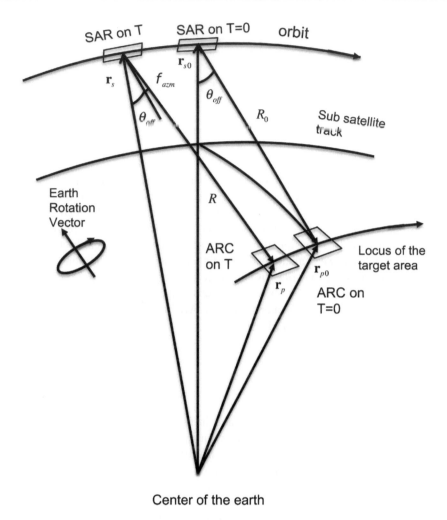

FIGURE 10-1 SAR and target positions in the inertial system fixed at Earth's center. The ARC and the background are fixed to the Earth and move with it; $\mathbf{r}_{p0}$ and $\mathbf{r}_{s0}$ are the satellite position and ARC positions at $T = 0$.

direction and the distance between ARC and SAR is R_0. A pulse, $\mathbf{S}_t$, transmitted to the ground at T is expressed by:

$$\mathbf{S}_t(t,T) = rect\left(\frac{t}{\tau}\right) \cdot e^{2\pi j\left(f_0 t + \frac{k}{2}t^2\right)} \tag{10.2a}$$

$$rect\left(\frac{t}{\tau}\right) = \left\{ \begin{array}{ll} 1 & |t| \le \tau/2 \\ 0 & else \end{array} \right\} \tag{10.2b}$$

where t is the time delay from T; f_0 is the transmission frequency; τ is the pulse width; and k is the chirp rate (Hz/s). Since a pulse is detected by the ARC, the ARC changes the received signal phase by the frequency of f_s (positive or negative), amplifies it, and retransmits it to the SAR until the SAR main beam is outbound from the ARC. Here, two phase modulations are initiated: one is the nonlinear phase modulation due to the time compression by the satellite–ARC relative movement; and the other is the active linear modulation by the ARC. The SAR–ARC relative

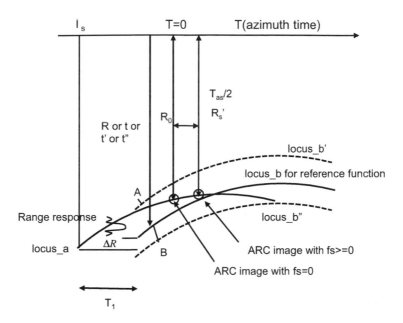

FIGURE 10-2 Geometry of the azimuth correlation. The range response for ARC is on the locus_a. The phase on A is propagated to Point B and correlated with the reference function along locus_b. The correlation on locus_b, therefore, maximizes the output and those on locus_b' and locus b" minimize it.

motion compresses the pulse width and the time delay from the ARC at the center of the pulse as follows:

$$\tau \to \frac{c+\dot{R}}{c-\dot{R}}\tau \tag{10.3a}$$

$$t = \frac{2R}{c} \to \frac{2R}{c-\dot{R}} \tag{10.3b}$$

where $\dot{R}$ is the time derivative of distance at T, and c is the speed of light. Therefore, the received signal, $S_r(t,T)$, is expressed as follows:

$$\mathbf{S}_r(t,T) \propto rect\left(\frac{t-\dfrac{2R}{c-\dot{R}}}{\dfrac{c+\dot{R}}{c-\dot{R}}\dfrac{\tau}{2}}\right) \cdot G_{ant}\left(t-\frac{2R}{c-\dot{R}},T\right) \cdot e^{2\pi j\left\{f_s(t+T+T_s)+f_0\left(\frac{c-\dot{R}}{c+\dot{R}}t-\frac{2R}{c-\dot{R}}\right)+\frac{k}{2}\left(\frac{c-\dot{R}}{c+\dot{R}}t-\frac{2R}{c-\dot{R}}\right)^2-f_0 t\right\}} \tag{10.4a}$$

$$G_{ant}(t,T_i) \cong G_{ele}(\theta_{off}) \cdot G_{azm}(\phi_{azm}) \tag{10.4b}$$

where f_s is the frequency shift added by ARC; T_s is the original time when the ARC starts the frequency modulation; G_{ant} is the one-way relative antenna pattern; G_{ele} is the elevation pattern; G_{azm} is the azimuth pattern; θ_{off} is the off nadir angle to ARC; and ϕ_{azm} is the azimuth angle to ARC (see Figure 10-1). Here, we do not discuss the amplitude decrease in the free space propagation. Under the maximum error of 2% within the azimuth correlation duration for spaceborne SARs, Equation (10.4a) is approximated by

$$\mathbf{S}_r(t,T) \cong rect\left(\frac{t-2R/c}{\tau}\right) \cdot G_{ant}\left(t-\frac{2R}{c},T\right) \cdot e^{2\pi j\left\{f_s(t+T+T_s)-\frac{2\dot{R}f_0}{c}t-\frac{2Rf_0}{c}+\frac{k}{2}\left(t-\frac{2R}{c}\right)^2\right\}}. \tag{10.5}$$

In the exponent of Equation (10.5), $f_s \cdot (t + T + T_s)$ is the phase shift introduced by ARC, $-2\dot{R}f_0 t / c$ the phase change (Doppler) within a pulse due to time compression, $-2Rf_0 / c$ the phase delay between ARC and SAR, and $k / 2 \cdot (t - 2R/c)^2$ the delayed pulse modulation. In the SAR antenna main beam, the following approximations are valid:

$$R \cong R_0 + \frac{1}{2}\ddot{R}_0 \cdot (T + \beta)^2 \qquad (10.6a)$$

$$\beta \equiv \frac{\dot{R}_0}{\ddot{R}_0} \qquad (10.6b)$$

$$t = \frac{2}{c}\left\{ R_0 + \frac{1}{2}\ddot{R}_0 (T + \beta)^2 \right\} \qquad (10.6c)$$

where $\dot{R}_0$ and $\ddot{R}_0$ are the first and second derivatives of slant range at $T = 0$.

10.2.3 RANGE CORRELATION

A range correlation output, $\mathbf{S}_{c,r}$, is given by

$$\mathbf{S}_{c,r}(t',T) = \int_{-\infty}^{\infty} \mathbf{S}_r(t,T) \cdot \mathbf{S}_{r,ref}^*(t - t')\,dt, \qquad (10.7a)$$

where

$$\mathbf{S}_{r,ref}(t - t') = rect\left(\frac{t - t' - 2R/c}{\tau} \right) e^{-2\pi j \frac{k}{2}\left(t - t' - \frac{2R}{c} \right)^2} \qquad (10.7b)$$

is the transmission pulse with unit amplitude, "*" the complex conjugate, and t' the new variable replacing t. Integration is performed over a pulse width ($-\tau / 2$ to $\tau / 2$) centered at $2R/c$. If the antenna elevation pattern does not change radically within a pulse width (within the main lobe), Equation (10.7a) can be well approximated by

$$\mathbf{S}_{c,r}(t',T) \cong e^{2\pi j F(t',T)} \tau \cdot \overline{G_{ele}}\left(t' - \frac{2R}{c},T \right) \cdot G_{azm}\left(t' - \frac{2R}{c},T \right) \cdot \frac{\sin\{E(t',T)\tau\pi\}}{E(t',T)\tau\pi}, \qquad (10.8)$$

where

$$F(t',T) = f_s T + f_s T_s - \frac{2Rf_0}{c} - \frac{k}{2}\left(t'^2 - \frac{4R^2}{c^2} \right) + E(t',T)\frac{2R}{c}, \qquad (10.9a)$$

$$E(t',T) = f_s - \frac{2\dot{R}f_0}{c} + k\left(t' - \frac{2R}{c} \right), \qquad (10.9b)$$

and $\overline{G_{ant}}$ is the averaged antenna pattern over a pulse width. The third term on the right side of Equation (10.9b) dominates the other two terms because k is huge. If we use δ as a deviation from R, the sinc function in Equation (10.8) becomes $\sin(2k\tau\pi\delta/c)/(2k\tau\pi\delta/c)$ and suppresses range resolution in several meters (for standard SAR with k approximately $1.0e^{11}$Hz/s and τ approximately 35 μs). Terms involving T in F cannot be ignored because they exist for a relatively longer duration and correlate with the azimuth reference function. Note that different existence areas of the two integrands in Equation (10.7a) does not produce a significant error (see Chapter 3).

10.2.4 Azimuth Correlation

The azimuth correlation is the same as the range correlation except that the integration is conducted on the range curvature. The range curvature is the locus of any target location on the azimuth time (T)—slant range (or t') coordinate system, and it is bounded by the antenna azimuth pattern and the processing frequency width. Let the locus_a in Figure 10-2 be the range curvature for the ARC response. Since the azimuth reference function is defined on the locus_b which is almost the same as locus_a shifted by the azimuth time delay T_1 in T (i.e., $T \to T - T_1$) and by the range shift ΔR in R (i.e., $R_0 \to R_0 + \Delta R$), the expressions for the locus_b in terms of R'' and t'' become:

$$R'' \cong R_0 + \Delta R + \frac{1}{2} \ddot{R}_0 \left(T - T_1 + \beta' \right)^2 \tag{10.10a}$$

$$t'' \cong \frac{2}{c} \left\{ R_0 + \Delta R + \frac{1}{2} \ddot{R}_0 \left(T - T_1 + \beta' \right)^2 \right\} \tag{10.10b}$$

where β' is an estimation based on knowledge of the orbit, attitude, and Doppler azimuth spectrum. Although erroneous information makes β' slightly different from β, we assume $\beta = \beta'$ for simplicity. On the locus_b, we do not consider the frequency shift at the ARC, so the azimuth reference function, $\mathbf{S}_{a,ref}$, should be

$$\mathbf{S}_{a,ref} \left(T - T_1 \right) = rect \left(\frac{T - T_1}{T_a} \right) e^{-\frac{2f_0}{c} R''}. \tag{10.11}$$

Then, the azimuth correlation output, $\mathbf{S}_{c,r,a}$, is given by

$$\mathbf{S}_{c,r,a} \left(t'', T_1 \right) = \int_{-\infty}^{\infty} \mathbf{S}_{c,r} \left(t'', T \right)_{locus\_b} \mathbf{S}_{a,ref}^{*} \left(T - T_1 \right)_{locus\_b} dT, \tag{10.12}$$

where T_a is the azimuth correlation duration. Since the azimuth antenna pattern in Equation (10.8) changes steeply within the azimuth correlation duration, it cannot be moved out from the integration of Equation (10.12). Integrand $\mathbf{S}_{c,r}$ on the locus_b can be interpolated by replacing t' of Equation (10.8) by t'' of Equation (10.10b). The Taylor expansion of the exponential terms in the two integrands and rearrangement of the dominant terms simplify Equation (10.12) to:

$$\mathbf{S}_{c,r,a} \left(t'', T_1 \right) \cong \overline{G}_{ele} \left(t'' - \frac{2R}{c} \right)$$

$$\cdot \int_{-\infty}^{\infty} rect \left(\frac{T - T_1}{T_a} \right) \cdot G_{azm} \left(t'' - \frac{2R}{c}, T \right) \frac{\sin \left(E(t'', T) \pi \tau \right)}{E(t'', T) \pi \tau} e^{2\pi j \left(f_s + f_{DD} \cdot T_1 \right) T} dT, \tag{10.13}$$

where

$$E \left(t'', T \right) \cong f_s + f_{DD} \beta + \frac{2k}{c} \left\{ \Delta R - \frac{1}{2} \ddot{R}_0 T_1 \left(-T_1 + 2\beta \right) \right\} - \left(-f_{DD} + \frac{2k}{c} \ddot{R}_0 T_1 \right) T, \tag{10.14}$$

and f_{DD} is the Doppler chirp rate given by $-2 f_0 \ddot{R}_0 / c$. The $\exp \left\{ 2\pi j T_i \left(f_s + f_{DD} T_1 \right) \right\}$ in Equation (10.13) provides a sinc-function-like output after the integration. The $\sin(\tau \pi E) / \tau \pi E$ maximizes the

TABLE 10-1

Parameters of the SAR Simulation

Parameter	Value
f_0	1.275 GHz
f_s	0–200 Hz
k	$-4.2857e^{11}$ Hz/s
c	300,000 km/s
τ	35 μs
Height	568 km
R_0	730 km
$T_a(s)$	1.8

Note: JERS-1's orbit data and ARC's actual locations were used for the simulation parameters.

integration if we select a ΔR that crosses locus_b and locus_a at their centers (see Figure 10-2). The integration is then maximized at:

$$T_1 \equiv T_{as} = -\frac{f_s}{f_{DD}} \tag{10.15a}$$

$$\Delta R = \frac{1}{4}\ddot{R}_0 T_{as}\left(\frac{T_{as}}{2} + 2\beta\right) \tag{10.15b}$$

where T_{as} is the azimuth time shift. Then, the azimuth location shift (x_a) and range location shift (x_r) are given by:

$$x_a = T_{as}V_g \tag{10.16a}$$

$$x_r = R_s' - R_0 = \ddot{R}_0\beta T_{as} = \dot{R}_0 T_{as} \tag{10.16b}$$

where V_g is the ground speed of the beam center along the sub-satellite track. Thus, IRF moves in the azimuth direction and its shift amount depends on the frequency shift (f_s) and Doppler chirp rate (f_{DD}); IRF also moves in the range direction if the product of the slant range velocity at the center of the beam and the time shift (T_{as}) is bigger than a range resolution; if the antenna beam is controlled to track $\dot{R}_0 = 0$ (i.e., yaw steering), range shift is eliminated; and the decreased overlapped area of two loci may lose the correlation gain (see Figure 10-2). A computation with parameters based on JERS-1 SAR (Table 10-1) evaluated the dependence of the impulse response peaks on the frequency shift and showed that the correlation gains decrease as f_s increases (Figure 10-3); and the correlation gain loss is predicted to be -10.1 dB at $f_s = 180$ Hz.

10.3 EXPERIMENTS

Calibration experiments for JERS-1 SAR were conducted at Japanese test sites using two L-band, frequency-tunable ARCs f_s varying from 0 to 180 Hz. The experiments sought to acquire data relating the frequency and location shifts, correlation gain loss, and the resolution broadening. The models developed in the previous section were verified and the applicability of such an ARC to the calibration was evaluated.

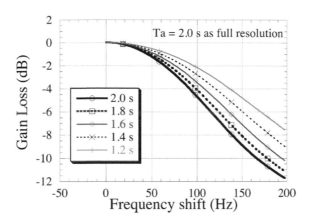

FIGURE 10-3 Frequency-shift dependency of the correlation gain loss. The number shows the azimuth integration time.

10.3.1 FREQUENCY-TUNABLE ARC

When a pulse is detected, the phase controller keeps mixing the signal (S_{ARC}) whose phase changes linearly with time after the pulse is first detected (a block diagram is shown in Chapter 9):

$$S_{ARC} = e^{2\pi j \cdot (f_s T + f_s T_s)} \tag{10.17}$$

10.3.2 SAR PRODUCTS

We use single look complex (SLC) data processed by a range-Doppler type SAR processor (Shimada 1999). Interpolation using the Fast Fourier Transform (FFT) (Freeman et al. 1988) was applied eight times to analyze the IRF. These SAR data are corrected for sensitivity variations by the antenna pattern, sensitivity time control (STC), and automatic gain control (AGC) (Shimada and Nakai 1994). A rectangular window function was selected when generating the SAR data in order to compare the data with the theory. We selected 42 dBm^2 as the ARC's σ, so as not saturate the SAR raw data but instead to achieve a signal-to-clutter ratio of 30 dB for the background σ^0 of −10 dB.

10.3.3 EXPERIMENTAL DESCRIPTION

The JERS-1 recurrence cycle is 44 days, and the path moves 49 km west every day at a latitude of around 35 degrees. A 26 km (75 km minus 49 km) area can be doubly observed over two consecutive days—the first day for the far range and the second day for the near range. We established the following experiment strategy: first, we selected test sites that can be observed by SAR over two consecutive days; second, we located two ARCs at the same slant ranges in order to exclude the SAR intensity dependency on the slant range; third, we always frequency-shifted ARC$_1$ but not ARC$_2$.

Because the Doppler chirp rate depends on the slant range, the IRF azimuth location shift slightly depends on the ARC's physical location. From an example for the JERS-1 restituted orbit of Path 66 on August 19, 1993, we estimated the differential location shift to be approximately 10m/Hz (e.g., 200 m for 20 Hz, 400 m for 40 Hz, 900 m for 80 Hz, and 2,000 m for 180 Hz, Figure 10-4).

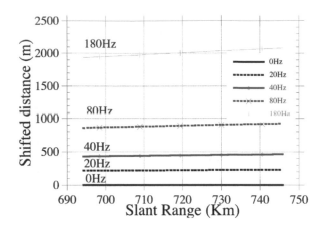

FIGURE 10-4   Slant range dependency of the ARC location shift for five different frequency shifts (0, 20, 40, 80, and 180 Hz).

Table 10-2 shows the history of the experiments and the parameter configuration for the two ARCs. Test sites were selected from NASDA (National Space Development Agency of Japan) calibration sites: Hatoyama site (N 35 deg 58'50.4", E 139 deg 23'14.4"), Kumagaya site (N 36 deg 08'00.5", E 139 deg 16'52.9"), and Niigata site (N 37 deg 54'26.3", E 139 deg 09'11.5"). Combinations of frequency shifts and the ARC locations were selected so that the IRF could appear on the dark background and be recognized clearly. Ideal backgrounds were rivers running nearby. However, those river widths were less than 500 m, and the effective river width along the satellite track was less than 1,000 m. A location shift of 2,000 m moved the IRF across the river. The physical ARC location as adjusted across the track so that the IRF could fall in rice fields and not in the nearby urban area.

Figure 10-5a–d shows the typical ARC image samples at the Niigata test site in response to the four different frequency shifts. Each image size is 3,520 m (azimuth) × 1,800 m (slant range).

TABLE 10-2

Experiments Using the ARC with Frequency Shift

	ARC$_1$		ARC$_2$			
No.	f_s (Hz)	x_m (m)	f_s (Hz)	x_m (m)	Date	Site
1.	40	462.5	0	0	11/16/'93	Niigata
2.	0	0	0	0	11/17/'93	Niigata
3.	0	0	0	0	2/11/'94	Oppe
4.	180	2,026	0	0	5/10/'94	Oppe
5.	80	862.5	0	0	5/11/'94	Kumagaya
6.	180	0	0	0	6/25/'94	Niigata
7.	0	0	0	0	8/6/'94	Oppe
8.	80	862.5	0	0	8/7/'94	Kumagaya
9.	80		0	0	9/20/'94	Niigata
10.	40	0	0	0	9/21/'94	Niigata
11.	0	0	0	0	10/21/'95	Niigata
12.	0	0	0	0	10/22/'95	Niigata
13.	20	223	0	0	7/25/'95	Oppe

Note: RCSs are selected as 42 dBm^2 for both ARCs. JERS-1 SAR is incorporated in this experiment; x_m is the measured location shift (m).

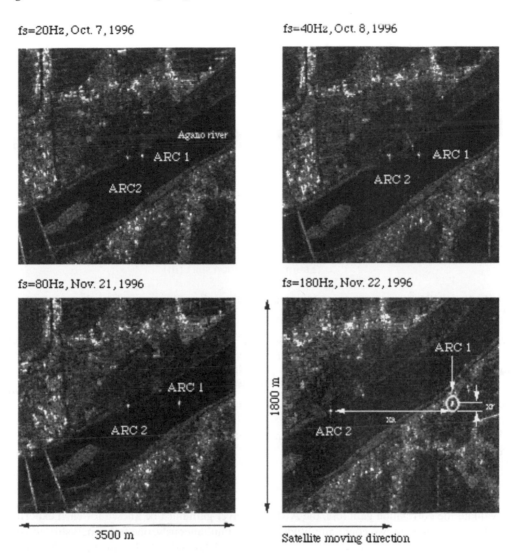

FIGURE 10-5 Sample images of the shifted ARC point targets, $f_s = 20, 40, 80,$ and 180 Hz (top left, top right, bottom left, and bottom right, respectively). In order to distinguish the ARC$_1$ image at $f_s = 180$ Hz from the bright targets on the land, it is surrounded by a white circle.

We confirmed that more frequency shift defocused the IRF and moves it in the azimuth and range directions. For these figures, ARC$_1$ and ARC$_2$ were deployed at almost the same location; their locations differ only a few meters.

10.4 ANALYSIS AND DISCUSSION

Analyses were conducted for location shift, correlation gain loss, resolution broadening, adaptability of the frequency tunable ARC to the calibration, and the frequency shift allowance.

10.4.1 IMAGE SHIFTS IN AZIMUTH AND RANGE

Figure 10-6 shows the location difference between the measured azimuth shift (x_m) and the theoretical azimuth shift (x_e) at five frequency shifts. The mean difference is −0.65 m, and the standard

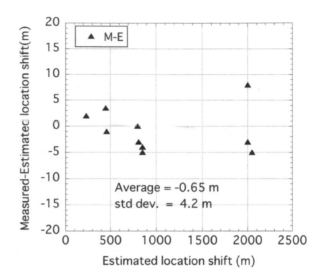

FIGURE 10-6   IRF location difference (M = measured; E = estimated) versus estimated location.

deviation is 4.2 m. These errors might arise from manual recognition of the IRF and satellite velocity determination error. The velocity error is around 15 cm/s for each component (Tsukuba 1991); it deviates from the Doppler chirp rate error of around only 0.007%. We then concluded that the orbital might not be the main error. A good estimation of the location shift requires accurate calculation of the Doppler chirp rate f_{DD} (Curlander and McDonough 1991). Range image shift has been confirmed from Figure 10-5d, for which the theoretical shift x_r at f_s = 180 Hz is 59 m ($\dot{R}_0$ = 0.22 km/s, and T_{as} = 0.28 s), and the measured shift in slant range is 63 m; therefore, both agree very well.

10.4.2   Correlation Gain Loss

To evaluate the correlation gain loss, the ARCs' receiver gain must be known for each experiment. If an ARC can transmit two equal-amplitude signals simultaneously, one frequency-shifted and the other not frequency-shifted the same as that of the time delay ARC (Daleman et al. 1990), the measurement loads for these ARC differences may be reduced. The ARCs used in these experiments were manufactured to stabilize within ±0.5 dB over two years. To ensure measurement credibility, however, those gains were monitored all through the experiments using calibrated signal generators and spectrum analyzers before and after the satellite passages. Results confirmed that the two receiver gains differed within 0.5 to 1.0 dB except at several points but it was necessary to correct them at the gain calculation.

The clutter and the less-focused SAR data also may prevent accurate detection of the impulse response peak. In turn, the integration of the impulse response is known to be well stabilized (Gray et al. 1990). We introduced the following two parameters (R_P and R_I) for the gain loss evaluation:

$$R_P = \frac{P_1(0,0) - P_{back,1}}{P_2(0,0) - P_{back,2}} \cdot \frac{G_2}{G_1} \tag{10.18a}$$

$$R_I = \frac{\iint\limits_{A_1}\left\{P_1(x,y) - P_{back,1}\right\}dxdy}{\iint\limits_{A_2}\left\{P_2(x,y) - P_{back,2}\right\}dxdy} \cdot \frac{G_2}{G_1} \tag{10.18b}$$

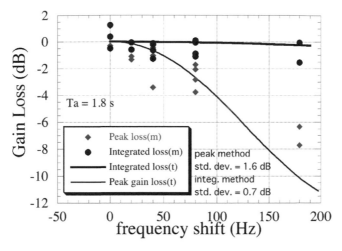

FIGURE 10-7 Peak power loss and integral power are plotted for the frequency-shifted ARC.

where R_P is the gain loss ratio using the peak method; R_I is that using the integral method; $P_1(x, y)$ and P_2 are the impulse response function of ARC_1 and ARC_2; G_1 and G_2 are the gain of ARC_1 and ARC_2; and $P_{back,1}$ and $P_{back,2}$ are the averaged background intensities for ARC_1 and ARC_2. We selected the integral area "A_1" (A_2) that surrounds the impulse response to be as large as possible but not to contain other brighter targets. Background intensity is estimated from the far sides of the aforementioned integral areas.

The evaluation results are depicted in Figure 10-7 and in Tables 10-3 and 10-4, where measured peak gain loss (R_P) is represented by black diamonds, measured integral gain loss (R_I) by white diamonds, theoretical (peak) gain loss by the solid thin line, and theoretical integral gain loss by the solid thick line. The azimuth integral time is 1.8 s. Integral gain loss shows a good agreement between theory and measurement with a standard deviation (SD) of 0.7 dB and a mean residue (MR) of −0.3 dB, but the peak gains do not agree with an SD of 1.6 dB and an MR of 0.1 dB. This tells us that the frequency shift-based gain/loss can be inferred theoretically, and the calibration factor could be corrected. In this evaluation, we corrected the gain difference of the two ARCs.

TABLE 10-3

Comparison of the Theoretical Gain Loss (Peak and Integral)

No.	f_s(Hz)	R_P(Peak)(dB)	P_I(Integral)(dB)
1	0	0	0
2	20	−0.15	−0.00
3	40	−0.65	−0.01
4	80	−2.67	−0.05
5	180	−10.14	−0.24

TABLE 10-4

Comparison of the Peak and Integral Method

	Peak Method	Integral Method
Mean residual	0.1 dB	−0.3 dB
Standard deviation	1.6 dB	0.7 dB

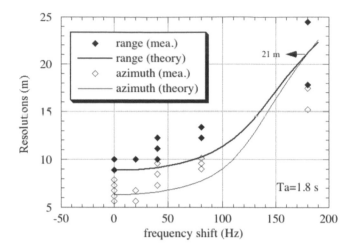

FIGURE 10-8   Frequency dependence of the range and azimuth resolutions for the measured data and the model.

10.4.3   RESOLUTIONS

Figure 10-8 shows the measured and theoretical azimuth and range resolutions (3 dB down width). This shows that both agree well with the theoretical values until $f_s = 20$ Hz; the theoretical values are lower than the measurements for f_s of 40 Hz and 80 Hz; and after that the theoretical values exceed the measurements. Disagreement at f_s less than or equal to 80 Hz is probably because the range-Doppler SAR processing algorithm used in this study is not perfectly matched to the data and needs some improvement. Simulations of the azimuth and range resolutions are discussed in the following subsections.

10.4.3.1   Azimuth Resolution

The azimuth cross sections of the IRF are calculated from Equation (10.13) by deviating the azimuth time (T_1) around the shifted time (T_{as}) and fixing ΔR as given by Equation (10.15b). Figure 10-9 shows the results for four frequency shifts (f_s: 0, 40, 80, and 180 Hz) with the azimuth distance ($V_g(T_1 - T_{as})$) instead of azimuth time. We confirmed that the azimuth resolution of

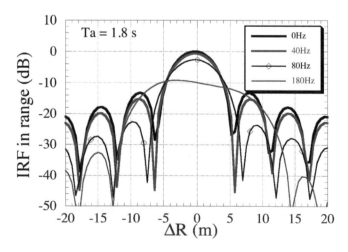

FIGURE 10-9   Azimuth cross section of the impulse response function centered at each location shift. The vertical axis is normalized by the peak at $f_s = 0$ Hz.

the single-look image is almost 6 m until f_s is less than 80 Hz; after that, it becomes broader quickly as f_s increases; the IRF for f_s = 180 Hz has wider Plato peaks than the lower frequency cases.

10.4.3.2   Range Resolution

A similar calculation is performed for the range resolution by changing ΔR in Equation (10.13) from −20 m to 20 m at $T_1 = T_{as}$. Figure 10-10 shows that the best resolution is confirmed to be 9 m for f_s of 40 Hz; but for f_s = 180 Hz, the resolution quickly increases to 22 m and asymmetry appears.

10.4.4   DISCUSSION ON THE CALIBRATION APPLICABILITY

The frequency-tunable ARC has several advantages compared with the standard ARC (non-frequency tunable). Shifting the impulse response location to a low background area or an area that does not receive radiometric interference from brighter man-made targets is the major advantage. Although the ARC can generate a larger radar cross section, the interference from other targets or medium bright clutter decreases the quality of the impulse response. This shifting capability is thus welcomed when searching for a test area that is easily accessible from the data processing/calibration center located in an industrialized area. In turn, the correlation gain loss, the resolution broadening, and some complexity of manufacturing the phase-shifting function are the disadvantages. The applicability of such an ARC to calibration is discussed here in terms of the improvement of the calibration coefficient.

Two representative SAR calibration methods using the point target responses are the peak method and the integral method. Ulander (1991) theoretically compared the accuracy of those two methods and summarized that the peak method achieves smaller error than the integral method if the image is well focused, but that the integral method is a robust method regardless of the SAR focus.

The calibration coefficient by the peak method depends on the peak value of the response and on the azimuth/range resolutions. As shown in Figures 10-3 and 10-10, those terms are significantly reduced at a higher frequency (e.g., the peak value by 10.1 dB at f_s = 180 Hz, the azimuth resolution to 22 m from 6 m, and the range resolution to 22 m from 9 m) and drop the calibration coefficient dramatically. An SAR image may be calibrated with an accuracy of 1.6 dB if no error is assumed for resolution estimation. However, we cannot recommend this method because the integral method gives better results.

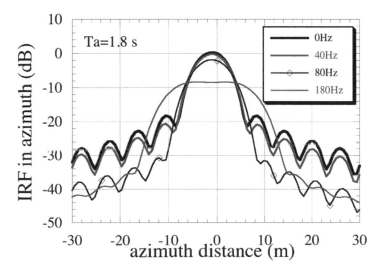

FIGURE 10-10   Range cross section of the IRF. The vertical axis is normalized by the peak of the referenced IRF with $\Delta R = f_s = 0$.

The calibration coefficient by the integral method depends only on the integration of the impulse response function—not on the resolutions. As shown in Figure 10-7, the theoretical integral gain loss agrees well with the measurement with an accuracy of 0.7 dB. This means that the integral method corrects the calibration coefficients. Although the hardware limitations do not allow evaluating the integral gain loss at an f_s higher than 180 Hz, we expect that the integral method can be applied to higher frequency cases.

Defocusing affects the resolution measurement, which is one item of the SAR image quality. From Figure 10-8, the azimuth and range resolutions become broader as f_s increases. The range cross section becomes asymmetric at $f_s = 180$ Hz (Figure 10-10), and the range and azimuth responses differ from the case of $f_s = 0$ if f_s exceeds 80 Hz. The frequency shift should therefore preferably be less than 40 Hz from the image quality evaluation point of view.

10.5   CONCLUSIONS

In this chapter, we have shown that the frequency tunable active radar calibrator suffers from correlation gain loss and resolution broadening as disadvantages; but it has a location shifting advantage. We have verified that the correlation gain loss can be estimated theoretically with an accuracy of 1.6 dB and the location shift within 4.2 m. We have proposed that the integral method is a robust calibration method for using frequency-tunable ARC; and the frequency shift should be less than 40 Hz for image quality evaluation for L-band spaceborne SAR.

REFERENCES

Curlander, J. C. and McDonough, R., 1991, *Synthetic Aperture Radar: Systems and Signal Processing*, Wiley, Hoboken, NJ.

Daleman, P. S., Hawkins, R. K., and Lukowski, T. I., 1990, "Experience with Active Radar Calibrators for Airborne SAR," *Proc. 10th Annual International Symposium on Geoscience and Remote Sensing*, College Park, MD, May 20–24, 1990, pp. 795–798.

Freeman, A., Curlander, J. C., Dubois, P. D., and Klein, J., 1988, "SIR-C Calibration Workshop Report," JPL Center for Radar Studies Publication, No. 88-003, Jet Propulsion Laboratory, California Institute of Technology, Pasadena, CA.

Gray, L. A., Vachon, P. W., Livingstone, C. E., and Lukowski, T. I., 1990, "Synthetic Aperture Radar Calibration Using Reference Reflectors," *IEEE T. Geosci. Remote*, Vol. 28, No. 3, pp. 374–383.

Jin, M. Y. and Wu, C., 1984, "A SAR Correlation Algorithm Which Accommodates Large-Range Migration," *IEEE T. Geosci. Remote*, Vol. GE-22, No. 6, pp. 592–597.

Raney, R. K., 1971, "Synthetic-Aperture Imaging Radar and Moving Targets," *IEEE T. Aero. Elec. Sys.*, Vol. AES-7, No. 3, pp. 499–505.

Shimada, M., 1999, "Verification processor for SAR calibration and interferometry," *Adv. Space Res.* vol. 23, no. 8, pp. 1477–1486, 1999.

Shimada, M., Oaku, H., and Nakai, M., 1999, "SAR Calibration Using Frequency-Tunable Active Radar Calibrators," *IEEE Trans. GRS*, Vol. 37, No. 1, pp. 564–573, Jan.

Shimada, M. and Nakai, M., 1994, "In-Flight Evaluation of L Band SAR of Japanese Earth Resources Satellite-1," *Adv. Space Res.*, Vol. 14, No. 3, pp. 231–240.

Tsukuba Space Center/NASDA, Personal Communication with Tracking Network Technology Department, 1991.

Ulander, L. M. H., 1991, "Accuracy of Using Point Targets for SAR Calibration," *IEEE T. Aero. Elec. Sys.*, Vol. 27, No. 1, pp. 139–148.

Van de Lindt, W. J., 1977, "Digital Technique for Generating Synthetic Aperture Radar Images," *IBM J. Res. Develop.*, Vol. 21, No. 5, pp. 415–432.

Wu, C., 1976, "A Digital Approach To Produce Imagery From SAR Data," presented at the *AIAA System Design Driven by Sensors Conference*, Paper No. 76-968, Pasadena, CA, October 18–20, 1976.

11 Mosaicking and Multi-Temporal SAR Imaging

11.1 INTRODUCTION

A high-resolution sensor helps in monitoring global environmental changes caused by human activities and nature. Repeated land observation enables the detection of various changes on Earth, such as deforestation and forest degradation, earthquakes, forest fires, and so on. The use of SAR images is promising for these purposes because of its all-weather observation capabilities while its sensitivity relies on radar frequency as such the L-band, which has more penetration in vegetation and less reflection from rough surfaces than higher frequency bands. SAR images will be more useful when compiled as time-series mosaics (Shimada and Ohtaki 2010).

Recent SAR systems routinely collect a huge amount of data and require a high computation performance for processing. Compilation of these data onto a large canvas as time-space co-registered data is an ideal process but is rather complicated especially with regard to radiometric and geometric issues. There are several mosaicking algorithms, and one is robust enough to cope with the long-strip processes. These algorithms were adopted to mosaic JERS-1 SAR (De Grandi et al. 2004; Shimada and Isoguchi 2002), ALOS/PALSAR, and ALOS-2/PALSAR-2. We discuss long-strip processing and mosaicking in this chapter.

11.2 LONG-STRIP SAR IMAGING

11.2.1 REQUIREMENTS AND DIFFICULTIES

When we utilize an SAR for long-term observation, hardware heating and data recording or downlinking are key issues. For example, ALOS/PALSAR was designed to provide 70 min of observation per orbit (duration of 99 min) and was verified at 40 min due to recording limitations. Two more issues also impact SAR imaging: temporal shifts of Doppler bandwidth and observation area variations.

11.2.1.1 Doppler Bandwidth Variation

SAR imaging is made up of signal compressions along a history of satellite-target relative motion on the range-Doppler plane (trend) associated with a ranging compressed FM signal (as discussed in Chapter 3). Because the SAR system flies on a circular orbit and the Earth rotates around the z-axis, the azimuth time dependency of trend varies slowly in time or latitude as does the distance: they appear as variations of Doppler bandwidth and the distance from the nadir track. Thus, imaging of the moving Earth requires an exact expression of the trend in time or frequency domain. To capture the varying Doppler bandwidth, the pulse repetition frequency (PRF) needs to be set large enough. As an example, ALOS-PALSAR experienced variation in the Doppler bandwidth around 2,000 Hz.

11.2.1.2 Variation of the SAR Nominal Track

The second issue impacting SAR imaging results from the along-orbit changes in distance between the satellite and the nominal imaging track. To ensure that SAR imaging contains the target of interest, the sampling window start time (SWST), which controls the delay time between the pulse transmission and the reception, should be correctly set and updated.

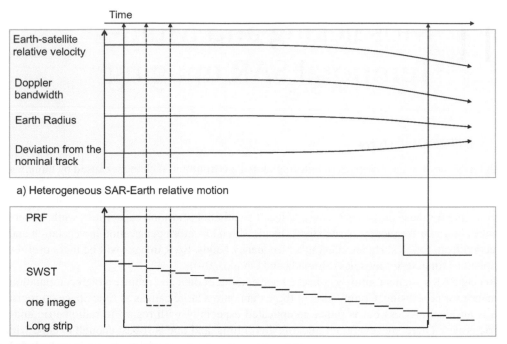

a) Heterogeneous SAR-Earth relative motion

b) Two parameter adjustments to solve the above mismatching

FIGURE 11-1   Schematic view of the temporal variation of the SAR–Earth relative motion, related parameter, and the two parameter adjustments that solve this gradual change.

Figure 11-1 shows the temporal variation of the satellite-Earth relative motion, the related Doppler frequency variation in Figure 11-1a, and the corresponding solution to the problem brought about by adaptively changing the PRF and SWST in a time-dependent fashion (Figure 11-1b). Two additional blocks in the bottom of the image show the duration of the signal SAR image and the long-strip data. Obviously, the difference is the number of the PRFs and SWSTs included in their windows. The issue concerning long-strip processing is how to adapt the SAR data acquired with different parameters and produce the image as if single PRF and SWST were used. This chapter expands on this theme using the PALSAR long-strip data that follows.

11.3   STRIP SAR IMAGING

The SAR imaging algorithm is robustly made up on a single scene base. Its expansion to long-strip data requires an appropriate update of the SWST, PRF, Doppler parameters, antenna elevation pattern, and radio frequency interference (RFI) filtering.

11.3.1.1   Basic Algorithm

SAR imaging is simply expressed by the two correlations:

$$f_{rc}(t-t_0) = F^{-1}\Big[F\{f_{rec}(t-t_0)\}\cdot F\{f_{ref,r}(t)\}\Big],\tag{11.1}$$

$$f_{ac}(t,T) = F^{-1}\Big[F\{f_{rc,I}(t,T)\}_C \cdot F\{f_{ref,A}(t,T)\}\Big].\tag{11.2}$$

where $f_{rc}(t-t_0)$ is the range-compression, $f_{ref,r}(t)$ the range reference, $f_{rec}(t)$ the raw data, $F(\ )$ and $F^{-1}(\ )$ the Fourier and inverse Fourier transforms, and t the range time; t_0 is reserved for a shifted sampling window (SW).

The azimuthally compressed data are represented by $f_{ac}(t,T)$; the suffix I is the interpolation operator to the range-compressed data by some PRF; the suffix c is the data collected along the range curvature in frequency domain; $f_{ref,A}(T)$ is the azimuth reference function, and T is the azimuth time or modified azimuth time. Compression (correlation) is performed in the frequency domain to obtain the processing speed.

11.3.1.2 Sampling Window (SW)

Within a pulse transmit-receive sequence, the SW is set to collect the scattered signal from the Region Of Interest (ROI) by specifying the start time, t_s, and the end time, t_e. They are updated every several tens of seconds to chase the nominal track on the Earth. Under long-strip processing, $t_{s\,min}$ is selected as the minimum of t_s as the original start time, and the maximum is to the end. For each image data, zeros are padded for void areas:

$$f_{rec}(t') = \begin{cases} f\{t-(t_s-t_{s\min})\} & (t_s-t_{s\min} \le t < t_e), \\ 0 & (t < t_s \ or \ t >= t_e), \end{cases} \quad (11.3)$$

where $f(t)$ is the received data at time (t) of each record, and $f_{rec}(t')$ is the arranged data originated at the time t'.

11.3.1.3 Long-Strip SAR

Figure 11-2 shows a process flow for the long-strip data associated with two PRFs where the reference PRF is selected at the beginning of the PRFs (Shimada 1999). Single-look complex data are multi-looked with M looks in range and MN looks in azimuth (i.e., M^2N looks in total) to suppress the speckle.

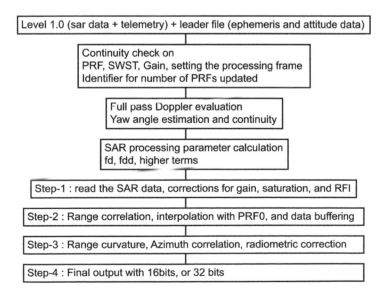

FIGURE 11-2 Flow chart outlining the SAR strip processing adopted for Sigma-SAR.

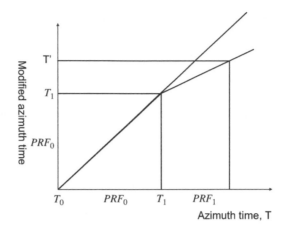

FIGURE 11-3 The relationship between the real azimuth time and the modified azimuth time. The PRF is shown as PRF_0 between azimuth times T_0 and T_1.

11.3.2 STRIP PROCESSING OVER VARIOUS PRFs

Although the SAR processing of the multiple PRFs can be performed by individually processing the data at each PRF and merging them, the amplitude and phase could be discontinuous at the order. To overcome this, it is preferable to reconstruct the raw data or the range compressed data in azimuth time as if the SAR were operated with a constant PRF, and this is performed by interpolation, such as in Equation (11.4), or by an FFT-based shift (Figure 11-3).

PALSAR updates PRF approximately every 2,000 km (Shimada et al. 2009; Shimada et al. 2010) and, thus, 10 min of data contains at least one PRF update:

$$f_{rc}(T') = \sum_{k=-N/2}^{N/2} f_{rc}(T_k) \frac{\sin(T_k - T')}{T_k - T'} \tag{11.4}$$

Here, $f_{rc}(T')$ is the range-compressed complex data interpolated at the target azimuth time T', N is the number of data used for the interpolation, and T is the real azimuth time.

11.3.3 DOPPLER PARAMETERS

While the target moves slowly, the SAR processing runs with a constant block size that Doppler parameters do not vary significantly. Doppler parameters (frequency, chirp rate, and the higher derivatives) should be precisely calculated and updated every segment. During the observation, the attitude is controlled either by the antenna bore site directing zero-Doppler or by the satellite's moving direction staying within the orbital plane leading non-zero Doppler. In these cases, the spectrum analysis provides the estimated yaw angle at each segment. The Doppler parameters are initiated from the following equation:

$$f_D = \frac{2 f_0}{c} (\mathbf{v}_s - \boldsymbol{\omega} \times \mathbf{r}_p) \cdot \frac{(\mathbf{r}_p - \mathbf{r}_s)}{|\mathbf{r}_p - \mathbf{r}_s|} \tag{11.5}$$

Here, f_D is Doppler frequency, f_0 the carrier frequency, c the speed of light, $\mathbf{v}_s$ the satellite velocity, $\boldsymbol{\omega}$ the Earth rotation vector, $\mathbf{r}_p$ the target position, and $\mathbf{r}_s$ the satellite position. The satellite state vectors at arbitrary time can be precisely interpolated from the 28 1-min. interval ephemeris. The Doppler chirp rate and higher terms are obtained from differentiation of the f_D using the geopotential function (Chapter 3).

11.3.4 Conditions for Image Generation

11.3.4.1 Secondary Range Compression

The range-Doppler algorithm requires secondary range compression (Jin and Wu 1984) to the non-zero Doppler signal with a wider bandwidth. If the signal is such that Equation (11.6) is valid, a hybrid range compression achieves fine focusing without algorithm complexity:

$$\left(\frac{\Delta K}{K} \right) B_w \tau < 2, \tag{11.6}$$

$$K = \frac{k_t}{1 + \dfrac{k_t}{f_{DD}} \left(\dfrac{\lambda \cdot f_D}{c} \right)^2}, \tag{11.7}$$

where K is the hybrid chirp rate, ΔK its variation within the image, B_w the bandwidth, τ the pulse width, k_t the chirp rate of the transmission signal, f_{DD} the Doppler chirp rate, f_D the Doppler frequency, λ the wavelength, and c the light speed. We name the left-side term of Equation (11.6) the secondary range compression parameter (SRCP). This effect is already included in Chirp Scaling Algorithm (CSA) processing.

11.3.4.2 Length of Segment

Long-strip processing is made up of a set of sub-scene processing. While the azimuth size is limited by the FFT size, the resultant truncation error and the possible discontinuity of two segments (sub-scenes) at the border remain issues to be addressed. The appropriate length of the segment needs to be determined by the simulation study to make sure no discontinuity or overlapping of the data occurs.

11.3.4.3 Backscattering Coefficient

The radar equation and the backscattering coefficients are summarized in Chapter 4. The output from the SAR processor is proportional to gamma-naught as follows:

$$P \propto \begin{pmatrix} \dfrac{G_R^2 \cos\theta}{R^2 \sin\theta} \gamma^0 \ strip \\[2em] \dfrac{G_R^2 \cos\theta}{R^4 \sin\theta} \gamma^0 \ Scan \end{pmatrix} \tag{11.8}$$

11.3.5 Common Themes in Standard Scene Processing

Other than the aforementioned, the following are applied as well:

- RFI notch filtering (Chapter 13)
- Converting the DN to sigma-zero (Chapter 9)
- Ortho-rectification and slope correction (Chapter 8)

11.4 MOSAICKING

11.4.1 General

Mosaicking of multi-strip data often faces banding or striping issues at the connecting area even when the data were acquired in the same season. Although this reflects the natural temporal variation of the intrinsic scattering properties, it certainly reduces the scientific value of the data and

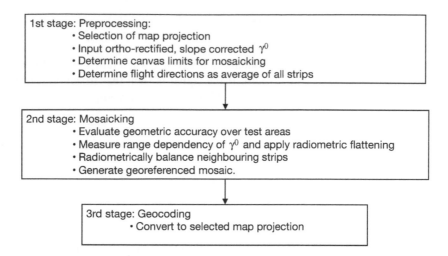

FIGURE 11-4   Flow chart outlining the stages of mosaic generation.

limits the application—for regional classification or retrieval of biophysical attributes. To over-come this, radiometric normalization is mandatory and needs to be implemented as a core part of mosaicking processes (Figure 11-4). The key elements are given here (Shimada and Isoguchi 2002):

Step-1: Step-1 prepares a set of long-strip imaging, slope-correction, and ortho-rectification for the given paths and selects a map out of the map-north or geo-reference coordinates. The former has the universal transverse Mercator, Lambert conformal conic, and equirect-angular projections, and the latter has global and regional coordinates (Figure 11-5).

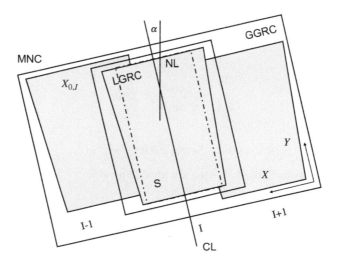

FIGURE 11-5   The coordinate system used for mosaic generation is shown where three strips are selected. The pass number increases from west to east as I − 1, I, and I + 1 (in the case of the ascending orbit). All three strips are inclined counterclockwise at an α angle from the north. Geometric and radiometric correction of each strip is conducted on each of the LGRC coordinates, and images are mosaicked using the exact co-registration process and related radiometric balancing on GGRC. The area in grey is selected to appear in the mosaic. CL represents the centerline of the LGRC image. Finally, the mosaic on GGRC is projected to MNC.

Step-2: The radiometric normalization (intensity normalization) is applied to all the strips, and all corrected strips are merged on the mosaic plane. Merging two strips follows the rule that the overlapped areas are divided into two along their center line (the determination of which will be discussed in Appendix 11A-2). Their intensities are corrected and connected. Recent SAR systems have accurate geometry, and no further geometric adjustment is necessary.

Step-3: Finally, the data will be projected onto the geo-code coordinate. The projection is a combination of the scaling and rotation between the input and the output plane as follows:

$$\mathbf{x}_I = \frac{1}{\Delta s} \mathbf{M}(\alpha) \cdot \left\{ \hat{\mathbf{X}}(\mathbf{q}) - \hat{\mathbf{X}}_C \right\} \tag{11.9}$$

$$\mathbf{X}_I = \mathbf{x}_I + \mathbf{X}_{0,I} \tag{11.10}$$

$$\mathbf{M}(\alpha) = \begin{pmatrix} \cos\alpha & \sin\alpha \\ -\sin\alpha & \cos\alpha \end{pmatrix} \tag{11.11}$$

$$\mathbf{x} = \left(pixel, line \right) \tag{11.12}$$

$$\mathbf{q} = (\varphi, \lambda) \tag{11.13}$$

$$\hat{\mathbf{X}} = \mathbf{f}(\mathbf{q}) \tag{11.14}$$

There are three coordinate systems associated in this projection: the local geo-referenced coordinate (LGRC), which is denoted by $\mathbf{x}$ (pixel, line) in each strip; the global geo-referenced coordinate (GGRC) denoted by $\mathbf{X}$; and map north coordinate (MNC) as the final mosaic plane, which is denoted by $\hat{\mathbf{X}}$. The first two have a shift relationship in Equation (11.10) where $\mathbf{X}_{0,I}$ is the origin of the Ith path in GGRC, and I is the path number. The latter two have the rotation and shift relationship using the matrix $\mathbf{M}(\alpha)$. Here, α is the rotation angle selected so that the angle between the satellite passage and the north direction is averaged; Δs is the spacing; $\mathbf{q}$ contains latitude (φ) and longitude (λ); $\mathbf{f}(\mathbf{q})$ is the mapping function; and $\hat{\mathbf{X}}_C$ is an offset of MNC.

11.4.2 Intensity Normalization

The simple and effective way to suppress banding is to equalize two averaged intensities of the image chips from the overlapped regions. In Appendix 11A-3, as shown in Figure 11A-1, the image square chips are piled along a path (X directions) and across a path (Y directions) in the overlapped zone at equal spacing and over all the associated strips. All the chips are processed to calculate the averaged intensity. At each point, the intensity correction factor is calculated referencing their geometric mean: $\gamma^0_{ref} = \sqrt{\gamma^0_a \cdot \gamma^0_b}$. The intensity connection factor is utilized in two ways in the X and Y directions. To the inter tie-point regions in the X direction, the factors are simply bi-linearly interpolated. In the Y direction, the polygonal curve is applied for the normalization (Figure 11-6) as shown in Equation (11.15). We assume the new gamma-zero $\left(\tilde{\gamma}^0 \right)$ becomes range-interpolated by applying the intensity correction function $G(Y)$ as range dependent:

$$\tilde{\gamma}^0 = G(Y) \cdot \gamma^0 \tag{11.15}$$

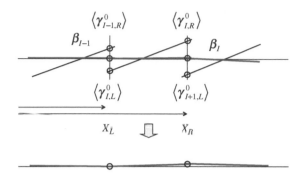

FIGURE 11-6 A simplified concept of radiometrically connecting two neighboring lines. This figure shows a range cross section of the mosaicked image with three associated paths, each of which has the gain offset at the border area. The range dependency of the three paths is similar because the range dependency of σ^0 and γ^0 is also similar. Gain offsets at the border points are minimized by changing the tilt angle of each cross section (i.e., range dependence of the gain offset). The final connected image obtained is shown at the bottom.

A factor (β_I) for the path I at the overlap is determined as

$$\frac{\left\langle \gamma_{I,R}^0 \right\rangle}{\beta_I} = \beta_I \left\langle \gamma_{I+1,L}^0 \right\rangle \tag{11.16}$$

$$\beta_I = \sqrt{\frac{\left\langle \gamma_{I,R}^0 \right\rangle}{\left\langle \gamma_{I+1,L}^0 \right\rangle}} \tag{11.17}$$

where $\left\langle \gamma_{I,R}^0 \right\rangle$ is the averaged γ^0 of the Ith strip at the overlap between I and $I+1$. Using β_{I+1} for the left side of the image and $1/\beta_I$ for the right, we can derive a gain correction function $G(Y)$:

$$G(Y_2) = \frac{\beta_{I+1} - \dfrac{1}{\beta_I}}{Y_R - Y_L}(Y_2 - Y_L) + \frac{1}{\beta_I} \tag{11.18}$$

where Y_L and Y_R are the Y coordinates of the image chips in LGRC. This equation is simply expanded to the Y direction.

In principle, this method does not suffer from error propagation because the correction simply pivots the range dependence of γ^0 around almost the center of the swath, which means that the intensity of each strip center is preserved. As seen from the principle, two strips are not exactly connected at both ends, but the gaps are as small as possible. We cannot recognize a gap through the experiment.

If the overlap occurs in an area of flat land, one near and the other far, two image chips can be used for intensity correction. If it occurs in a mountainous area, their direct comparison could not be easily performed because of possible inclusion of layover or shadowing at the chip. An F-distribution test can ascertain whether the two chips can be used for the gain calculation. If we assume two images are well co-registered, intensities of their pixels follow the chi-squared distribution with the degree of freedom being double the number of looks. Then, their intensity ratio follows the F-distribution. These two chips and their intensity correction factor can be used

TABLE 11-1
List of Metadata

No.	Contents	Expression
1	Total days from launch	Unsigned short integer, unit = 1 day
2	Local incidence angle	Unsigned character: 0–90; unit: 1 degree in positive
3	Mask	Unsigned character: 255; normal 0: outside the image; 50: ocean; 100: layover; 150: shadowing

if the averaged intensity ratio is less than a criterion of 6.59 (confidence of 95%) for two-look SAR images (10-m ALOS PALSAR mosaics), while a value of 1.55 is noted for 64-look SAR images (50-m mosaics):

$$\left(\frac{\gamma_1^0}{\gamma_2^0} \right)_N < a_{95} \tag{11.19}$$

Here, γ_1^0 is the pixel intensity (gamma-naught) of Image 1 and γ_2^0 is that for Image 2. Suffix N is the number of looks, and a_{95} is the confidence level of 95%.

11.4.3 META DATA

To assist the interpretation of the SAR data and time-series analysis of the mosaic purpose, meta data are so important. They are represented by total days from the launch, local incidence angle, mask for normal area, outside of image, ocean, layover, and shadowing. Table 11-1 shows the meta data example.

11.5 EVALUATION

Three data sets were used to evaluate the method described earlier: (1) a 5.5 min PALSAR strip passing from Indonesia to the Philippines, which had changed PRF once; (2) three strips for Riau Province in Indonesia; and (3) five strips covering Tasmania to Australia's northern coast. The first data set is used to validate different PRFs, and the latter two are used to evaluate the mosaicking and banding suppression. These three strips are shown in Figure 11-7 and listed in Table 11-2.

11.5.1 LONG-STRIP IMAGING

In Figure 11-8a, the long-strip has 77 segments covering a ground length of 2,220 km with SWST and PRF variations. The satellite approaches the Equator from the south, and then the local Earth radius increases. The Earth-satellite relative velocity and the Doppler bandwidth at the IFOV increase mainly because of the Earth's rotation. Thus, PALSAR changed the PRF from 2,141.32 Hz to 2,169.19 Hz. The SWST maintains the relative position of the imaging swath to the sub-satellite track (SST) and, in 1 μs, can change steps every 30 s. Equation (11.8) shows the relationship between σ^0 and range, incidence angle, and antenna elevation pattern; therefore, it is important to understand how these parameters change. The SWST decreases stepwise independently from the PRF change. The one-way antenna elevation pattern within five representative segments is given in Figure 11-8b and indicates a variation of 0.7 dB and 1 dB at the near and far range, respectively.

Azimuth time dependence of Doppler parameters (Figure 11-9a and b) show that although the Doppler chirp rate is almost constant over the 77 segments, the Doppler frequency varies from

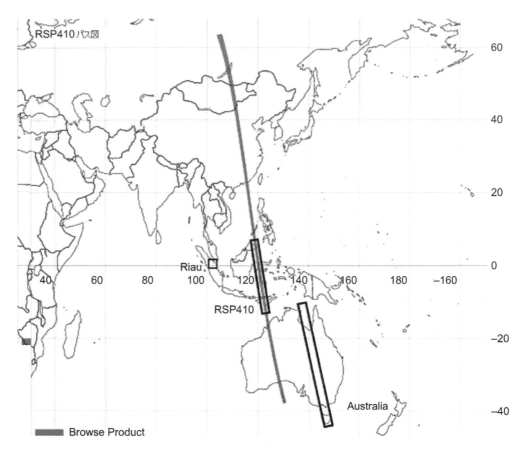

FIGURE 11-7 Ground track of the PALSAR long-strip of RSP410 (blue-green line) and coverage of the PALSAR images considered in this study.

TABLE 11-2

A List of PALSAR Data Sets and Evaluation of the Proposed Algorithm

No.	Area	Sensor and Mode	Objectives and Description
1	Australia to Siberia passing through Sulawesi Island, Indonesia	PALSAR-FBD	Long-strip processing and ortho-rectification Acquisition date: August 9, 2008 Ascending node Pass number: RSP410 DEM: SRTM3 Off-nadir angle: 34.3° Start latitude is −12.1 degrees, and end latitude is 7.84 degrees
2	Riau Province, Indonesia	PALSAR FBS	Off-nadir angle: 34.3° RSP443 (March 30, 2008), RSP444 (April 16, 2008), RSP445 (March 18, 2008)
3	Five paths covering Tasmania to the north of Australia	PALSAR FBD	Off-nadir angle: 34.3° RSP385 (June 2, 2007), RSP384 (July 7, 2007), RSP383 (June 20, 2007), RSP382 (July 19, 2007), RSP381 (July 2, 2007)

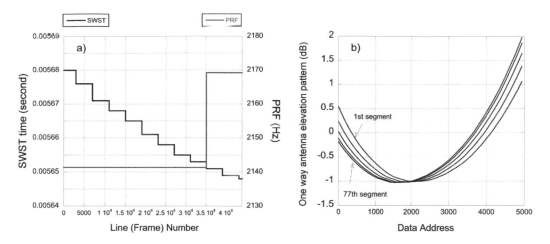

FIGURE 11-8   (a) Temporal variation of SWST and PRF associated in this long-strip processing of 77 total segments as a function of the raw data number, (b) temporal variation of antenna elevation pattern versus data address for five cases. The line number of the horizontal axis in the left image (a) is the relative frame number and the data address of the horizontal axis in the right image (b) is 16-look summed data. SWST is expressed in seconds and PRF is in Hertz.

38 Hz to 56 Hz at the near range and at an average of 0.23 Hz/segment in this case. While an RFI correction process is always required, this strip is not seriously affected by noise—except at the very end of the path where 1% of the spectrum is contaminated by RFI. The latitude dependence of the secondary range compression was evaluated; however, the PALSAR normally was operated with the yaw-steering mode and the secondary range compression is necessary. Even the hybrid correlation that is applied to the almost zero Doppler and Doppler chirp rate may not degrade the image quality without decreasing the processing speed.

11.5.2   ORTHO-RECTIFICATION

Look at the processed images that are ortho-rectified georeferenced sigma-zero in Figure 11-10a, slope corrected and ortho-georeferenced gamma-zero in Figure 11-10b, the local incidence angle generated using the SRTM3 DEM (Shuttle Radar Topography Mission with 3 acrsec DEM) in

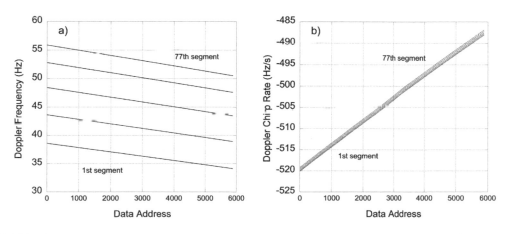

FIGURE 11-9   (a) Doppler frequencies versus data address in range at five selected segments, and (b) Doppler chirp rate versus data address in range at five selected segments.

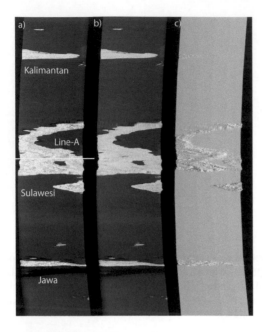

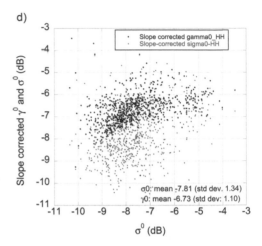

FIGURE 11-10 Comparison of (a) the ortho-rectified georeferenced image, (b) the slope corrected and ortho-rectified γ^0 image, (c) the local incidence angle image generated using the SRTM DEM (Kalimantan, Sulawesi, and Jawa Islands in Indonesia, acquired on August 9, 2009), and (d) comparison of σ^0 (dB) and slope-corrected γ^0 (dB) as well as slope-corrected σ^0 (dB) along Line A of the first image (a).

Figure 11-10c, and the comparison between the sigma-zero and gamma-zero in Figure 11-10d. The image is largely occupied by the ocean except for Sulawesi Island at the center. The elevation ranges from 0 m to 2,500 m. The vertical axis of the image is compressed and the horizontal axis expanded to fit this page. Figure 11-10 indicates that the slope-corrected γ^0 has lower variation than the normal σ^0 and slope-corrected σ^0 and may be attributed to valid dependence of the cosine of the local incidence angle on σ^0 in forested and mountainous areas. The geolocation accuracy was measured as 11.925 m, refererencing the corner reflectors deployed globally (Shimada 2010).

11.5.3 MOSAICKED IMAGES

The proposed mosaicking method (M-1) was evaluated using the data at Riau Province and eastern Australia in comparison with a non-correction method (M-2) that pastes the images without gain corrections.

11.5.3.1 The Riau Case

a. **Continuity Evaluation:** Riau Province is located in northeastern Sumatra facing Malaysia across the Melaka Strait. Significant deforestation has taken place in Riau over the past few decades, which has resulted in a loss of biomass and the release of carbon dioxide into the atmosphere (Uryu et al. 2001). Although the province mosaic was produced by using the normal mosaic algorithm complied with Method M-1, we focus on the ground range mosaic as a subset in Figure 11-11a because of page limitations. The subset image using Method M-2 is shown in Figure 11-11b, and the two evaluation sites are outlined in white

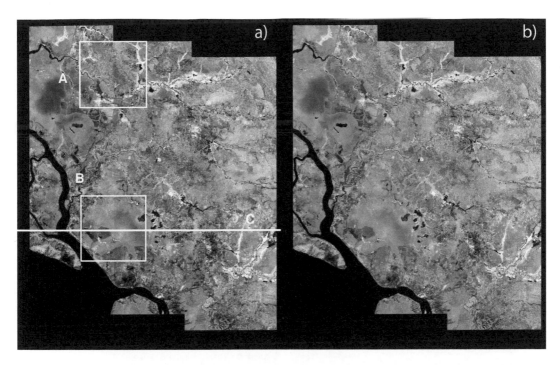

FIGURE 11-11   Subsets of mosaics generated using (a) Method M-1 and (b) Method M-2 (no correction) and showing the location of subsets A and B (see Figure 11-12) and line C.

squares (A and B) and lines (C). A comparison indicates that Method M-1 succeeded in making corrections at even slight changes of image intensity. However, slight radiometric differences are evident between strips in the bottom center of Figure 11-12a and b, which suggests that Method M-1 corrects the intensity offset but not local changes such as inundation, rainfall, and so on. Such differences are perhaps best reduced by selecting scenes acquired under similar environmental conditions.

Figure 11-13 compares two γ^0 (dB) using both methods along Line C. The two intensities are the same (Figure 11-13a), but within the two black circles denoted by intersections 1 and 2, γ^0 from Method M-1 is more continuous than that of Method M-2 (Figure 11-13b). The 0.3-dB drop in Method M-2 was corrected and made smooth at the curve by Method M-1. The average values for Methods M-1 and M-2 are −7.16 dB and −7.16 dB, respectively, while the standard deviation is slightly smaller using Method M-1 (0.081 dB compared to 0.093 dB). No clear intensive differences were observed at intersection 1; therefore, a closelup was not prepared. The two-dimensional correction map at the GGRC coordinate in Figure 11-14 shows that the gain deviates from 0.94 to 1.06 and exists widely across and along the strip

b. **Co-Registration Accuracy:** Co-registration accuracies of two neighbor strips were measured by using the area matching to all image chips selected from the overlapped area; 500 were selected within a strip (Appendix 11A-3). The co-registration errors were measured as −0.42 to approximately −0.44 pixel in the x-direction (approximately in range) and 0.14 to approximately 0.28 pixel in the y-direction (azimuth). Since PALSAR ortho-rectification has a geometric accuracy of 12 m, this residual might have occurred because the orbital data used are of lower accuracy in this case. (Note that Figure 11-15 shows almost a zero-offset for the two co-registration errors.)

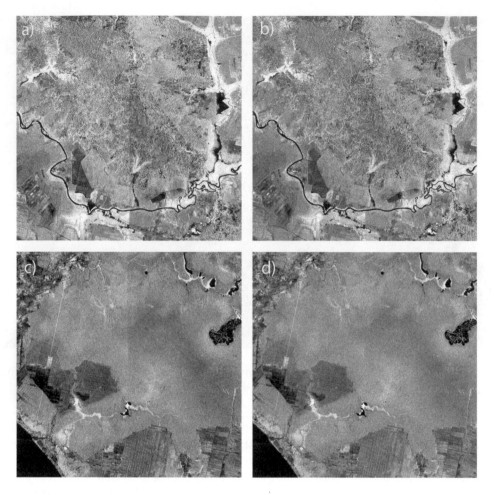

FIGURE 11-12   (a, c) Subsets A and B before and (b, d) following correction using Method-1. All images were acquired in FBS mode (HH polarization).

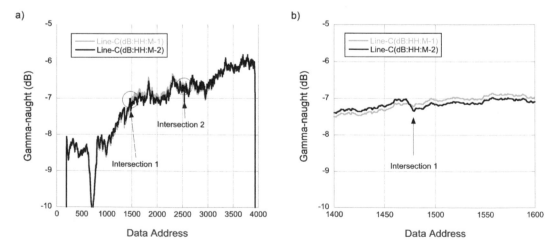

FIGURE 11-13   (a) Range dependence of gamma-naught along Line-C of two images corrected by Method M-1 and Method M-2, and (b) a close-up for the range address between 1400 and 1600.

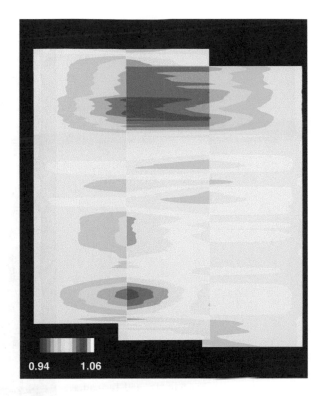

FIGURE 11-14   Area distribution of the additive correction gain.

11.5.3.2   The Australia Case

a. **Visual Evaluation of the Destriping Correction:** The five-strip mosaic of Australia covering 4,000 km was evaluated (Figure 11-16). Northern Queensland and the Northern Territory are low lying, relatively flat, and primarily associated with wooded savannas, while Tasmania is more variable in terrain with higher biomass forests. Banding within this mosaic is particularly prominent; hence, the method for removing banding between

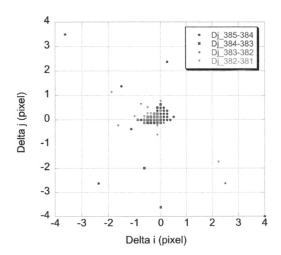

FIGURE 11-15   Geometric location error observed for five consecutive paths in the Australia mosaic (see Figure 11-17).

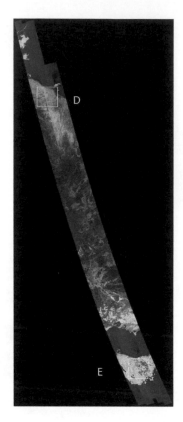

FIGURE 11-16 Mosaic of five ALOS PALSAR HH polarization strips extending from northern to southern Australia. Subsets for Northern Queensland and the Northern Territory are shown in "D" and Tasmania in "E".

strips was evaluated for area "D" in the north and "E" in the south. In region "D", the banding was attributed to precipitation differences at the time of image acquisition, but this intensity variation was largely suppressed following application of Method M-1. The approach was, however, more successful for region "E" where the land cover is more homogeneous and large variations in surface moisture between dates are less evident (see Figure 11-17).

b. **Geometric Co-Registration:** A total of 1,060 samples was taken from RSPs 385–384 ($n = 261$), 384–383 ($n = 273$), 383–382 ($n = 266$), and 382–381 ($n = 260$), where "n" is the number of samples. Here, RSP (Reference System for Planning) expresses the unique orbit number for observing the Earth by ALOS. The averaged shift in x and y was -0.172 pixels and 0.065 pixels, with a standard deviation of 0.261 and 0.277 pixels, respectively. Delta-x/delta-y relationships indicated that most of the points locate at zero values and confirmed that neighboring strips can be accurately co-registered. The mosaic was supported by the ALOS precision orbit data and processing parameter update by matching DEM-based Simulated SAR Image (DSSI); SAR data was not necessary in this case.

11.6 GENERATION OF LARGE-SCALE MOSAICS AND THEIR INTERPRETATIONS

Under the Kyoto and Carbon Initiative (Rosenqvist et al. 2007), JAXA produced PALSAR mosaic data sets operationally. The available mosaics are listed in Table 11-3 and are uploaded routinely to the JAXA/Earth Observation Research Center (EORC) Web site (http://www.eorc.jaxa.jp/ALOS/en/kc_mosaic/kc_mosaic.htm). The data sets consist of 2-byte unsigned short integer-type PALSAR mosaic raw data and JPG images. They are now available for download and can be used for scientific research purposes. Since the data sets preserve the geometric and radiometric accuracy, comparison

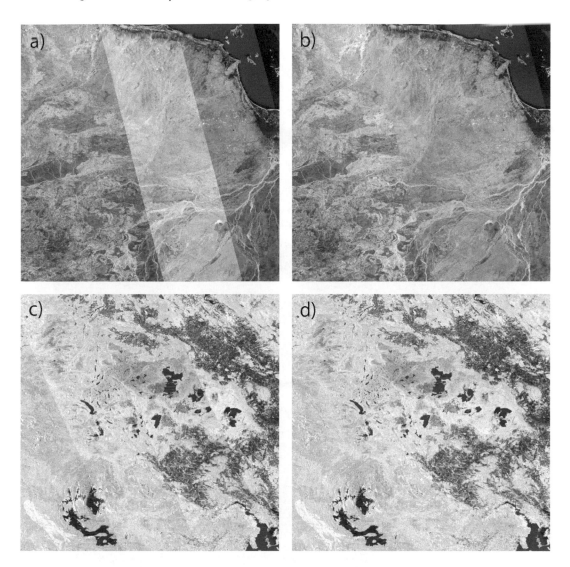

FIGURE 11-17 Comparison of before and after corrections (a) and (b) for Area D and (c) and (d) for Area E. Here, Method M-2 is applied to (a) and (c), while Method M-1 is applied to (b) and (d).

TABLE 11-3

List of Mosaic Datasets Available to the Public

No.	Area	Year	Polarization
1	Indonesia, Malaysia, and Philippines	2007, 2008, 2009	HH+HV
2	Indochina Peninsula	2007, 2008, 2009	HH+HV
3	Indonesia and Malaysia	2007, 2008, 2009	HH+HV and HH
4	Japan	2007, 2008, 2009	HH+HV
5	Central Africa	2008, 2009	HH+HV
6	Australia	2009	HH+HV

http://www.eorc.jaxa.jp/ALOS/en/kc_mosaic/kc_mosaic.htm

FIGURE 11-18   PALSAR mosaics for (a) Africa in 2008 and (b) Australia in 2009.

of the data sets supports, for example, the detection of land cover change including deforestation. Two mosaics are presented in Figure 11-18 for Australia (2009) and Africa (2008). The PALSAR mosaic data sets are generated with 25-m spacing from the slant range long-strip amplitude data, which are also produced with 16 looks in the azimuth direction: 54 m; four looks in the slant range direction: approximately 30 m for FBS and approximately 60 m for FBD; with bilinear interpolation, and 25-m spaced ortho-rectifications.

11.6.1   Comparison between PALSAR and Landsat Mosaics

The United States Geological Survey (USGS) provides calibrated, 30-m, cloud-free Landsat mosaics; 1,393 ground control points (GCPs) were selected to assess the geometric accuracy of PALSAR

TABLE 11-4
Summary of Geolocation RMSE of the JAXA PALSAR Mosaic

Area	Northing RMSE (m)	Easting RMSE (m)	Total RMSE (m)	No. of GCPs
Japan (2007)	22.81(−112.9,43.8)	34.20(−114.2,69.9)	41.11(0.0,119.3)	104
Borneo-Jawa (2007)	23.13(−76.7,71.1)	32.15(−94.5,49.4)	39.61(0.0,98.0)	104
Sumatra (2007)	27.98(−96.9,65.8)	30.03(−86.3,60.7)	41.05(0.0,129.8)	70
Philippines (2007)	17.19(−35.67,35.66)	16.86(−26.89,33.23)	24.08(0.48,43.56)	49
Philippines (2009)	22.83(−54.90,74.90)	29.34(−75.18,39.54)	37.17(0.02,98.39)	101
Borneo-Jawa (2009)	24.79(−62.75,71.95)	30.23(−79.32,26.33)	39.09(0.0,85.42)	83
Sumatra (2009)	26.42(−50.9,67.1)	32.99(−131.9,39.7)	42.26(0.0,131.9)	83
Japan (2009)	26.46(−55.8,52.3)	33.26(−90.0,61.3)	42.50(0.0,99.8)	69
Indochina (2009)	27.96(−52.5,72.9)	30.60(−92.8,75.5)	41.45(0.0,118.0)	89
Central Africa (2008)	24.30(−46.7,47.4)	21.16(−48.2,42.3)	32.22(2.9,63.0)	131
Central Africa (2009)	16.52(−35.17,30.81)	16.20(−39.16,35.88)	23.14(2.73,44.36)	147
Sulawesi (2007)	17.01(−35.14,31.79)	15.44(−30.68,37.59)	22.98(2.30,43.27)	68
Sulawesi (2009)	15.38(−33.76,33.74)	16.21(−41.20,34.76)	22.35(0.85,45.16)	67
Australia (2009)	19.66(−44.41,30.90)	18.91(−41.28,48.26)	27.28(2.35,58.44)	218
All	22.35	25.81	34.14	1383

Note: The numbers in brackets represent the minimum and maximum values, respectively.

and Landsat data sets. The GCPs were selected manually with the condition that two sensors satisfy; the same land feature was identified; that feature was mainly from coastal and riverside areas; and that the feature was flat with limited texture and distributed uniformly at the site listed in Table 11-4. The RMSE was calculated using:

$$\varepsilon = \sqrt{\frac{1}{N} \sum_{i=0}^{N-1} \left(x_{SAR,i} - x_{Landsat,i} \right)^2 + \left(y_{SAR,i} - y_{Landsat,i} \right)^2} \qquad (11.20)$$

where x is the easting and y the northing; the suffixes are the sensors; i is the sample number; and N is the total number of samples. The RMSEs are 22.35 m, 25.81 m, and 34.14 m for the northing, easting, and distance (Table 11-4).

Possible reasons for differences are:

a. PALSAR observes in the squint geometry while Landsat observes in the nadir.
b. PALSAR and Landsat have different visibilities from the target, and this could lead to mis-identification of GCPs. For example, the PALSAR identifies the corner reflectors clearly but Landsat does not.
c. The bi-linear interpolation of the 64-look slant range image reduces the original geometric accuracy.

11.6.2 GEOMETRIC VALIDATION USING THE CORNER REFLECTOR

A 10-m two-look PALSAR mosaic of Japan was generated to quantify geometric accuracy using corner reflectors at three calibration sites: Tomakomai-Hokkaido, Sawara-Chiba, and Yoshimi-Saitama. The RMSE, average northing, and easting errors were 13.19 m, −3.12 m, and 12.82 m with standard deviations being 7.0 m, 3.6 m, and 6.0 m, respectively. More rigorous measurements of the geometric accuracy based on the corner reflectors were reported in Shimada et al. (2009) and Shimada (2010), with these averaging 9.7 m with the single-look slant range image and 12.103 m using the two-look ortho-rectified image (using the SRTM). At the time of publication, JAXA is generating 10-m resolution global mosaic data using the same algorithm applied to the 50-m mosaic data.

11.7 DISCUSSIONS

11.7.1 Intensity Normalization

The PALSAR image swath is 70 km, and the incidence angle varies by 6 degrees. The antenna gain offset at this incidence angle variation (e.g., off-nadirs at the far and near ranges) is less than 1 dB and is sufficiently large to cause the strip. In northern Australia, strips of brighter intensity were paired with those of lower intensity, which was attributed to heavy precipitation and a local increase in the backscattering characteristics. In Tasmania (as with Riau Province), intensity changes were evident but less extreme. The radiometric normalization method successfully removed the variation between strips. It should be noted that the intensity values were modified such that they were more similar to those associated with the majority of values occurring in the region of interest.

11.7.2 Long-Strip Mosaicking Method

Mosaics can be produced by combining many scenes (such as Type-A [Eidenshink 2006]). However, the preferred option is to combine a limited number of long-strip SAR data (Type B [Shimada and Isoguchi 2002]). The typical image width is 70 km, but there is no limitation. As an illustration, if the region of interest is 4,000 km^2 by 4,000 km^2, 3,300 scenes or 60 strips are required if Type-A and Type-B methods are used. Processing using both methods is now fully automated.

One SAR scene can be generated by a SAR processor but, to process long strips, the Doppler chirp rate, frequency, and radar parameters (SWST, antenna elevation pattern, and so on) need to be updated. Both image quality and geometric/radiometric performance are the same. When we consider mosaic processing using these two types of data sets, there are three differences: co-registration, destriping correction, and memory allocation.

Both methods provide almost the same co-registrations because geometry is very accurate. For the intensity normalization, Type-A requires the radiometry to be continuous between the target scene and the surrounding scenes. This radiometric continuity should be confirmed for each of the scenes used, but this is a significant task. However, if the sensor is stable and the target has the same intensity, the processing can be achieved more easily. The use of the Type-B method is better for destriping and reduces the number of unknown parameters because the discontinuity problem appears only at the border of neighboring strips. In terms of memory allocation, requirements are greater for Type B.

11.7.3 Geolocation Accuracies

Geolocation errors were measured in two ways: co-registration errors of image chips and comparison of the Landsat and SAR mosaics. In the first, the RMSE was 20 m (0.4 pixels) for Riau and almost zero for Australia. The latter was measured as 34.14 m for all 50-m mosaics. Since the geolocation accuracies of the three products: the slant range, ortho-rectified images, and the 10-m mosaic, were measured as 9.7 m (Shimada et al. 2009), 11.25 m (Shimada 2010), and 13.19 m, respectively, with regard to CRs, this value might be raised from an error in the co-registration of the SAR and optical systems.

11.8 SUMMARY

This chapter introduced a mosaic method that consists of SAR long-strip data, ortho-rectification, slope correction using the DEM, intensity normalization, and strip integration to the global mosaic. The DEMs have become available globally by the SRTM (2005) and ASTER Global DEM

(GDEM) (2009). GPS improvement makes the state vector more accurate, and then the importance of the long-strip images is slightly reduced when generating a mosaic that contains islands. Since the intensity normalization is so important, long-strip mosaicking makes the process more simplified than that from the individual scenes. Thus, the importance of long-strip processing is the simplification of the overall processing. Using the ALOS/PALSAR archive, we demonstrated the performance of the proposed method: (a) the method succeeded in reducing striping artifacts; (b) the inter-geometric error with Landsat is 34.14 m; and (c) slope-correction using the SRTM canceled the topographic modification effect of the backscatter. JAXA operationally generates 25-m-resolution PALSAR and JERS-1 SAR mosaics. SAR mosaics allow increases in application field monitoring: for land use, land cover change, map generation, coastlines, desertification, wetlands, disasters, and Antarctic glaciers.

APPENDIX 11A-1: EDGE EXPRESSION

A long-strip image is impacted by the stepwise intensity variation at the near and far ends because zeros are padded following the SWST along a path. On the other hand, the ortho-image has a zigzag border due to the layover effect—especially for mountainous areas viewed by smaller off-nadir views. Determination of the center line of two overlapped regions becomes necessary further along in the process. Although there are several complex methods, polynomial approximation for these borders can be used to determine the center of an overlapped region.

APPENDIX 11A-2: IMAGE SELECTION AT THE OVERLAPPED REGION AND CENTERLINE DETERMINATION

There are four ways to use the overlapped images: (1) prioritize the far-range image over the near-range image, (2) process the near-range image, (3) average both images, and (4) connect both halves from the far and near ranges at the border. In general, the near- and far-range images are degraded, and the first two are not recommended. Averaging may have intensity variation due to temporal changes in the target area. The fourth option seems the best alternative, although the calculation process is complex. Thus, our algorithm adopts the fourth method.

The key step is to determine the centerline (CL) of the overlapped region. From Figure 11A-1, the CL is determined as follows:

$$b_{i,f,I} = \frac{a_{i,f,I} + a_{i,n,I+1}}{2}, \left(b_{i,n,I} = \frac{a_{i,n,I-1} + a_{i,n,I}}{2} \right) \tag{11A-1.1}$$

$$X_{C,f,I} = \sum_{i=0}^{N-1} b_{i,f,I} Y_I^i, \left(X_{C,n,I+1} = \sum_{i=0}^{N-1} b_{i,n,I+1} Y_{I+1}^i \right) \tag{11A-1.2}$$

Here, i is the order exponent, and b is the additive mean coefficients of a. This edge information is also used as the border of valid/invalid information at the mosaicking.

APPENDIX 11A-3: TEST AREA DETERMINATION WITHIN THE OVERLAPPED REGION

Since each image pixel has polynomial expressions on latitude/longitude, we may know a pixel/line of the second image overlapped with the first image of pixel/line as

$$(\varphi, \lambda) = \left\{ f_R(x_R, y_R), g_R(x_R, y_R) \right\} \tag{11A-1.3}$$

$$(\varphi, \lambda) = \left\{ f_L(x_L, y_L), g_L(x_L, y_L) \right\} \tag{11A-1.4}$$

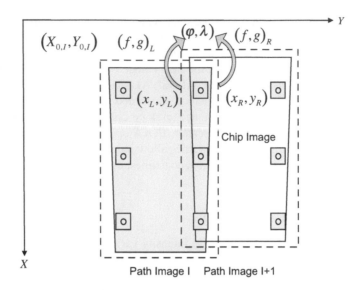

FIGURE 11-A1   Comparing the image location difference between the model and the image compression.

where suffix R and L means the right and left images; x and y are the LGRCs, $f(\cdot)$ and $g(\cdot)$ are polynomials of pixel/line; φ is the latitude; λ is the longitude; and (x_L, y_L) are determined for a given (x_R, y_R) iteratively or vice versa.

REFERENCES

De Grandi, J. F., Spirolazzi, G., Rauste, V., Curto, Y. A., Rosenqvist, A., and Shimada, M., 2004, "The GBFM Radar Mosaic of the Eurasian Taiga: Selected Topics on Geo-Location and Preliminary Thematic Products," *Proc. International Geoscience and Remote Sensing Symposium, IGARSS 2004*, Anchorage, AK, September 20–24, 2004, http://dx.doi.org/10.1109/IGARSS.2004.1369075

Eidenshink, J., 2006, "A 16-Year Time Series of 1 km AVHRR Satellite Data of the Conterminous United States and Alaska," *Photogramm. Eng. Remote S.*, Vol. 72, No. 9, pp. 1027–1035.

Jin, M. Y. and Wu, C., 1984, "A SAR Correlation Algorithm Which Accommodates Large-Range Migration," *IEEE T. Geosci. Remote*, Vol. GE-22, No. 6, pp. 592–597.

Rosenqvist, A., Shimada, M., Itoh, N., and Watanabe, M., 2007, "ALOS PALSAR: A Pathfinder Mission for Global-Scale Monitoring of Environment," *IEEE T. Geosci. Remote*, Vol. 45, No. 11, pp. 3307–3316.

Shimada, M., 1999, "Verification Processor for SAR Calibration and Interferometry," *Adv. Space Res.*, Vol. 23, No. 8, pp. 1477–1486.

Shimada, M., 2010, "Ortho-Rectification of the SAR Data Using the DEM-Based Simulated Data and Its Accuracy Evaluation," *IEEE J-STARS Special Issue on Kyoto and Carbon Initiative*, Vol. 3, No. 4, pp. 657–671.

Shimada, M. and Isoguchi, O., 2002, "JERS-1 SAR Mosaics of Southeast Asia Using Calibrated Path Images," *Int. J. Remote Sens.*, Vol. 23, No. 7, pp. 1507–1526.

Shimada, M., Isoguchi, O., Tadono, T., and Isono, K., 2009. "PALSAR Radiometric and Geometric Calibration," *IEEE T. Geosci. Remote*, Vol. 47, No. 12, pp. 3915–3932.

Shimada, M. and Ohtaki, T., 2010, "Generating Continent-Scale High-Quality SAR Mosaic Datasets: Application to PALSAR Data for Global Monitoring," *IEEE J-STARS Special Issue on Kyoto and Carbon Initiative*, Vol. 3, No. 4, pp. 637–656.

Shimada, M., Tadono, T., and Rosenqvist, A., 2010, "Advanced Land Observation Satellite (ALOS) and Applications for Monitoring Water, Carbon, and Global Climate Change," *Proc. IEEE*, Vol. 98, No. 5, pp. 780–799.

Uryu, Y., Mott, C., Foead, N., Yulianto, K., Budiman, A., Setiabudi, Takakai, F., et al., 2001, "Deforestation, Forest Degradation, Biodiversity Loss and CO2 Emissions in Riau, Sumatra, Indonesia," *WWF Indonesia Technical Report*, Jakarta, Indonesia.

12 SAR Interferometry

12.1 INTRODUCTION

SAR interferometry (InSAR) measures the phase difference of two co-registered SAR complex images and converts it to geophysical parameters: topography, surface deformation, forest cover, forest height, moving target (as the along-track interferometry), and so on (Massonnet et al. 1993). Recent progress in space technology, such as position determination within several centimeters and orbit maintenance within a hundred meters or so of the nominal orbit, enables the differential or time-series InSAR to monitor the surface deformation precisely. Differential interferometric synthetic aperture radar (DInSAR), persistent scatterer interferometric synthetic aperture radar (PSIn-SAR), and small baseline subset (SBAS) are the representing methods. The accuracy of InSAR is related to the characteristics of the target and the signal propagation media: (1) spatio-temporal and scattering property, (2) SAR signal quality, (3) inter-orbit property: baseline and accuracy, (4) signal propagation property: ionosphere and troposphere, and (5) processing property: co-registration of two images, filtering, unwrapping, and so on.

The first three characteristics (target property, SAR signal quality, and baseline distance) have been extensively covered in the literature (Zebker and Villasenor 1992; Zebker et al. 1992; Zebker and Goldstein 1986; Moccia and Vetrella 1992; Prati et al. 1993; Monti-Guarnieri et al. 1993; Rodriguez and Martin 1992; Bamler and Just 1993; Gray and Farris-Manning 1993; Li and Goldstein 1990). Phase unwrapping has also been discussed quite a lot (Goldstein et al. 1988; Spagnolini 1995; Hunt 1979; Takajo and Takahashi 1988; Ghiglia and Pritt 1998), although there remain some issues. The others characteristics have not been discussed as much; the co-registration problem may depend on the SAR imaging algorithm.

12.2 SAR INTERFEROMETRY PRINCIPLE

12.2.1 Scenario

As shown in Figure 12-1, a target area with height (z) over the Earth is observed repeatedly at least two times from almost the same orbits with the same SAR but different timing, or from a single orbit with two identical SARs simultaneously, both of which are separated with a cross track baseline. The former configuration is called a "repeat pass InSAR" and the latter is a "single pass InSAR." For various reasons, including hardware manufacturing difficulties, measurement sensitivity, and budget issues, the single-pass InSAR is utilized only at National Aeronautics and Space Administration/Shuttle Radar Topography Mission (NASA/SRTM), Deutsches Zentrum fur Luft- unt Raumfahrt/TerraSAR-X and TanDem X (NASA/SRTM, DLR/TSX, and TDX) for space-borne purposes but it has been used more often for airborne purposes. Repeat pass SAR interferometry is used most often for space but faces some challenges for airborne use.

The general observation concept is shown in Figure 12-2, where two images, the master image and the slave image, are obtained from almost parallel orbits, and one target is observed twice at slightly different incidence angles (b–c). The target is included in both images, and it is also included in a pixel as shown in the close-up image (a). We assume that two orbits are precisely determined and that a one-to-one relationship with regard to the images can be obtained as well. In reality, however, two orbits are seldom parallel and are less accurate. Building an accurate co-registration frame relies on image correlation, which is difficult to attain, thereby causing another error source.

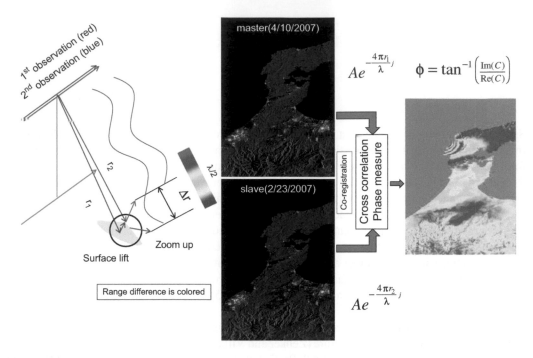

FIGURE 12-1 Differential SAR interferometry and principle for detecting the deformation (and differential land change speed).

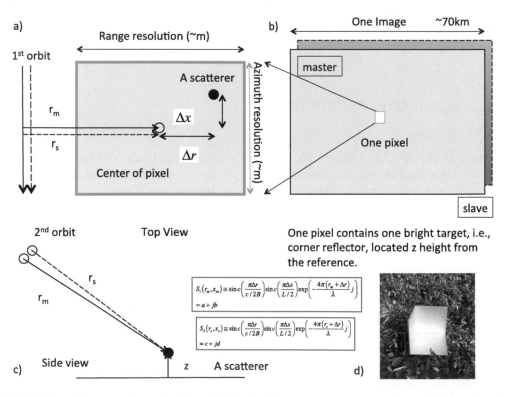

FIGURE 12-2 Schematic views of collocated SAR images in top and horizontal dimensions. (a) Top view of the two co-registered pixels. (b) View of the two-image allocation. (c) Side view. (d) One representative scatterer: corner reflector.

12.2.2 Expression of the Pixel

Two pixel expressions are approximated by a point target and by distributed targets:

a. In point target approximation, an SAR complex image pixel can be expressed by

$$S_m(r_m, x_m) = A(r_m, x_m) \, \text{sin} \, c\left\{\frac{\pi(r - r_m)}{c/2B}\right\} \text{sin} \, c\left\{\frac{\pi(x - x_m)}{L/2}\right\} \exp\left(-\frac{4\pi r_m}{\lambda}j\right) \qquad (12.1)$$

where r, r_m, x, and x_m are the range to the point target, to the pixel center, the azimuth distance to the point target, and the distance to the pixel center, respectively. Equation (12.1) considers a general condition that a point target (scatterer) does not necessary locate at the pixel center: $x \neq x_m, r \neq r_m$.

b. For distributed targets, each pixel is constructed by a linear summation of all the responses from the point targets that exist within a pixel and whose gain is distance-weighted as sinc functions in range and azimuth. So, a pixel, $S_m(r_m, x_m)$, at the location of r_m and x_m as the slant range and azimuth location, can be expressed by (Figure 12-3)

$$S_m(r_m, x_m) = e^{-2\pi j \frac{2r_m}{\lambda}} \sum_{i=0}^{N} A_{m,i} e^{-2\pi \Phi_{m,i}} W_{m,i} + A_{m,n} e^{j\phi_{m,n}} \qquad (12.2)$$

$$= a + bj,$$

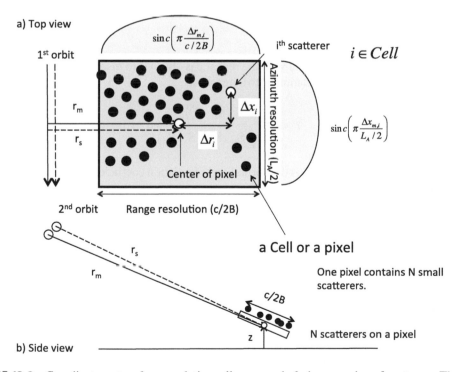

FIGURE 12-3 Coordinate system for a resolution cell composed of a huge number of scatterers. The amplitude of the correlated scatter signal decreases as the scatterer is further from the center of the pixel. Image (a) is the top view of a pixel structure filled by all the scatterers in a pixel or a resolution cell, and (b) is the side view of such a pixel in which all the scatterers are involved.

$$\Phi_{m,i} = \delta_{m,i} - \frac{2 \cdot \Delta r_{m,i}}{\lambda}, \tag{12.3}$$

$$\Delta r_{m,i} = r_{m,i} - r_m \tag{12.4}$$

$$W_{m,i} = \tau T_a \sin c \left(\pi \frac{\Delta r_{m,i}}{c / 2B} \right) \cdot \sin c \left(\pi \frac{\Delta x_{m,i}}{L_A / 2} \right), \tag{12.5}$$

where suffix m stands for the master; $A_{m,i}$ is the amplitude of the signal scattered from the ith scatterer in a master pixel; $\Phi_{m,i}$ is the balance of two phases defined at the pixel center: the instantaneous phase change ($\delta_{m,i}$) and the range dislocation ($\Delta r_{m,i}$); Δx is the azimuth pixel spacing; $W_{m,i}$ is the impulse response function; B is the bandwidth in range; L_A is the antenna length in azimuth; $\tau \cdot T_a$ is the pulse width $\times$ azimuth synthetic aperture time; V_g is the satellite ground speed; and A_n and ϕ_n are the amplitude and phase of the associated noise, respectively (the thermal noise at a receiver, AD redundancy noise, and AD saturation noise).

It should be noted that the scatterers can be numbered one dimensionally for simplicity—even they are distributed two dimensionally.

In summary,

1. The truncated phase accumulated for all the scatterers in a resolution cell is preserved.
2. The correlation amplitude decreases as ith scatterer dislocates from the pixel center.
3. Several noises degrade the cross correlation.

12.2.3 Cross Correlation

Among the InSAR scenarios, we deal with the repeat pass InSAR to manipulate the InSAR formulation and consider the cross correlation of two images as

$$\gamma \cdot e^{j\phi} = \frac{\langle \mathbf{S}_m \cdot \mathbf{S}_s^* \rangle}{\sqrt{\langle \mathbf{S}_m \cdot \mathbf{S}_m^* \rangle \langle \mathbf{S}_s \cdot \mathbf{S}_s^* \rangle}}, \tag{12.6}$$

where $\langle \cdot \rangle$ is the ensemble average, γ is the normalized correlation coefficient ranging from 0.0 to 1.0, and ϕ is the phase ranging from $-\pi$ to π.

Inserting Equations (12.3) and (12.5) into Equation (12.6), the numerator becomes

$$\langle \mathbf{S}_m \cdot \mathbf{S}_s^* \rangle = \left\langle \left(e^{-4\pi j \frac{r_m}{\lambda}} \sum_i A_{m,i} e^{2\pi j \Phi_{m,i}} W_{m,i} + A_{m,n} e^{j\phi_{m,n}} \right) \left(e^{-4\pi j \frac{r_s}{\lambda}} \sum_k A_{s,k} e^{2\pi j \Phi_{s,k}} W_{s,k} + A_{s,n} e^{j\phi_{s,n}} \right)^* \right\rangle$$

$$= e^{-4\pi j \frac{r_m - r_s}{\lambda}} \left\langle \sum_i \sum_k A_{m,i} A_{s,k} e^{2\pi j (\Phi_{m,i} - \Phi_{s,k})} W_{m,i} W_{s,k} \right\rangle + e^{-4\pi j \frac{r_m}{\lambda}} \left\langle \sum_i A_{m,i} e^{2\pi j \Phi_{m,i}} W_{m,i} A_{s,n} e^{-j\phi_{s,n}} \right\rangle$$

$$+ e^{4\pi j \frac{r_s}{\lambda}} \left\langle A_{m,n} e^{j\phi_{m,n}} \sum_k A_{s,k} e^{-2\pi j \Phi_{s,k}} W_{s,k} \right\rangle + \left\langle A_{m,n} e^{j\phi_{m,n}} A_{s,n} e^{-j\phi_{s,n}} \right\rangle$$

$$= e^{-4\pi j \frac{r_m - r_s}{\lambda}} \left\langle \sum_i \sum_k A_{m,i} A_{s,k} e^{2\pi j (\Phi_{m,i} - \Phi_{s,k})} W_{m,i} W_{s,k} \right\rangle. \tag{12.7}$$

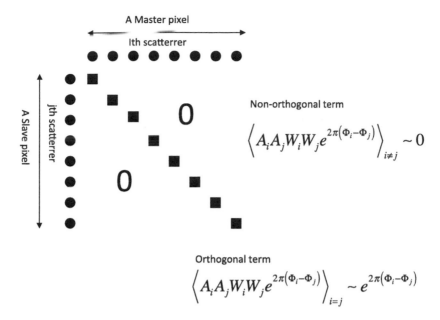

FIGURE 12-4   Ensemble average of cross correlation of two pixels, each of which consists of N scatterers.

Since the ensemble average of the correlation with noise is zero, the latter three terms at the second equivalence become zero, and the first term of the second line could be nonzero. This term is the cross correlation of the two pixels, each of which consists of N scatterers; its expectation can be the covariance of the two partially independent pixels as shown in Figure 12-4.

This figure schematically explains how cross correlation is conducted and how ensemble averaging is performed. We pick up one pixel of master and slave images, both of which are co-registered. All the scatterers involved in this pixel are numbered from 1 to N and aligned in the horizontal direction from left to right. In a similar fashion, the slave pixel is directed to align the related scatterers but in the vertical direction. As Figure 12-4 shows, the master consists of all the scatterers numbered as i for all N in horizontal, and the scatterers the slave pixel has aligned vertically.

Covariance of two such pixels is performed as an ensemble and has nonzero components only at the orthogonal components with ensemble-zero at the non-orthogonal terms. In this way, the summation of the orthogonal components can be expressed by the phase rotation representing the averaged dislocation of two pixels contaminated by the variation of the scatterers.

The co-registered pixels dislocate slightly in range and azimuth (Figure 12-3); therefore, the term contained in $\langle \cdot \rangle$ in Equation (12.6) becomes a complex constant:

$$\left\langle \sum_i \sum_k A_{m,i} A_{s,k} e^{2\pi j(\Phi_{m,i} - \Phi_{s,k})} W_{m,i} W_{s,k} \right\rangle = f \cdot \exp(\delta j) \qquad (12.8)$$

where δ is the averaged excess phase difference rather than the master-slave range difference defined at the center of the pixel. If the scatterers uniformly distribute in a pixel, δ must be zero. Usually, the natural case nearly equals zero δ.

On the other hand, as shown in the second equivalent of Equation (12.2), each pixel is a complex value. Cross correlation of the two complex values is calculated by

$$\begin{aligned} \langle \mathbf{S}_m \cdot \mathbf{S}_s^* \rangle &= (a + jb)(c - jd) \\ &= (ac + bd) + j(bc - ad) \\ &= ABe^{j\phi} \end{aligned} \qquad (12.9)$$

where a, b, c, d, A, B, and ϕ are the real values; ϕ is simply obtained by the manipulation of Equation (12.9):

$$\phi = \tan^{-1}\left(\frac{bc-ad}{ac+bd}\right) = \tan^{-1}\left\{\frac{\Im\left(\langle \mathbf{S}_m \cdot \mathbf{S}_s^* \rangle\right)}{\Re\left(\langle \mathbf{S}_m \cdot \mathbf{S}_s^* \rangle\right)}\right\}. \tag{12.10}$$

From this relation, the numerator of Equation (12.6) is rewritten as

$$\langle \mathbf{S}_m \cdot \mathbf{S}_s^* \rangle = A\exp\left(\phi j\right)$$
$$= e^{-4\pi j\frac{r_m-r_s}{\lambda}} e^{j\delta} \tag{12.11}$$

Comparing the exponent terms, we finally have an important equation describing the measured phase and the differential slant ranges:

$$\phi = -\frac{4\pi}{\lambda}(r_m - r_s) + \delta$$
$$= \tan^{-1}\frac{\Im\left(\langle \mathbf{S}_m \cdot \mathbf{S}_s^* \rangle\right)}{\Re\left(\langle \mathbf{S}_m \cdot \mathbf{S}_s^* \rangle\right)} \tag{12.12}$$

where $\Im(\cdot)$ and $\Re(\cdot)$ are the imaginary part and the real part. The ensemble calculation is performed over several pixels.

12.2.4 Arrangement of $r_m - r_s$ and Theoretical InSAR Phase

In order to analyze the differential range of $r_m - r_s$ in Equation (12.12), we adopt the coordinate system of Figure 12-5: The target P locates at height Z over the Earth's surface (sphere); two orbits, m and s, observe P as the side-looking geometry, and two orbits are separated in B and dh that are parallel and normal to the Earth's surface, respectively. We make the following assumptions: During two observations, no decorrelation (temporal, thermal, or spatial) occurs; a point P was shifted to P' in dx (parallel to the Earth's surface) and vertically in dz as a surface deformation event; and SAR characteristics do not change.

For Equation (12.12), we have the following:

$$\phi = -\frac{4\pi}{\lambda}(r_m - r_s)$$
$$= -\frac{4\pi}{\lambda}(r_s' - r_s) + \frac{4\pi}{\lambda}(r_m' - r_s'). \tag{12.13}$$

From the figure, it can be seen that $r_m' = r_m$, and we determine six equations as follows:

$$r_m = \left\{(R_1+z)^2 + R_m^2 - 2(R_1+z)R_m\cos(\varphi_1 + \Delta\varphi_0)\right\}^{1/2} \tag{12.14}$$

$$r_s = \left\{(R_1+z)^2 + (R_m+dh)^2 - 2(R_1+z+dz)(R_m+dh)\cos\left(\varphi_1 + \Delta\varphi_0 + \frac{B}{R_m} + \frac{dx}{R_1}\right)\right\}^{1/2} \tag{12.15}$$

$$r_m' = \left(R_1^2 + R_m^2 - 2R_1 R_m \cos\varphi_1\right)^{1/2} \tag{12.16}$$

$$r_s' = \left\{R_1^2 + (R_m+dh)^2 - 2R_1(R_m+dh)\cos\left(\varphi_1 + \frac{B}{R_m}\right)\right\}^{1/2} \tag{12.17}$$

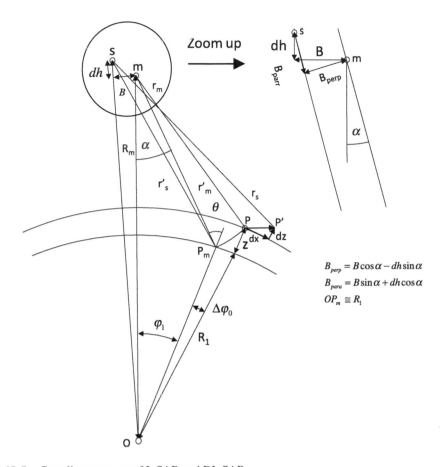

$$B_{perp} = B\cos\alpha - dh\sin\alpha$$
$$B_{para} = B\sin\alpha + dh\cos\alpha$$
$$OP_m \cong R_1$$

FIGURE 12-5 Coordinate system of InSAR and DInSAR.

$$\varphi_1 = \cos^{-1}\left(\frac{R_1^2 + R_m^2 - r_m^2}{2R_1R_m}\right) \tag{12.18}$$

$$\Delta\varphi_0 = \frac{z}{R_1 + z}\frac{R_m^2 - r_m^2 - R_1^2}{2R_1R_m\sin\varphi_1} \tag{12.19}$$

where r_m is the slant range to the point P observed by a master orbit (m); r_s is that of P' observed by a slave orbit (s); R_1 is the Earth's radius at P; R_m is the altitude of the master orbit; φ_1 is the Earth's center angle between the master orbit and P_m; dh is the distance of the surface's perpendicular orbit; z is the height of P; and B is the orbit distance of the parallel surface. The point P is projected on the reference Earth as P_m; in keeping with r_m, we have two more slant ranges, r_m' and r_s', for P_m.

From Equation (12.17), r_s' can be expanded by using the Taylor series and arranged by renaming the first two terms as:

$$\begin{aligned}
r_s' &\cong \left\{R_1^2 + R_m^2 + 2R_mdh - 2R_1(R_m + dh)\cos\varphi_1 + 2R_1(B\sin\varphi_1)\right\}^{1/2}\\
&= \left\{r_m'^2 + 2R_mdh - 2R_1dh\cos\varphi_1 + 2R_1(B\sin\varphi_1)\right\}^{1/2} \tag{12.20}\\
&= r_m' + \frac{1}{r_m'}\left\{dh(R_m - R_1\cos\varphi_1) + R_1(B\sin\varphi_1)\right\}
\end{aligned}$$

There are two more equations in Figure 12-5:

$$\cos\alpha = \frac{R_m - R_1\cos\varphi_1}{r'_m},\tag{12.21}$$

$$\sin\alpha = \frac{R_1\sin\varphi_1}{r'_m}\tag{12.22}$$

Then, we have the parallel baseline distance, B_{para}, as follows:

$$r'_s - r'_m \cong dh\cos\alpha + B\sin\alpha$$
$$\equiv B_{para}\tag{12.23}$$

Inserting Equations (12.15) and (12.20) into Equation (12.13), the first term at the second line of Equation (12.13) can be rearranged as

$$r'_s - r_s \cong \frac{zR_1}{r_mR_m}\left\{\frac{B\cdot\left(r_m^2 + R_m^2 - R_1^2\right)}{\left(-r_m^4 + 2r_m^2R_m^2 - R_m^4 + 2r_m^2R_1^2 + 2R_m^2R_1^2 - R_1^4\right)^{1/2}} - dh\right\}$$
$$+ \left[dz\frac{r_m^2 - R_m^2 + R_1^2}{2r_mR_1} + dx\frac{R_m}{r_m}\left\{1 - \frac{\left(-r_m^2 + R_m^2 + R_1^2\right)^2}{4R_m^2R_1^2}\right\}^{1/2}\right].\tag{12.24}$$

When combined with the following relations:

$$\cos(\pi - \alpha) = \frac{r_m^2 - R_m^2 + R_1^2}{2r_mR_1}$$
$$\cos\varphi_1 = \frac{-r_m^2 + R_m^2 + R_1^2}{2R_1R_m},\tag{12.25}$$

$$\left(-r_m^4 + 2r_m^2R_m^2 - R_m^4 + 2r_m^2R_1^2 + 2R_m^2R_1^2 - R_1^4\right)$$
$$= 4r_m^2R_m^2 - \left(r_m^2 + R_m^2 - R_1^2\right)^2\tag{12.26}$$
$$= 4r_m^2R_m^2\sin^2\alpha,$$

and

$$\frac{r_m^2 - R_m^2 + R_1^2}{2r_mR_1} = \cos(\pi - \theta)$$
$$r_m\sin\theta = R_m\sin\varphi_1,\tag{12.27}$$

Equation (12.24) is rearranged as

$$r'_s - r_s \cong \frac{zR_1}{r_mR_m}\left\{\frac{B\left(r_m^2 + R_m^2 - R_1^2\right)}{2r_mR_m\sin\alpha} - dh\right\} + (-dz\cdot\cos\theta + dx\cdot\sin\theta)$$
$$= \frac{zR_1}{r_mR_m}\left(\frac{B\cos\alpha}{\sin\alpha} - dh\right) + (-dz\cdot\cos\theta + dx\cdot\sin\theta)\tag{12.28}$$

Defining the perpendicular baseline, B_{perp}, using B and dh as Equation (12.29) and using the relation between R_1 and R_m in Equation (12.30):

$$B_{perp} = B\cos\alpha - dh\sin\alpha, \tag{12.29}$$

$$R_m\sin\alpha = R_1\sin(\pi - \alpha), \tag{12.30}$$

we have:

$$r_s' - r_s \cong \frac{zB_p}{r_m\sin\theta} - dz\cdot\cos\theta + dx\cdot\sin\theta. \tag{12.31}$$

Inserting Equations (12.23) and (12.31) into Equation (12.13), we finally have a very important equation for the interferometric phase and the two-image differentiation:

$$\phi = -\frac{4\pi}{\lambda}(r_m - r_s)$$
$$\cong -\frac{4\pi}{\lambda}\left(\frac{B_{perp}z}{r_m\sin\theta} + dD + B_{para}\right) \tag{12.32}$$

$$dD = -dz\cdot\cos\theta + dx\cdot\sin\theta. \tag{12.33}$$

Here, dD is the deformation value measured in the line of sight.

12.2.5 ALONG TRACK INTERFEROMETRY (ATI)

ATI is designed to detect the moving target and its slant range speed, which may appear in the InSAR phase and the coherence drop. The measurement principle is the same as the cross-track InSAR. If the InSAR is based on the single-pass InSAR, the processing difficulty can be significantly reduced. Otherwise, the repeat-pass InSAR based on the airborne systems may be quite difficult. Most spaceborne ATI adopts the single-pass InSAR; the rigorous co-registration frame is always prepared without question and the troublesome corrections for orbit error or the terrain corrections are not necessary. The results are obtained by simply correlating the two-image data.

Among the three ATI configurations (Stiefvater Consultants 2006)—single transmit and dual receives, the ping-pong operation of the two transmit and dual receive system, and the dual individual transmit/receive mode as the repeat pass InSAR—we select the first case. It easily detects the phase variation during the time that two antennas are moving, and the InSAR phase can be obtained from Equation (12.32), rewriting it as

$$\phi = -\frac{2\pi}{\lambda}\left(\frac{B_{perp}z}{r_m\sin\theta} + dD + B_{para}\right)$$
$$= -\frac{2\pi}{\lambda}dD \tag{12.34}$$
$$= -\frac{2\pi}{\lambda}\frac{B}{V_g}U$$

Here, B_{perp} and B_{para} are zero and dD is the line of sight displacement, B is the antenna baseline in azimuth, Vg is the satellite ground speed, and U is the speed of the target in the line of sight.

Inserting 2π at the left-hand side of Equation (12.34), we have the maximum ambiguous speed as

$$U = \frac{\lambda}{B}V_g. \tag{12.35}$$

12.2.6   INSAR COHERENCE

The phase measured with Equation (12.12) is statistical, and its standard deviation (σ_Φ) has been derived by Rodriguez and Martin (1992) and Madsen et al. (1995) using the Cramer-Rao bound as:

$$\sigma_\Phi = \frac{1}{\sqrt{2N}} \frac{\sqrt{1-\gamma^2}}{\gamma}, \tag{12.36}$$

where γ is the correlation coefficient of both images defined by Zebker and Villasenor (1992), which is decomposed to three multiplicative components from Equation (12.6):

$$\gamma = \frac{\left|\langle \mathbf{S}^m \cdot \mathbf{S}^{s*}\rangle\right|}{\sqrt{\langle \mathbf{S}^m \cdot \mathbf{S}^{m*}\rangle}\sqrt{\langle \mathbf{S}^s \cdot \mathbf{S}^{s*}\rangle}}$$

$$= \frac{P_{cl}}{P_c} \cdot \frac{P_c}{P_c + P_d} \cdot \frac{1}{1 + SNR^{-1}} \tag{12.37}$$

$$\rightarrow \gamma_{temp} \cdot \gamma_{spatial} \cdot \gamma_{thermal}$$

where P_{cl} is the power of the correlative component without temporal decorrelation, P_c the power of the correlative component including temporal decorrelation, P_d the power of the non-correlative component, SNR the signal to noise ratio, γ_{temp} the temporal decorrelation, $\gamma_{spatial}$ the spatial decorrelation, and $\gamma_{thermal}$ the thermal decorrelation. The total correlation coefficient is modeled after Zebker and Villasenor (1992):

$$\gamma = \exp\left\{-\frac{1}{2}\left(\frac{4\pi}{\lambda}\right)^2\left(\sigma_y^2\sin^2\theta + \sigma_z^2\cos^2\theta\right)\right\} \cdot \left(1 - \frac{2|B|R_y\cos^2\theta}{\lambda r}\right) \cdot \frac{1}{1 + SNR^{-1}} \tag{12.38}$$

where σ_y (σ_x) are the RMSE motions in the across track (and vertical) directions, θ the local incidence angle, ρ_r the slant range resolution, r the slant range, B the horizontal baseline distance, and R_y the range resolution. In earlier chapters, we discussed the fact that the JERS-1 SAR data sometimes are seriously saturated. We rank the decorrelation by thermal noise and saturation noise. If SNR and saturation rate are represented by 5 dB and 5% (see Tables 9-3, 9-4, and 9-5), respectively, SNR gives $\gamma_{thermal}$ of 0.75 and saturation gives 0.95% of decorrelation. PALSAR had an averaged SNR of 8 dB providing $\gamma_{thermal}$ of 0.86, and PALSAR-2 had an averaged SNR of 12 dB providing $\gamma_{thermal}$ of 0.94. Therefore, SNR affects more than saturation for the decorrelation.

12.2.7   GEOPHYSICAL PARAMETERS

Here, we summarize the geophysical parameters that could be obtained from the InSAR.

12.2.7.1   Height

Height z can be obtained by manipulating Equation (12.32) by letting $dD = 0$ as follows:

$$z = -\phi_{uw}\frac{\lambda}{4\pi}\frac{r_m\sin\theta}{B_{perp}} - \phi_a + z_b, \tag{12.39}$$

and the standard deviation (σ_z) can be given by

$$\sigma_z = \sqrt{\frac{1-\gamma^2}{2N\gamma^2}}\frac{\lambda}{4\pi}\frac{r_m\cdot\sin\theta}{B_{perp}}. \tag{12.40}$$

Here, ϕ_{UW} is the unwrapped phase (see Section 12.8), ϕ_a is the atmospheric correction, and z_b is the bias of the height.

12.2.7.2 Deformation

Arranging Equation (12.32) for the deformation dD, we have the following:

$$dD = -\frac{\lambda}{4\pi}\phi - \left(\frac{B_{perp}\cdot z}{r_m\cdot\sin\theta} + B_{para}\right) \tag{12.41}$$

and its standard deviation

$$\sigma_{dD}^2 = \left(\frac{\lambda}{4\pi}\right)^2\sigma_\phi^2 + \left(\frac{B_{perp}}{r_m\sin\theta}\right)^2\sigma_Z^2 + \left(\frac{z}{r_m\sin\theta}\right)^2\sigma_{B_{perp}}^2 + \sigma_{B_{para}}^2. \tag{12.42}$$

12.2.7.3 Local Normal Vector

Local normal vector ($\mathbf{n}_l$), which is the parameter that describes the surface normal vector, can be given by:

$$n_l = \frac{\left(-\dfrac{\partial z}{\partial x}, -\dfrac{\partial z}{\partial y}, 1\right)^t}{\sqrt{\left(\dfrac{\partial z}{\partial x}\right)^2 + \left(\dfrac{\partial z}{\partial y}\right)^2 + 1}} \tag{12.43}$$

$$\begin{pmatrix} \dfrac{\partial z}{\partial x} \\ \dfrac{\partial z}{\partial y} \end{pmatrix} = -\frac{\lambda}{4\pi}\frac{r_m\sin\theta}{B_{perp}}\begin{pmatrix} \dfrac{\partial\phi}{\partial x} \\ \dfrac{\partial\phi}{\partial y} \end{pmatrix}. \tag{12.44}$$

12.3 PROCESS FLOW

The InSAR process is summarized as follows and shown in Figure 12-6.

1. The SLC images (master and slave) are generated using almost the same Doppler center frequency. (This is not necessarily required, but the same Doppler band is preferable for making the following co-registration easier.)
2. The co-registration frame is built.
3. For InSAR and DInSAR, including a raw fringe generation, the following steps are undertaken: co-registration to the ground surface, correction of the (digital elevation model) (DEM)-based fringe, correction of the orbit, and correction of the atmosphere and ionosphere.
4. The final product is generated and projected onto the map.

InSAR, and especially DInSAR, require that the SAR image be well associated and co-registered with the Earth's topography. As such, precise co-registration of the SAR image and DEM are the mandatory issue, for which accurate SAR geolocation is a must.

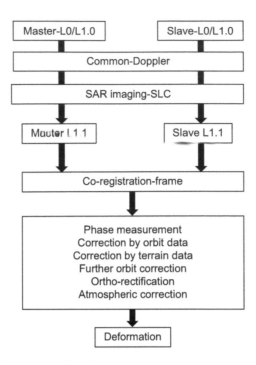

FIGURE 12-6   InSAR-DInSAR process flowchart.

12.4   SPATIAL DECORRELATION AND CRITICAL BASELINE

The InSAR coherence mainly depends on three decorrelations, which are presented in Equation (12.37): thermal decorrelation, spatial decorrelation, and temporal decorrelation. By the term "decorrelation," we mean to decrease the correlation in negative way (such as correlation = 1 − decorrelation. Some additional types include vegetation decorrelation, height-related decorrelation, and ionospheric decorrelation. One of the more common, spatial decorrelation, can be derived theoretically when considering how the geometric relationship of wavelengths projected on the same surface from two adjacent observations (and incidence angles) differ slightly each other and that the wave lengths match by shifting one or the other observation. Here, we assume that the interference occurs from two wavelengths whose waves are the same by observing the series of particles located on the surface (Gatelli et al. 1994). Figure 12-7 shows the geometry of two observations to the pixel A on the surface, where the master orbit is θ, the slave orbit follows $\theta' = \theta - \Delta\theta$, and both separate with B_{perp} as a perpendicular baseline. The interferometry correlates with the scatterers locating on the surface with the same wavelength components from the two observations, and we consider the wavelength projected on the surface from this satellite. Here, the wavelengths on the surface are given by Λ (master) and Λ' (slave) as follows:

$$\Lambda = \frac{\lambda}{\sin\theta} \qquad (12.45)$$

$$\Lambda' = \frac{\lambda}{\sin\theta'} \qquad (12.46)$$

Here,

$$\theta' = \theta - \Delta\theta; \qquad (12.47)$$

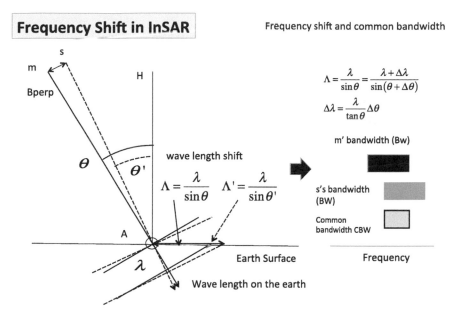

FIGURE 12-7 Frequency shift that has occurred in cross track InSAR.

then, the wavelength should change with $\Delta\lambda$ in such a way that

$$\Lambda = \frac{\lambda}{\sin\theta} = \frac{\lambda + \Delta\lambda}{\sin(\theta - \Delta\theta)} \tag{12.48}$$

$$\Delta\lambda = -\frac{\lambda}{\tan\theta}\Delta\theta. \tag{12.49}$$

Differentiating

$$\lambda = \frac{c}{f}, \tag{12.50}$$

we have

$$\Delta f = -\frac{f^2}{c}\Delta\lambda. \tag{12.51}$$

Thus, the master and slave orbits have a mutual wavelength shift depending on the incidence angle difference. When the resultant $\Delta\lambda$ exceeds the total bandwidth (B_w), the interferometry collapses. The spatial decorrelation can be expressed by

$$\gamma_{spatial} = \frac{B_w + \dfrac{f^2}{c}\Delta\theta}{B_w}. \tag{12.52}$$

At the critical baseline, the correlation becomes zero.

Figure 12-8 shows one example of the ALOS-PALSAR for γ and the baseline. Bandwidths of 14, 28, 42, and 84 MHz SAR at the height of 628 km have 7, 15, 23, and 45 km of the critical baselines in theory. From experience, we know that 10% of the critical baseline will provide successful InSAR results.

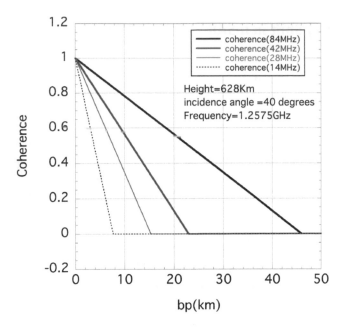

FIGURE 12-8   Gamma and critical baseline versus B_{perp}.

12.5   ACCURACY REQUIREMENTS

12.5.1   Height Accuracy Requirement for DInSAR

We show the equality of Equation (12.42) as

$$\sigma_{dD}^2 = \left(\frac{\lambda}{4\pi}\right)^2 \sigma_\phi^2 + \left(\frac{B_p}{r_m \sin\theta}\right)^2 \sigma_Z^2. \tag{12.53}$$

By letting the second term on the right side be zero, we can estimate the surface deformation error. Normally, the radar system works with σ_Φ of about 30 degrees, so σ_{dD} is 0.97 cm in the L-band case. It increases as σ_Φ increases. If the second term on the right side of Equation (12.53) is less than the first term, we have the following inequality:

$$\sigma_z \leq \frac{\lambda}{4\pi} \frac{r_m \sin\theta}{B_{perp}} \sigma_\phi. \tag{12.54}$$

A sample result of Figure 12-9 shows that the DEM-induced error may be up to 20 m for $B_p = 0.5$ km, $r_m = 710$ km, and $\theta = 38$ degrees. This may be the normal condition.

12.5.2   Orbit Determination Accuracy Requirement

Letting ΔB_{perp} and ΔB_{para} have the bias errors in orbit data, expanding Equation (12.41) by the bias errors, and ignoring the phase measurement error, we have the height error (dz):

$$\Delta z = r_m \sin\theta \frac{B_{para}}{B_{perp}} \frac{\Delta B_{para} - \Delta B_{perp}}{B_{perp}}. \tag{12.55}$$

It should be noted that this equation is valid regardless of the wavelength. If B_{perp} (B_{para}) is 200 m, ΔB_{para} 10 m, ΔB_{perp} 20 m, r_m 710,000 m, and $\sin\theta$ 0.5, Δz is 18 km and unusable. If we desire Δz to be less than 3 m under these conditions, ΔB_{para} (ΔB_{perp}) should be less than 1 cm.

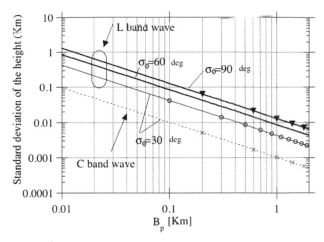

FIGURE 12-9 Relationship between the perpendicular baseline distance (B_{perp}) and the height standard deviation for three L-band waves and one C-band wave.

12.6 ERROR ANALYSIS

All the InSAR error sources are listed in Table 12-1. These errors are only related to InSAR processing; these random and/or bias errors can be minimized by selecting the co-registration parameters and correcting the orbit parameters.

TABLE 12-1
Summary of Error Sources

No.	Item	Type	Non/Improvable	InSAR/DInSAR
1	Decorrelation			
	Spatial decorrelation	R	N	B
	Thermal decorrelation	R	N	B
	Temporal decorrelation	R	N	B
2	Co-registration of master/slave			
	Co-registration frame	R	I	B
	Co-registration coefficient	R	I	B
	Co-registration criteria	R	I	B
	Interpolation	R	I	B
3	Co-registration of the master/SAR simulated image			
	Generation of the SAR simulated image	R	I	DInSAR
4	DEM			
	Height accuracy	R/B	N	DInSAR
	Variation of the height	R	N	DInSAR
5	Orbital error			
	Along track (shift)	B	I	B
	Cross track (shift)	B	I	B
	Non-parallel component	B	I	B
6	Miscellaneous			
	Water vapor	B	I	B
	Foreshortening		N	B
	Layover		N	B
	Focusing	R	I	B

Note: B = Both InSAR/DInSAR. R = Random error. B = Bias error. I = Improved. N = Not improved.

12.6.1 Co-Registration

Perfect co-registration of the two images is always desired but essentially is difficult to attain because the orbital properties—orientation, position, and speed—vary temporally. However, producing both images under the same Doppler centroid and rearranging the zero-baseline relaxes the co-registration issues (Massonnet and Rabaute 1993; Zebker et al. 1994). This section deals with co-registration versus accuracy: the parameter, the frame, the accuracy requirement, and the interpolation.

12.6.1.1 Co-Registration Parameters

There are two parameters that dictate the quality of the co-registration—the complex correlation (γ in Equation [12.56]) and the amplitude covariance (β in Equation [12.57]):

$$\gamma = \frac{\left|\langle \mathbf{C}_m \cdot \mathbf{C}_s^* \rangle\right|}{\sqrt{\langle \mathbf{C}_m \cdot \mathbf{C}_m^* \rangle}\sqrt{\langle \mathbf{C}_s \cdot \mathbf{C}_s^* \rangle}} \tag{12.56}$$

$$\beta = \frac{\left|\langle \left(A_m - \overline{A_m}\right) \cdot \left(A_s - \overline{A_s}\right) \rangle\right|}{\sqrt{\langle \left(A_m - \overline{A_m}\right)^2 \rangle}\sqrt{\langle \left(A_s - \overline{A_s}\right)^2 \rangle}} \tag{12.57}$$

where A is the amplitude.

We performed temporal decorrelation simulations with various SNRs. In the simulations, two sets of the simulated signal, each consisting of $\varepsilon\%$ (power ratio) of the correlative component and $(1-\varepsilon)\%$ (power ratio) of the non-correlative component, are prepared. The former is weighted by a sinc function centered at a given address, and the latter by the thermal noise and the non-correlative signal component. Thus, the simulated signals for the master and slave are expressed by:

$$\mathbf{C} = \mathbf{S}_c + \mathbf{S}_{uc} + \mathbf{N} \tag{12.58}$$

$$\left|\mathbf{S}_c\right|^2 = \varepsilon\left|\mathbf{S}\right|^2 \tag{12.59}$$

$$\left|\mathbf{S}_{uc}\right|^2 = (1-\varepsilon)\left|\mathbf{S}\right|^2 \tag{12.60}$$

$$SNR = \frac{\left|\mathbf{S}_c + \mathbf{S}_{uc}\right|^2}{\left|\mathbf{N}\right|^2} = \frac{\left|\mathbf{S}\right|^2}{\left|\mathbf{N}\right|^2} \tag{12.61}$$

$$\gamma \cong \frac{\varepsilon\left|\mathbf{S}\right|^2}{\left|\mathbf{S}_c + \mathbf{S}_{uc}\right|^2 + \left|\mathbf{N}\right|^2} = \frac{\varepsilon}{\left(1 + SNR^{-1}\right)} \tag{12.62}$$

where $\mathbf{C}$ is the simulated complex image, $\mathbf{S}_c$ the correlative signal component, $\mathbf{S}_{uc}$ the non-correlative signal component, and $\mathbf{N}$ the thermal noise. Suffixes for the master and the slave images are omitted; $\mathbf{S}_c$ is introduced as a function that contains the texture of the target (i.e., the sinc function). In this simulation, to find the tie-point successfully when using β, the point target-like texture is adopted.

Figure 12-10a shows the responses of γ and β around the tie-point for three cases ($\gamma = 0.3$, 0.5, and 0.7). Since β is the same in these cases, only one curve is shown. If γ shows a significantly higher peak at the tie-point than the neighbors, the tie-point can be correctly selected. If not, miss-selection might occur. Figure 12-10b shows the relationship between γ and the success rate: the ratio of the number of successes to the total trials. It shows that γ selects the correct tie-point of 100% for $\gamma \geq 0.6$, while a smaller γ may result in selection failures. β always selects the correct tie-point regardless of γ.

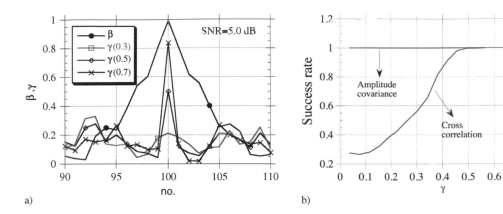

a)   b)

FIGURE 12-10   Co-registration simulation of β and γ for finding the tie points: (a) behavior around the exact tie points; (b) success selection rate versus the cross-correlation parameter (γ) for both amplitude covariance and cross correlation.

Simulation results are summarized here.

i. Complex correlation (γ): Complex correlation has several features. (1) It shows the steep response with a large value around the tie-point as $γ ≥ 0.5$ and SNR $≥ 5$ dB. (2) A less favorable correlation ($γ < 0.5$) does not select the correct tie-point. (3) Complex correlation, γ, does not require the target texture, which means that snow-covered ice, an ice sheet, and a glacier may be matched up. (4) The SNR for γ should exceed 0.5 (5.0 dB).

ii. Amplitude correlation (β): This does not consider the phase, so a large β does not assure a good fringe. The advantage is that β provides information on the co-registration regardless of γ if the images contain texture. The amplitude covariance has two disadvantages. (1) It does not show the fringe information. (2) It needs some textures in both images, which means that featureless areas, such as ice sheets and glaciers with the same colors, cannot be matched up.

iii. Other parameters
Information about other effective parameters is available and outlined in Table 12-2.

12.6.1.2   Co-Registration Frame

A co-registration frame plays a very important role in InSAR processing because it connects the master and slave images geometrically. Thus, building an accurate co-registration frame is at the core of the InSAR and every attempt should be made to ensure that this frame and its preparation are as accurate as possible.

TABLE 12-2

Comparison of the Co-Registration Parameters

No.	Parameter	Features	For Sensor
1	Complex correlation γ	a) Needs good SNR and scene correlation b) Indicates fringe quality	ERS-1/2, SIR-C, Seasat, AIRSAR
2	Amplitude covariance β	a) Needs texture b) Does not depend on SNR so much c) Does not indicate fringe quality	JERS-1 SAR
3	Frequency spectrum of complex correlation	a) Needs good SNR and scene correlation b) Estimates fringe quality	ERS-1/2, SIR-C, Seasat, AIRSAR

As long as the two orbits are aligned close each other, the following pseudo affine equation is the most stable and robust frame:

$$x = ax' + by' + cx'y' + d$$
$$y = a'x' + b'y' + c'x'y' + d',$$
(12.63)

where x, y are the address of the master pixel and x', y' are the corresponding address of the slave pixel; a, b, c, d, a', b', c', d' are the unknown constants. These coefficients are determined by collecting the tie points from the master and slave images using the parameters of α or β and screening them. Tie point selection is performed by the correlation of two chip images and the density of the piling points; the correlation area also is important. Although Equation (12.63) is experimentally confirmed as robust and stable, a high-order polynomial of the range dependence could be considered when the image size and the B_{perp} both become large. The following simulation investigates the range dependence and the co-registration accuracy.

Because the two images are generated in the same orientation, azimuth co-registration could be made by an image shift (if the image size is small). A range co-registration, serving as the slant range function, is affected by the local topography, B_{perp} and B_{para}. We consider how the local topography affects the accuracy of the range co-registration frame.

In this case, we will ignore the surface deformation and the offsets. We also will ignore the local deviation of the atmosphere. The range distance between two images, $\Delta r = r_s - r_m$, is given by rearranging Equation (12.63); its primary term, B_{para}, can be approximated by the second-order expansion around $r_m = r_{m0}$:

$$\Delta r = B_h - \frac{z \cdot B_p}{r_m \sin\theta}$$
$$\cong B_{h0} + B_{h1} \cdot (r_m - r_{m0}) + \frac{B_{h2}}{2} \cdot (r_m - r_{m0})^2 - \frac{z \cdot B_p}{r_m \sin\theta}$$
(12.64)

$$B_{h0} = B \cdot \sin\alpha_0 + dh \cdot \cos\alpha_0$$

$$B_{h1} = (B \cdot \cos\alpha_0 - dh \cdot \sin\alpha_0) \cdot \frac{1}{r_{m0}\tan\theta_0}$$

$$B_{h2} = -\frac{B}{r_{m0}^2 \tan^2\theta_0}\left(\sin\alpha_0 + \tan\theta_0 \cos\alpha_0 + \frac{\cos^2\alpha_0}{\cos^2\theta_0 \sin\alpha_0}\right)$$
(12.65)
$$-\frac{dh}{r_{m0}^2 \tan^2\theta_0}\left(\cos\alpha_0 - \tan\theta_0 \sin\alpha_0 - \frac{\cos\alpha_0}{\cos^2\theta_0}\right).$$

A co-registration error primarily is caused by z, and its maximum error, Δr_{error}, is

$$\Delta r_{error} \leq \frac{\Delta z_{max} \cdot B_p}{2 \cdot r_{m.min} \cdot \sin\theta_{min}}$$
(12.66)

where Δz_{max} is the maximum height difference within an image, $r_{m.min}$ the minimum master slant range, and θ_{min} the minimum incidence angle. We conducted a simulation on the dependence of the Δr on r_m for three cases. For Case 1, B_{perp} (B_{para}) is 0.5 km (0.1 km) at $r_{m0} = 690$ km. The image has the constant height of 4 km over a swath. Case 2 differs from Case 1 in that a sinusoidal mountain (a peak height of 4 km over a 20-km span) is included. Case 3 differs from Case 1 in that B_{perp} is 1.0 km. Two approximation models, a linear model and a second-order power model, are examined to evaluate the residual of Δr. Each case (1, 2, and 3) is approximated by two models, and their residuals are calculated (i.e., Case 1-1 means that case is approximated by a liner model, and Case 1-2 means that Case 1 is approximated by a second-order power model). From Figure 12-11, we determine that the second-order power model minimizes the co-registration frame error and co-registers the

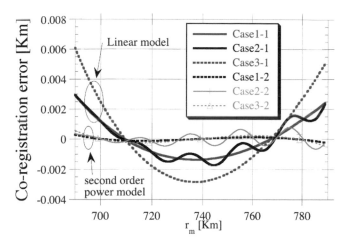

FIGURE 12-11 Co-registration error comparison for three patterns in two models: Case 1 with B_{perp} of 0.5 km at 4 km height, Case 2 with B_{perp} of 0.5 km at sinusoidal mountain, and Case 3 with B_{perp} of 1 km at 4 km height.

mountain area that is 4 km in height within an accuracy of 1 m. In reality, most InSAR is conducted at B_{perp} of less than 0.5 km and an image size of less than 50 km in the slant range. In which case, Equation (12.63) is well accepted. The only issue is the ScanSAR-InSAR case where the wide range of the swath is prepared. The registration frame of the ScanInSAR will be discussed in Section 12.11.

12.6.1.3 Co-Registration Accuracy Requirement

It is obvious that a more accurate co-registration gives a higher coherence and vice versa. However, it is difficult to clarify the co-registration requirement from the simulations. Therefore, an experimental study was conducted. Assuming a co-registration frame made for a sample image is error-free, in this study, a relationship between the fringe quality and co-registration accuracy was measured by adding an offset of 0 to 2 pixels in 0.2 step in range and azimuth to the aforementioned co-registration frame. The JERS-1 SAR image pair over Mt. Kuju in Kyushu, Japan, was used as a sample. Figure 12-12a shows a hundred fringe quality patterns, and Figure 12-12b shows

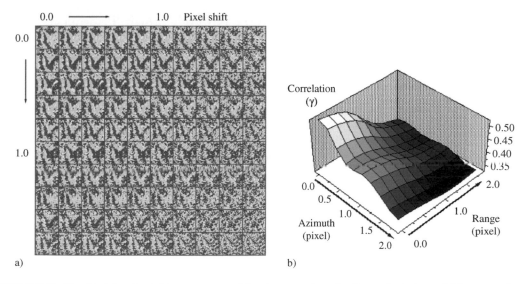

FIGURE 12-12 (a) A hundred fringes generated for the co-registration frame deviation, which ranges from 0 to 2 pixels in 0.2 steps in range and azimuth direction. The allowed deviation is 0.4-pixel, (b) is the corresponding coherence distribution.

TABLE 12-3

Comparison of the Interpolation Method

No.	Interpolation	Fringe Quality	Computation Time Relative to Nearest Neighbor (NN)
1	Nearest neighbor (NN)	Satisfied (0.37)	1.0
2	Bilinear (BL)	Satisficd (0.4)	1.14
3	Cubic convolution (CC)	Satisfied (0.4)	3.00

the corresponding coherence as the three-dimensional demonstration, where the highest coherence is prepared at the zero displacement in range and azimuth. A co-registration error of 0.4 pixel may be the maximum allowance.

12.6.1.4   Interpolation

The dependence of fringe quality on interpolation was evaluated for three methods—nearest neighbor (NN), bilinear (BL), and cubic convolution (CC)—using the Mt. Kuju image. The results are: (a) NN gives the smallest correlation and a little worse fringe quality than the others. (b) BL shows better fringe quality than NN, especially at the fringe edge. And (c) CC provides the best fringe quality, though it requires the longest computation time. The γs for an area are 0.44 (NN), 0.53 (BL), and 0.55 (CC), and those for the whole images are 0.37, 0.4, and 0.4. Finally, we recommend BL as the appropriate interpolation because it provides satisfactory fringe quality in an acceptable computation time (Table 12-3). This result agrees with the comments of Massonnet and Rabaute (1993) and Zebker et al. (1994) that the more accurate interpolation does not improve the fringe quality enough to justify the greater computation time.

A co-registered slave image is often contaminated by the weak wavy intensity pattern on the amplitude image as well as the InSAR phase. This occurs when an SAR image is created by a nonzero Doppler center and the interpolation does not correct the nonzero frequency. To prevent this, the slave SAR image needs to be zero-frequency shifted before the interpolation and then converted back to the original frequency. JERS-1 SAR and ALOS/PALSAR also need this preprocessing.

12.6.2   Orbital Error Correction (Cross Track and Vertical Speed Component)

The error model for the appropriate orbit model is discussed in Section 12.10.

12.7   PHASE DELAY DUE TO THE REFRACTIVE INDEX VARIATION

In the previous discussion, we assumed that two observations were made under the same atmospheric and/or ionospheric conditions. SAR measures the round-trip time to the target, and then the distance depends on the light speed. The ratio of the real light speed to the ideal one is defined as the refractive index, n. The repeat pass InSAR becomes more complex, using the refractive index as follows:

$$
\begin{aligned}
\phi &= -\frac{4\pi}{\lambda_0}\left(r_m n_m - r_s n_s\right) \\
&= -\frac{4\pi}{\lambda_0}\left(r_m n_m - r_s n_m + r_s n_m - r_s n_s\right) \\
&= -\frac{4\pi}{\lambda_0} n_m \left(r_m - r_s\right) - \frac{4\pi}{\lambda_0} r_s \left(n_m - n_s\right) \\
&= -\frac{4\pi}{\lambda_0} n_m \left(\frac{B_{perp} z}{r_m \sin\theta} + B_{para} + dr\right) - \frac{4\pi}{\lambda_0} r_s \left(n_m - n_s\right).
\end{aligned}
\tag{12.67}
$$

Here, the second term on the right-hand side is the distance change due to the two-day difference in the retractive index. The troposphere and ionosphere impact the refractive index daily. The former affects the phase deviation largely in area, and the latter limits the affecting areas. Mountainous and volcanic areas often are problematic.

12.7.1 Atmospheric Phase Delay

The refractive index (n) is defined as the ratio of the speed of light in space to that in the media:

$$n = \sqrt{\frac{\varepsilon \mu}{\varepsilon_s \mu_s}} \tag{12.68}$$

where ε_s is the permittivity of free space, μ_s is the magnetic permeability, ε is the permittivity of the target, and μ is the magnetic permeability. The observation path goes through the troposphere (between the surface and 10 to 15 km in height), the stratosphere (from 10 to 100 km), and the ionosphere (above 100 km). The stratosphere is stable in temperature and probably does not affect the refractive index. The ionospheric variation in daily phenomena and solar activity greatly impacts the refractive index. The refractive index in the troposphere depends on three major weather components— the atmospheric pressure (p), the temperature (T), and the humidity (e). The experimental model obtained by Kerr (1951) and Mushiake (1961) is:

$$n = 1 + \left(\frac{77.6}{T} p \times 10^{-6} + \frac{0.373}{T^2} e \right). \tag{12.69}$$

Because p and e depend on the height, so does n. The transmitted signal propagates within the atmosphere in a curve. Even if the refractive index is completely different over two days, if its vertical structure is almost the same over an image for each day, the geodetic quantities have only the offset over an image and will be corrected. If the refractive index changes locally within an image, the geodetic quantities change locally and are very hard to correct. Atmospheric pressure distribution does not change locally within the SAR image size, so this may not cause any error. Water vapor normally is supplied from the Earth's surface (i.e., volcanoes, wetland, or grassland, whose scale ranges from small to large). This may cause errors in the geodetic quantities. We will estimate the phase difference based on local variations in the refractive index.

The U.S. standard temperature, atmospheric pressure, and water vapor pressure models are used for the vertical structures, such as in Ulaby et al. (1982):

1. Temperature distribution

$$T(z) = \begin{cases} T_0 - a \cdot z & 0Km \leq z \leq 11Km \\ T(11) & 11Km \leq z \leq 20Km \\ T(11) + (z-20) & 20Km \leq z \leq 32Km \end{cases} \tag{12.70}$$

2. Atmospheric pressure distribution

$$P_a(z) = 2.87 \cdot 1.225 e^{-z/H_1} \cdot T [mbar] \tag{12.71}$$

3. Water vapor pressure distribution

$$P_w(z) = 4.59 \cdot \rho_0 e^{-z/H_4} \cdot T [mbar] \tag{12.72}$$

where $a = 6.5 \, Km^{-1}$; z is height; H_1 is density scale height (9.5 km) of the atmosphere; ρ_0 is water vapor density at the sea level [mgm^{-3}], which ranges from 0.01 gm^{-3} in a very cold, dry climate to as much as 30 gm^{-3} in hot, humid climates; and H_4 is the density scale height of water vapor, which ranges from 2.0 km to 2.5 km.

TABLE 12-4
Simulation Condition for the Atmospheric Uncertainty

		Simulation-1				Simulation-2	
Case	Path	T_0(K)	ρ_0($e^{-3}kg^{-3}$)	Case	Path	T_0(K)	Δ_0($e^{-3}kg^{-3}$)
1-1	1	300	7.5	2-1	1	290	7.5
	2	290 to 310	7.5		2	290	0 to 21
1-2	1	300	0	2-2	1	300	7.5
	2	290 to 310	0		2	300	0 to 21
1-3	1	300	30	2-3	1	273	7.5
	2	290 to 310	30		2	273	0 to 21

Note: Where H_4 is 4.0 (km) and H_1 is 9.5 km

We approximate the atmosphere as ten 1-km-width layers from the ground to a height of 10 km, with each layer having a different refractive index. Free space conditions are applied from 10 km to satellite altitude. The one-way phase change ($\Delta\phi$) between two paths is obtained by:

$$\Delta\phi = \frac{2\pi}{\lambda} \sum_{i=1}^{11} \left(n_i^m r_i^m - n_i^s r_i^s \right) \tag{12.73}$$

where n_i^m is the refractive index at the ith layer in the master image; n_i^s is the refractive index at the ith layer in the slave image; r_i^m is the path length in the ith layer in the master image; and r_i^s is the path length in the ith layer in the slave image. Two simulations of the phase dependencies on temperature and water vapor and based on U.S. air (Table 12-4) show that the temperature change at ten degrees makes the phase difference only 10 degrees, which is smaller than the receiver thermal noise. But the water vapor changes the phase error 360 degrees at maximum. This confirms the importance of the water vapor correction for deriving deformation information from DInSAR.

12.8   UNWRAPPING

Since the measured InSAR phase is the principal value of a wrapped true phase that ranges from $-\pi$ to π, the true phase needs to be derived accurately to measure the DEM or surface deformation. When doing so, unwrapping is an issue (Figure 12-13). The unwrapping phase has been defined

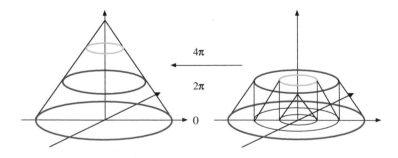

Unwrapping
Continues the wrapped phase ($-\pi \sim \pi$) to unwrapped phase.

method : the branch cut , the least square method, etc.

FIGURE 12-13   Conceptual explanation of unwrapping.

as follows: The "true phase difference," S_{ij}, between two adjacent pixels basically can be estimated from the "wrapped phase difference," $\Delta\varphi_{ij}$, between two adjacent wrapped phases (φ_{pi} and φ_{pj}). Let the wrapped phase difference be expressed as follows:

$$\Delta\varphi_{ij} = \varphi_{p,i} - \varphi_{p,j}. \tag{12.74}$$

Since the phase difference between the adjacent samples cannot be smaller than $-\pi$ or greater than π, we have the following expression for the phase differences:

$$\left(\begin{array}{ll} if\ |\Delta\varphi_{ij}| < \pi & then\ S_{ij} = \Delta\varphi_{ij} \\ if\ \Delta\varphi_{ij} \leq -\pi & then\ S_{ij} = \Delta\varphi_{ij} + 2\pi \\ if\ \Delta\varphi_{ij} \geq \pi & then\ S_{ij} = \Delta\varphi_{ij} - 2\pi \end{array} \right). \tag{12.75}$$

The phase can then easily be unwrapped by integrating the S_{ij} along any arbitrary path. However, SAR's several noise sources (thermal noise and speckle noise) may impact this method. These noise sources disturb the phase continuity of the target, so phase integration without considering phase discontinuity propagates errors over an image. The phase unwrapping (phase reconstruction) problem has been investigated by laser researchers since the 1970s and by radar researchers since the 1980s. Two representative unwrapping methods have been developed—the branch-cut method and the least square estimation method.

12.8.1 Branch-Cut Method

The branch-cut method (Goldstein et al. 1988; Prati et al. 1990) is based on the aforementioned premise and builds a number of branch-cuts in the image across which the phase difference S_{ij} should not be integrated. This branch-cut prevents the propagation of phase error, which is generated for several reasons. First, the thermal and speckle noise violate the sampling theorem. Second, the layover area is not properly sampled (under sampled). As a result, the rotation of the phase gradient is not zero because of these noise sources. The phase discontinuity in the integration was discussed in Takajo and Takahashi (1988). The integration along any enclosure, C, that does not cross the ghost line, is zero:

$$\sum_C S_{i,j} = 0. \tag{12.76}$$

If the enclosure crosses the ghost lines, the integration becomes

$$\sum_C S_{i,j} = \pm 2n\pi \tag{12.77}$$

where "n" is an integer (0, 1, 2,..) and the sign depends on the crossing direction. Residue is defined as a point that satisfies Equation (12.76) with nonzero "n," and its value is calculated by summing S_{ij} around four neighboring pixels. The residue with a plus value is called plus residue; residue with a minus value is called minus residue. The branch-cuts should be selected as a connected line (region) of the residues along which summation of the residue values is zero. Ideally, the number of plus and minus residues should be the same over an image. However, the actual image shows some imbalance. This may be due to layover and the lower SNR. Goldstein et al. (1988) proposed masking the lower SNR region as the algorithm-give-up area. For these reasons, this method is ambiguous, although the principle is easy to understand.

12.8.2 Least Square Phase Estimation Method

The least square estimation method solves the partial differential equation (elliptic type) that governs the unwrapped phase (φ) and the gradient of the phase difference (Δφ). The equation is

$$\frac{\partial^2 \varphi}{\partial x^2} + \frac{\partial^2 \varphi}{\partial y^2} = \Delta\phi \tag{12.78}$$

where x and y are the coordinate system, and the right-side term is the gradient of the measured phase difference. The right-side term is estimated from the measured phase difference, although this is wrapped and affected by the noise. The Neumann condition is also required as the boundary condition. Several methods have been proposed to estimate the local gradient: the multi-resolution phase gradient (local frequency) estimation method by Bamler et al. (1996), and the maximum likelihood method, principal value finite difference (PVFD), and complex signal phase derivative (CSPD) by Spagnolini (1995). This method was originally developed for the laser phase reconstruction problem (Hunt 1979; Takajo and Takahashi 1988). Once the conditions are prepared, Equation (12.77) can be solved by an iteration method (Hunt 1979) or a faster iteration method in space and frequency domain (Press et al. 1989). After that, several methods were developed (Ghiglia and Pritt 1998).

12.9   COHERENCE ANALYSIS

The coherence is a good indicator of how easily the InSAR phase can be obtained. A strong InSAR phase needs fewer samples for averaging, and a lesser quality phase needs more data for spatial averaging. Spatial and thermal decorrelation can be theoretically derived. But temporal decorrelation is only derived through experiments.

Sensitivity of L-band SAR to deforestation was analyzed using time series amplitude data and interferometric coherence data that ALOS/PALSAR collected for the Rio Branco region of the Amazon and the province of Riau in Indonesia over a period of five years and four months. Two different test sites in Brazil and Indonesia were selected because their deforestation processes are different from each other. Indonesia was logged for oil palms and acacia plantations without flattening after the process, whereas Brazil's crop plantations have resulted in land flattening. Interferometric coherence and amplitude are used to evaluate the temporal variation of the deforestation areas. Here, 20 scenes and 16 scenes were collected and analyzed for Riau in Indonesia and Rio Branco in Brazil, respectively. Figure 12-14 shows the temporal variation in the coherence of these areas. Both of them have decayed with time. Riau has three test sites: acacia, peat forest, and clear-cut. Rio Branco has two sites: forest and clear-cut. Most of them decay simply with time, while the decay rate is forest dependent. From the figures, the temporal coherence simply decreases within 300 days as the temporal baseline. The samples are shown only for the forest-covered area; the urban area has a different temporal decay.

12.10   CORRECTION OF ATMOSPHERIC EXCESS PATH DELAY AND THE ORBIT ERROR

The orbital error poses the fault fringe component on the true fringe which would be obtained by using the error-free orbit. This additional fringe can be removed by determining the orbital errors in a way so as to minimize the difference of the measured phase and the orbit-error-model by the least square method. Atmospheric excess path delay (AEPD) can be also corrected using the numerical model (Shimada 1999; Shimada 2000).

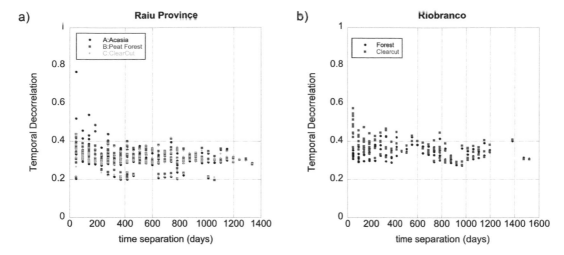

FIGURE 12-14   Temporal variation of the coherence of the natural forest and the clear-cut areaz: a) Riau state of the Indonesia, and b) Rio Branco area of Brazil.

12.10.1   Phase Model

We assume that a small area with natural relief is deformed by earthquakes and/or volcanic activities, the coherence of two repeat-pass SAR images is high enough to make this determination. The InSAR phase (ϕ) is expressed by

$$\phi = -4\pi \left(\int_{r_p}^{r_m} \frac{dr}{\lambda_m} - \int_{r'_p}^{r_s} \frac{dr}{\lambda_s} \right),$$ (12.79)

where ϕ is the unwrapped (Goldstein et al. 1988) phase; r_m and r_s are the position vectors of the master and slave orbits, r_p and r'_p are the position vectors of the targets observed from the master and slave orbits (with no orbit error, both orbits are the same); λ_m and λs are the wavelengths of the master and slave orbits (see Figure 12-15). The wavelength varies with the refractive index of the

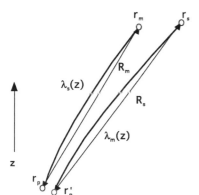

FIGURE 12-15   Coordinate system used in this analysis. Z represents height. Signals from the scatterers propagate on the curved lines.

propagation media along the integration path in Equation (12.79). By exchanging the integral ends, Equation (12.79) can be modified to

$$\phi = -4\pi \left(\int_{r_p}^{r_m} \frac{dr}{\lambda_s} - \int_{r_p}^{r_s} \frac{dr}{\lambda_s} \right) - 4\pi \left(\int_{r_p}^{r_s} \frac{dr}{\lambda_s} - \int_{r_{p'}}^{r_s} \frac{dr}{\lambda_s} \right) - 4\pi \left(\int_{r_p}^{r_m} \frac{dr}{\lambda_m} - \int_{r_p}^{r_m} \frac{dr}{\lambda_s} \right),$$

(12.80)

where the first term is the geometric phase difference of the target–satellite, the second term is the phase change due to the surface deformation, and the third term is the atmospheric excess path delay (AEPD). The first two terms can be simplified to

$$\phi_1 = -\frac{4\pi}{\lambda_0}(R_m - R_s)$$

(12.81)

$$\phi_2 = -\frac{4\pi}{\lambda_0}\Delta R_{ms},$$

(12.82)

where R_m and R_s are the distances of the satellite to the targets, λ_0 is the wavelength in a vacuum, and ΔR_{ms} is the surface deformation component in the line of sight. We can assume that the atmosphere (troposphere) is represented by several layers parallel to the Earth's surface up to 30 km, and each layer has components of temperature, pressure, and water vapor's partial pressure. Since these three parameters relate to the refractive index of the troposphere, the third term can be expressed by

$$\phi_3 = -4\pi \frac{1}{\lambda_0 \cos\Theta_0} \sum_i \left(n_{m,i} - n_{s,i} \right) \Delta r_i$$

(12.83)

$$n = 1 + \frac{77.6}{T} p \cdot 10^{-6} + \frac{0.373}{T^2} e.$$

(12.69)

(See Figure 12A-1 and the appendix for the derivation of Equation [12.83]). Here, Θ_0 is the off-nadir angle at the satellite; $n_{m,i}$ and $n_{s,i}$ are the refractive index at the ith layer of the master and slave images, and Δr_i is each layer's vertical thickness. Equation (12.69) is an empirical model for the refractive index of the troposphere (Kerr 1951; Mushiake 1961), and P, T, and e are the pressure (hPa), temperature (K), and the water vapor's partial pressure (hPa), respectively.

12.10.2 ORBITAL ERROR MODEL

Denoting the errors of the satellite state vector with Δr (a surface perpendicular component), ΔB (a surface parallel component), and Δv (a surface perpendicular velocity), the correct vector is expressed by the following equations (see Figure 12-16). As the Earth model also contains some errors, we define this orbit error as containing the error of the Earth model:

$$\mathbf{r} = \left(\mathbf{r}_0 + \mathbf{v}_0 t + \frac{1}{2}\mathbf{a}_0 t^2 \right) + (\Delta \mathbf{B} + \Delta \mathbf{r}) + (\Delta \mathbf{v})t$$

(12.84)

$$\Delta \mathbf{r} = \Delta r \cdot \mathbf{n}_r$$

(12.85)

$$\Delta \mathbf{B} = \Delta B \cdot \mathbf{n}_n$$

(12.86)

$$\Delta \mathbf{v} = \Delta v \cdot \mathbf{n}_r$$

(12.87)

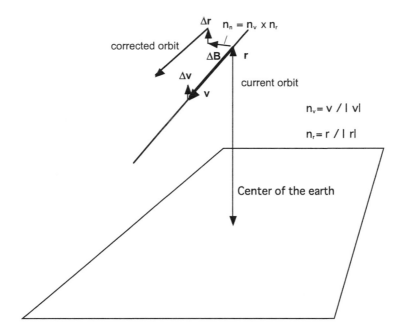

FIGURE 12-16 Coordinate system for the orbit. The solid line on the right-hand side represents the current state vector provided by the tracking center, and the thin solid line on the left-hand side is the corrected state vector.

$$\mathbf{x} \equiv (\Delta B, \Delta r, \Delta v) \qquad (12.88)$$

Here, $\mathbf{n}_r$ and $\mathbf{n}_n$ are the unit position vector of the satellite and that perpendicular to both $\mathbf{n}_r$ and the velocity vector; r_0 is a position vector, v_0 is a velocity vector, and a_0 is an acceleration vector at $t = 0$. Time t originates at the beginning of the image. The velocity error parallel to the surface can be negligible because it becomes much smaller than the vertical value. As a result, the error vector is expressed by both the vertical and horizontal position errors and one vertical velocity error. Therefore, the phase error, $\hat{\phi}$, is modeled by

$$\hat{\phi} = -\frac{4\pi}{\lambda_0}(R_m - R_s) \qquad (12.89)$$

$$R_s(t \mid \mathbf{x}) = \left|\mathbf{r}_s(t \mid \mathbf{x}) - \mathbf{r}_{GCP}\right| \qquad (12.90)$$

$$R_m(t \mid \mathbf{x}) = \left|\mathbf{r}_m(t \mid \mathbf{x}) - \mathbf{r}_{GCP}\right|. \qquad (12.91)$$

Here, $\mathbf{r}_m(t \mid \mathbf{x})$ and $\mathbf{r}_s(t \mid \mathbf{x})$ are the master and slave orbit expressions that, as the unknowns, contain the error, and $\mathbf{r}_{GCP}$ is the position vector of the GCP. GCPs are selected from the successfully unwrapped image area.

12.10.3 DETERMINATION OF THE ORBITAL ERROR AND THE PARAMETER ACCURACY

The InSAR phase (ϕ) deviates with a standard deviation of $\sqrt{(1-\gamma^2)/2N}/\gamma$, where γ is the coherence and N is the number of the samples (Rodriguez and Martin 1992). An AEPD-corrected phase

$(\phi - \phi_3)$ also deviates around $\hat{\phi}$ with the same standard deviation as noted earlier. Thus, a suitable orbital error should minimize the expectation of $\phi - \phi_3$:

$$E = \frac{1}{M} \sum_k \left\{ \frac{\phi_k - \hat{\phi}_k(x)}{\sigma_k} \right\}^2 \rightarrow \min. \tag{12.92}$$

Here, M is the number of GCPs, and σ_k is the standard deviation of the kth phase measurement. This equation is equivalent to

$$\frac{\partial}{\partial x_m} E = 0 (m = 0,1,2). \tag{12.93}$$

Taylor expansion of Equation (12.93) determines the unknowns as

$$x_m = (a_{mn})^{-1} \cdot b_n \tag{12.94}$$

$$a_{mn} = \sum_k \frac{1}{\sigma_k^2} \left\{ \frac{\partial^2 \hat{\phi}_k}{\partial x_m \partial x_n} \left(\phi_k - \hat{\phi}_k \right) - \frac{\partial \hat{\phi}_k}{\partial x_m} \frac{\partial \hat{\phi}_k}{\partial x_n} \right\} \tag{12.95}$$

$$b_n = -\sum_k \frac{1}{\sigma_k^2} \frac{\partial \hat{\phi}_k}{\partial x_n} \left(\phi_k - \hat{\phi}_k \right). \tag{12.96}$$

The differential coefficients of the slave vector are

$$\frac{\lambda_0}{4\pi} \frac{\partial \hat{\phi}}{\partial x_m} = \left(\frac{\partial R_s}{\partial \Delta B}, \frac{\partial R_s}{\partial \Delta r}, \frac{\partial R_s}{\partial \Delta v} \right)^t$$

$$= \begin{pmatrix} R_s^{-1}\left(1-C_n^2\hat{r}_s^{-2}\right) & C_nR_s^{-3}\left(r_s-C_r\right) & C_nR_s^{-3}\left(r_s-C_r\right)t \\ C_nR_s^{-3}\left(r_s-C_r\right) & \left(1-\left(r_s-C_r\right)^2 R_s^{-2}\right)R_s^{-1} & \left(1-\left(r_s-C_r\right)^2 R_s^{-2}\right)R_s^{-1}t \\ C_nR_s^{-3}\left(r_s-C_r\right)t & \left(1-\left(r_s-C_r\right)^2 R_s^{-2}\right)R_s^{-1}t & \left(1-\left(r_s-C_r\right)^2 R_s^{-2}\right)R_s^{-1}t^2 \end{pmatrix} \begin{pmatrix} \Delta B \\ \Delta r \\ \Delta v \end{pmatrix} + \begin{pmatrix} -R_s^{-1}C_n \\ R_s^{-1}\left(r_s-C_r\right) \\ R_s^{-1}\left(r_s-C_r\right)t \end{pmatrix}$$

$$\tag{12.97}$$

$$C_n \equiv \mathbf{r}_{GCP} \cdot \mathbf{n}_r \tag{12.98}$$

$$C_r \equiv \mathbf{r}_{GCP} \cdot \mathbf{n}_n. \tag{12.99}$$

Errors (standard deviations) of the estimated parameters are

$$\sigma_{\Delta B}^2 = \frac{a_{22}a_{33} - a_{23}^2}{\det(a_{mn})}$$

$$\sigma_{\Delta r}^2 = \frac{a_{11}a_{33} - a_{13}^2}{\det(a_{mn})} \tag{12.100}$$

$$\sigma_{\Delta v}^2 = \frac{a_{11}a_{22} - a_{12}^2}{\det(a_{mn})}$$

where $det\,(a_{mn})$ means the determinant of the matrix a_{mn}. Since the InSAR phase arises from a differential distance between the master-target and the slave-target, the unknowns should be determined from either the master minus the slave or either the master or the slave. Here, we assume that the slave orbit has some error but not the master. The iteration continues until the absolute ratio of the fluctuation divided by the converged value becomes less than 1.0e-5.

12.10.4   HIGHER CORRECTION

After the aforementioned AEPD and orbit corrections, we sometimes see that a phase changes as a second-order function of the slant range and azimuth coordinates. There are two possible causes. First, the interferometric coherence of the mountainous area tends to be lower than that of the flat land. The flat land is unwrapped successfully, but the mountainous area sometimes is not. Mt. Iwate, with a bigger baseline, is a good example. The Morioka flat plain is successfully unwrapped and the GCPs are only selected from the center of the image (nonuniformity). Second, the refraction values based on the global analysis data (GANAL) from the Japanese Meteorological Agency from the actual ones over a wide area. In the first case, the problem may be solved by selecting the GCPs uniformly over the image. This requires successful unwrapping over an image. In the second case, the GANAL must be improved. Here, we adopt a method that determines the second-order function of the phase difference in azimuth and range from the data and then subtracts it from the data.

12.10.5   VALIDATION

We evaluated the surface deformation derived from the proposed method. We selected Mt. Fuji and its vicinity as a non-deformation sample and the Mt. Iwate area as a deformation sample. All the evaluated images are listed in Table 12-5. Results with AEPD correction are shown in Table 12-6a and those without AEPD correction in Table 12-6b. These tables show the standard deviation of the residual phase at GCPs (σ_1), the three unknowns (orbital corrections) and their standard deviations, standard deviation of the residuals over an image (σ_2), phase error of the 10 km × 10 km area at the top of the mountain (σ_3), and base line lengths (B_{perp}). Several institutions, represented by the

---

TABLE 12-5

A List of the SAR Image Pairs Used in This Analysis

No.	Master	Slave	B$_p$(m)	Result	Target	Condition	Season
1	8/20/1993	7/7/1993	502	△	Mt. Fuji	NSD	Summer
2	5/15/1997	4/1/1997	154	△	Mt. Fuji	NSD	Spring
3	8/11/1997	6/28/1997	157	△	Mt. Fuji	NSD	Summer
4	10/21/1995	9/7/1995	350	○	Mt. Fuji	NSD	Fall
5	3/19/1998	2/3/1998	1300	○	Mt. Fuji	NSD	Winter
6	9/11/1998	3/19/1998	790	△	Mt. Fuji	NSD	S-W
7	9/9/1998	11/5/1997	120	○	Mt. Iwate	SD	
8	9/9/1998	6/13/1998	1191	○	Mt. Iwate	SD	
9	7/27/1998	6/13/1998	1440	○	Mt. Iwate	SD	
10	6/13/1998	4/30/1998	800	○	Mt. Iwate	SD	

Note:

○   Fairly well corrected.

△   Almost corrected.

NSD   No surface deformation.

SD   Surface deformation occurred during a period.

B$_p$ Perpendicular baseline distance (m).

TABLE 12-6A
Results with the Atmospheric Excess Path Delay Correction Using GANAL

No.	σ_1(cm)	ΔB(m)	(mm)	Δr(m)	(mm)	Δv(cm/s)	(mm/s)	σ_2(cm)	σ_3(cm)	B_p(m)
1	1.19	7.40	(1.9)	5.79	(2.9)	−4.3	(0.027)	1.96	1.70	−508.51
2	1.73	21.6	(5.9)	15.5	(4.2)	−2.8	(0.049)	1.05	1.47	−151.20
3	1.33	−1.3	(6.4)	−1.2	(4.6)	0.99	(0.055)	1.60	1.92	−157.22
4	0.39	50.0	(3.8)	38.0	(2.8)	−7.9	(0.039)	1.21	2.41	−347.24
5	0.75	27.6	(5.2)	19.4	(3.7)	−2.5	(0.043)	1.28	2.25	−168.96
6	1.34	50.1	(3.4)	38.1	(2.5)	−6.9	(0.046)	2.16	2.48	−788.22
7	1.19	55.1	(6.2)	38.1	(4.4)	−8.1	(0.087)	2.23	−	−145.48
8	0.42	20.0	(10.1)	14.6	(7.0)	−7.27	(0.051)	2.74	−	−1,178.64
9	4.43	−12.4	(1.8)	−8.39	(3.7)	−3.9	(0.025)	2.59	2.42	1436.56
10	1.36	−10.2	(7.1)	−6.87	(4.9)	3.9	(0.046)	2.28	1.68	−571.22
mean	1.41							1.91	2.04	

Japan Meteorological Agency, thoroughly investigated the deformation possibility of the Mt. Fuji region. Their work indicates that this area has been very stable, except for a medium-scale earthquake (M 5.3) in 1996 (Sourifu 1997), and volcanic activity is very low (Takahashi and Kobayashi 1998). Thus, we assume that there are no earthquakes around this area, and this area can be treated as a non-deformation area.

12.10.5.1 Correction Procedure
First, two SLC images were generated from the JERS-1 SAR raw signal data, and the InSAR phase was unwrapped (Shimada 1999). Second, simulated amplitude SAR images were generated using the DEM and master/slave state vectors, and a geometric relationship between these simulated and

TABLE 12-6B
Results without Atmospheric Excess Path Delay Correction

No.	σ_1(cm)	ΔB(m)	(mm)	Δr(m)	(mm)	Δv(cm/s)	(mm/s)	σ_2(cm)	σ_3(cm)	B_p(m)
1	1.50	7.42	(4.0)	5.86	(2.9)	−4.69	(0.027)	2.10	4.44	−508.57
2	2.09	21.3	(5.9)	15.1	(4.2)	−3.4	(0.049)	1.96	6.80	−150.77
3	0.62	−1.4	(6.4)	−1.2	(4.6)	0.67	(0.055)	1.63	3.20	−157.22
4	1.08	52.0	(3.8)	37.1	(2.8)	−7.2	(0.039)	1.38	5.04	−347.75
5	0.71	27.6	(5.2)	19.5	(3.7)	−2.5	(0.044)	1.25	1.60	−169.00
6	0.72	50.0	(3.4)	37.9	(2.5)	−8.4	(0.046)	2.25	2.58	−788.10
7	0.92	55.0	(6.2)	38.1	(4.4)	−8.5	(0.087)	2.32	−	−145.55
8	0.42	19.7	(10.1)	14.3	(7.0)	−7.3	(0.051)	2.72	−	−1178.20
9	4.42	−6.57	(5.4)	−13.5	(3.7)	−4.9	(0.026)	2.25	5.73	1436.23
10	1.44	−13.61	(7.1)	−4.5	(4.9)	3.8	(0.046)	2.45	4.38	−571.39
mean	1.39							2.03	4.22	

Note: B_p is represented in the nearest range. Values in brackets are standard deviations of previous values. The units are mm for ΔB and Δr and mm/s for Δv. σ_1 is the standard deviation of the observed phase-phase model, σ_2 is the standard deviation of all the phase differences in an image, and σ_3 is the standard deviation of the phase difference within a 10 km × 10 km square area centered at the mountain peak.

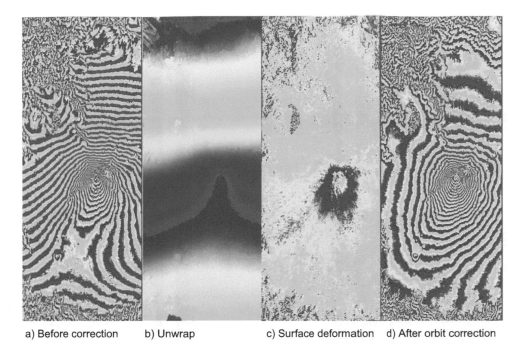

a) Before correction   b) Unwrap   c) Surface deformation   d) After orbit correction

FIGURE 12-17 Process of the orbital error correction: (a) orbit-corrected fringe, (b) orbit-corrected unwrapped phase, (c) orbit–terrain-corrected fringe, and (d) orbit-corrected final fringe. In this way, the orbital error is eliminated.

actual images was obtained. These orbits were roughly corrected by comparing the actual image and the simulated images. Third, AEPDs (ϕ_3) on the master and slave dates were calculated using the GANAL data. Fourth, we selected more than ten GCPs from the successfully unwrapped area and calculated the height of the GCPs using the geometric relationship obtained in the second step. Fifth, we estimated the errors of the slave vectors using selected GCPs. Finally, we reconstituted the slave orbit and calculated the surface deformation by correcting the AEPD. Figure 12-17 shows four image combinations: (a) an un-corrected DTM, (b) an unwrapped un-corrected DTM, (c) orbital error correct final deformation, and (d) orbital error-corrected DTM.

12.10.5.2 Water Vapor

The vertical structure of the water vapor's partial pressure, total pressure, and temperature was calculated from the GANAL data. Radiosonde data (JMA 1997) from Hamamatsu at 9:00 were used for validating the GANAL at Mt. Fuji. These data sets are made of 17 geopotentials (heights), wind directions, and wind speeds to cover the atmosphere from the surface to a height of 30,000 m (pressure, 10 hPa), and nine relative humidities from the surface to the height of 10,000 m (pressure, 300 hPa). They are generated at a latitude and longitude span of 1.25 degrees and a time span of 6 hours. These data are bilinearly interpolated in time and horizontal space. (They are interpolated vertically once converted in logarithmic space.) Since the evaluation area is as wide as 100 km, we assumed that the atmosphere is the same over a scene. GANAL is generated using the surface terrain averaged over 100 km × 100 km. Thus, it may not express a local inhomogeneous variation of the water vapor's partial pressure that may be caused by a possible wind perturbation at the peak of the mountains. Regional objective analysis data (RANAL) with a spatial averaging size of 20 km will be used for this type of correction.

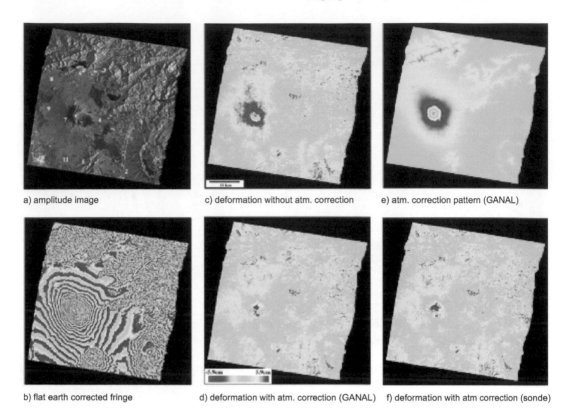

a) amplitude image   c) deformation without atm. correction   e) atm. correction pattern (GANAL)

b) flat earth corrected fringe   d) deformation with atm. correction (GANAL) f) deformation with atm correction (sonde)

FIGURE 12-18   Analysis of Mt. Fuji for image pairs acquired on October 21, 1997, and September 7, 1997: (a) Amplitude image, (b) flat-Earth-corrected fringe, (c) terrain-corrected fringe, (d) terrain and atmospheric excess path delay (GANAL)-corrected fringe, (e) atmospheric excess path delay correction pattern using GANAL, and (f) terrain and atmospheric excess path delay (radiosonde)-corrected fringe. The image dimensions are 45 km on the horizontal axis and 47 km on the vertical axis.

12.10.5.3   Test Sites (No Deformation Case)

Mt. Fuji and its vicinity are suitable for validating the proposed method since the mountain peak is 3,776 m high, the water vapor's partial pressure may vary due to the mountain features and the seasonal change, and no major surface deformation has occurred in this region (Figure 12-18a. Thus, six pairs of JERS-1 SAR data were selected from the data archive (summer, spring, summer, autumn, winter and autumn, and spring and autumn—Cases 1 to 6 of Table 12-6).

a. **Autumn and the Other Pairs**

In Case 4 of Table 12-5, the finally obtained flat Earth-corrected fringe (meaning the topography, wrapped between $-\pi$ and π) is given in Figure 12-18b; the surface deformation that was not corrected for AEPD $(\phi - \hat{\phi})$ is shown in Figure 12-18c; the final surface deformation $(\phi - \phi_3 - \hat{\phi})$ corrected for AEPD using GANAL is in Figure 12-18d; the AEPD pattern (ϕ_3) is in Figure 12-18e; and the surface deformation pattern corrected for AEPD from radiosonde data $(\phi - \phi_3 - \hat{\phi})$ is in Figure 12-18f. Here, the data are expressed as the change of one-way distance from –5.9 cm to 5.9 cm. The other examples for no surface deformation are shown in Figures 12-19a to 12-19e.

b. **Vertical Structure of the Atmospheric Excess Path Delay (GANAL and Radiosonde)**

AEPD (ϕ_3) values obtained for Cases 1, 4, 7, and 8 are shown in Figure 12-20a. GANAL and radiosonde related parameters, as well as AEPD (ϕ_3) and water vapor vertical distributions for two days of Case 3, are compared in Figure 12-20b.

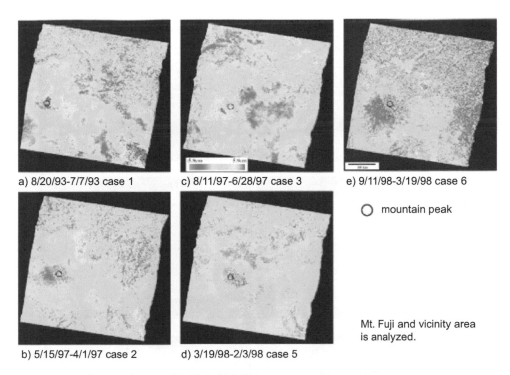

a) 8/20/93-7/7/93 case 1   c) 8/11/97-6/28/97 case 3   e) 9/11/98-3/19/98 case 6

○ mountain peak

b) 5/15/97-4/1/97 case 2   d) 3/19/98-2/3/98 case 5

Mt. Fuji and vicinity area
is analyzed.

FIGURE 12-19 Other examples of Mt. Fuji: (a) August 20, 1993–July 7, 1993 (Case 1); (b) May 15, 1997–April 1, 1997 (Case 2); (c) August 11, 1997–June 28, 1997 (Case 3); (d) March 19, 1998–February 3, 1998 (Case 5); and (e) September 11, 1998–March 19, 1998 (Case 6).

c. **Surface Deformation Cases**

Mt. Iwate data were used to validate the surface deformation cases. Mt. Iwate, 2038 m at its peak, was active in 1995 and later in April 1998. The volcanic activity reached a peak in July 1998, and a M. 6.1 earthquake occurred on September 3, 1998, after which Mt. Iwate became calm. Table 12-5 shows the data pairs processed by the proposed method. Temporal changes in the surface deformation are shown in Figure 12-21a–d, and the results without AEPD correction are shown in Figure 12-21e–h. This figure confirms the effectiveness of the proposed method.

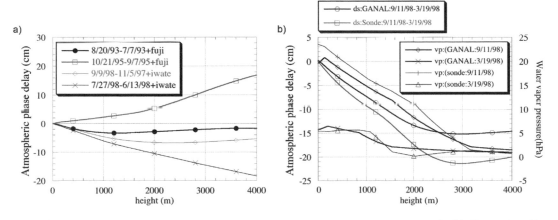

FIGURE 12-20 (a) Vertical profile of the atmospheric excess path delay (using GANAL. (b) Vertical profile of the atmospheric excess path delay and the water vapor pressure in comparison using GANAL and radiosonde data (Case 3).

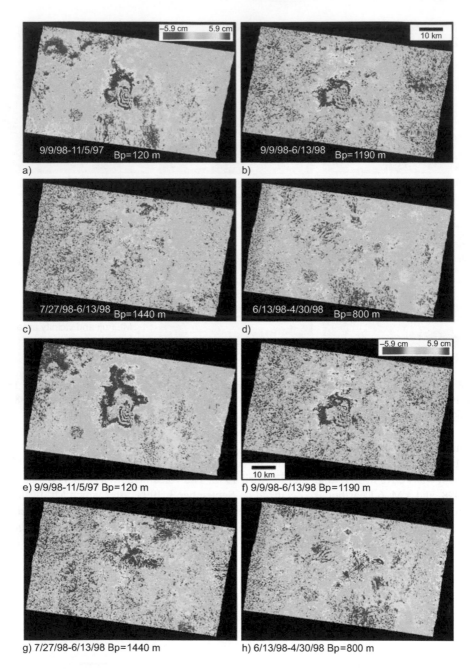

FIGURE 12-21 Surface deformation patterns of Mt. Iwate and vicinity during the volcanic and co-seismic activity period were estimated by the InSAR method with atmospheric excess path delay correction using GANAL. Images (a) through (d) show the results with atmospheric excess path delay correction and Images (e) through (h) show those without the delay correction. The image pairs are: (a) 9/9/1998–11/5/1997 (Case 7); (b) 9/9/1998–6/13/1998 (Case 8); (c) 7/27/1998–6/13/1998 (Case 9); (d) 6/13/1998–4/30/1998 (Case 10); (e) 9/9/1998–11/5/1997 (Case 7); (f) 9/9/1998–6/13/1998 (Case 8); (g) 7/27/1998–6/13/1998 (Case 9); and (h) 6/13/1998–4/30/1998 (Case 10). All the images are geocoded so the vertical axis corresponds to the north direction. Each image is 60 km vertically and 90 km horizontally. The atmospheric excess path delay that clearly appeared in (g) and (h) was eliminated by this proposed correction as shown in (c) and (d). Image pairs (c) and (d) were acquired during a less active period of surface deformation. Thus, the results in (c) and (d) might be consistent. From this, results (a) and (b) might be accurate.

12.10.6 Discussion

12.10.6.1 AEPD Correction

As shown in Figures 12-18 and 12-19, AEPD correction with GANAL is very effective in detecting surface deformation. Figure 12-18c and e indicates that the AEPD model agrees with the observations. The color variation at the top of the mountain in Figure 12-18d may give an incorrect result unless the mountain is deformed. AEPD varies with the off-nadir angle of each pixel. Equation (12.83) and its correction display more confidence in the result. The aforementioned color variation, which is incorrectly interpreted as the north side approaching the satellite and the south side removed, can be explained by possible inhomogeneous water vapor distribution due to wind perturbation around the mountain. Averaged residual phase error over a scene is 1.2 cm, which satisfies the phase measurement requirement of 20 to 30 degrees (and which corresponds to 1.5 cm for the one-way distance). A small (0.5 cm) dip at the top of the mountain is within the error allowance but can be seen as a systematic error. GANAL accuracy must be improved.

The proposed method also provided good results for the other data pairs from different seasons (Figure 12-19). The mountain peak is indicated by a circle. There are no additional colors at the top of the mountain except for Figure 12-19a, which looks similar to the mountain peak of Figure 12-18. GANAL data after 1995 have a time span of 6 hours and a spatial span of 1.25 degrees, and data for 1993 have a time span of 12 hours and a spatial span of 2 degrees. This greater course span may have caused the color change. Perpendicular baselines of the other examples in Figure 12-19 range from 154 m to 1,300 m. The bigger baseline gives better height resolution. Nonetheless, residual color does not appear at the top of the mountain. This may be because the water vapor distribution is small and homogeneous.

From the previous analysis, we see that the use of GANAL for AEPD correction is very effective for eliminating errors in estimating surface deformation. Moreover, color variation in the foothills of Mt. Fuji in Figure 12-18e may indicate residual water vapor distribution from the GANAL and may show a way to improve GANAL. GANAL is generated with a medium relief terrain that is averaged over a 100 km × 100 km region. This may cause a slight phase error at the top of the mountain. AEPD differs with GANAL and radiosonde data; AEPD using radiosonde data generated more errors. This may be due to the different locations of Mt. Fuji and Hamamatsu (radiosonde data) and the existence of a larger mountainous region between them. Regardless, GANAL is a reliable source for correcting the AEPD of the InSAR data.

12.10.6.2 Vertical Distribution of the Atmospheric Excess Path Delay

The two-day difference of AEPD varies significantly with time and space (Figure 12-20a. The AEPD of Mt. Iwate (between September 9, 1998 and November 5, 1997) decreases with height to 2,000 m then and increases again. The AEPD of Mt. Fuji (August 20, 1993 and July 7, 1993) decreases with height to 1000 m then increases thereafter. AEPD (October 21, 1995 and September 7, 1995) increased with height and became 12 cm at a height difference of 2,776 m (from the peak of Mt. Fuji at 3,776 m to 1,000 m, which is the altitude of Lake Yamanaka). This corresponds to one period of an L-band signal and causes a large error in deformation estimation. These two examples are for isolated mountain cases. Because the mountainous areas around the peaks generate updrafts and downdrafts, GANAL, which does not consider this, may express the phase delay almost exactly.

As shown in Figure 12-20b, radiosonde data capture more details of the vertical properties of the water vapor's partial pressure than GANAL. Although the water vapor pressures are almost the same on March 19, 1998, they differ significantly on September 11, 1998, at heights below 3,000 m. This causes a larger phase difference between the final products using GANAL and radiosonde. (Assuming 0 m as a reference, the AEPD of radiosonde data at 4,000 m is −20 cm and that of GANAL is −15 cm). It should be noted that the radiosonde and the GANAL data were collected 100 km apart for a mountain series. The difference in AEPD may not be discussed in the same table.

12.10.6.3   Combining Orbit Correction and Atmospheric Correction

The orbit estimation converged within three to four iterations when the error ratio became less than 1.0e-5. All the examples showed similar convergence, and the proposed method seems to work well. The sixth column (second from the right) shows the residual of the phase difference over a scene (standard deviation of an image). The standard deviation average is 1.91 cm after AEPD correction, while the phase residual without AEPD correction is 2.00 cm. We calculated the phase residual of the mountain peaks at Mt. Fuji and Mt. Iwate over a 10 km × 10 km area for eight cases (excluding Cases 7 and 8, which are crustal movement cases). The phase residual was improved from 4.22 cm, which is the average of the non-corrected case (6.8 cm max. to 1.6 cm min.), to 2.04 cm (2.4 cm max. to 1.7 cm min.), as the AEPD case. The estimated orbit elements and baseline distances change on the order of several tens of centimeters. Estimated accuracies of the error vectors are several millimeters for ΔB, Δr and several hundred millimeters per second for Δv. This means that the atmospheric correction changes the orbit error but does not affect the Bp estimation as much. Moreover, the final surface deformation depends on the accuracy of the atmospheric correction component. Delacourt (1998) corrected the Mt. Etna data with an accuracy of 6 ± 3 cm, but in this study, we improved the correction to an accuracy of 2.04 cm.

12.10.6.4   Miscellaneous on Orbit Corrections

Several parameters should be discussed in this estimation routine. These include the number of the estimation parameters and the terms to be estimated. In this study, we estimated three parameters in the slave orbit. It is possible to increase the number of parameters from three to six or eight by parameterizing the master orbit. Since interferometry is based on the relative position of two orbits, the parameterization should be fixed to one orbit. From a stability point of view, the recommended parameters are ΔB, Δr, and Δv.

On average, the actual SAR image does not co-register with the simulated image that is made from the DEM and orbit data in slant range direction. Due to this error, the phase difference behaves as a squared power of the slant range. One way to correct this is to cancel out the additional phase by applying a suitably selected phase function.

Successful unwrapping of the phase difference throughout an image is necessary because uniform and widely selected GCP over an image can only accurately estimate ΔB and Δr and the other parameters. If unwrapping is conducted at a part of the image, the unwrapped area must be enlarged as much as possible by increasing the number of samples of the data and then increasing the coherence. In the future, parameter estimation from a mixture of wrapped and unwrapped phases should be considered.

12.10.6.5   Application to Surface Deformation Monitoring

A global view of the surface deformation between September 9, 1998, and November 5, 1997, is shown in Figure 12-21a. The deformation caused by the earthquake of September 3, 1998, can be seen at the center of the image. Other deformation caused by volcanic activity is visible northwest of Mt. Iwate. The volcanic activity peaked on April 30, 1998. In Figure 12-21c (July 27, 1998 and June 13, 1998) and Figure 12-21d (June 13, 1998–April 30, 1998), a moss green-colored part is visible west of Mt. Iwate. This may indicate a small inflation of the mountain due to the magma uplift. Figure 12-21b is provided from September 9, 1998, and June 13, 1998, data as a bigger baseline case. This result looks similar to Figure 12-21a except for a change along a mountain ridge. This may be related to deformation before April 30, 1998, or to a local change in the water vapor.

12.10.7   Summary

First, we corrected the atmospheric excess path delay using global objective analysis (GANAL) data generated by the Japanese Meteorological Agency. Next, we estimated the error of the orbit vectors (ΔB, Δr, and Δv) using ground control points (GCPs) and unwrapped phase information in a least square sense. As a result, we were able to estimate the surface deformation with an accuracy of 2.04 cm. This value is significantly improved from the non-atmospheric correction value of 4.04 cm. We have also confirmed the effectiveness of GANAL data for this purpose.

It should be noted that this GANAL, as the reanalysis data for atmospheric behavior assimilated for the theory and validation data, was provided by Japan's Meteorological Research Institute in the 1990s and that organization no longer exists. New generation of the reanalysis data, called the global spectral model (GSM), spaced every 25 km, for 6 hours, and having 17 layers (10 hPa to 1,000 hPa, in Gridded Binary defined by the World Meteorological Organization GRIB format), has been provided routinely beginning in 2018.

12.11 ScanSAR-ScanSAR INTERFEROMETRY

12.11.1 Introduction

In principle, ScanSAR-ScanSAR interferometry (Scan-InSAR) is more appropriate for detecting the wide areal deformation rather than strip InSAR. However, the correct co-registration frames for all strips are difficult to prepare. Among two imaging algorithms, SPECAN is advantageous as it is quicker (Shimada 2009; Cumming and Wong 2005) than the full aperture SAR (FA), but it is disadvantageous due to difficulties at phase continuity across the full swath (Shimada 2008). In this section, a modified full aperture ScanSAR (MFA) is introduced to virtually integrate all the individual FA-SLC to a single large SLC and to improve the co-registration accuracy.

12.11.2 Processing

The FA preserves the InSAR phase and is often used for Scan-InSAR (Bamler et al. 1999). If the target area is texture enriched, all the beams are co-registered well. Otherwise, some may fail, the phase might collapse, and discontinuity could appear across the swath. FA is simply operated, but the difference in azimuth length is an issue. For building a robust and unified co-registration frame, the equally azimuth-spaced SLC is ideal.

Such a SLC is produced in a way so as to interpolate all the strips with a reference PRF after range correlation before azimuth correlation. Since the start of the timing of each burst and PRF is rigorously scheduled, the resampling of the range compressed data at the reference PRF and azimuth compression can generate the full-length SLC data. A reference PRF is selected from the data or assigned to the ideal value. Here, we selected the data at Scan #1. The timing offset is selected from the timing control table (Table 12-7), which is the sampled from ALOS/PALSAR.

12.11.3 Processing Parameters

12.11.3.1 Doppler Parameter
Doppler frequency is always measured and reported as less than 100 Hz. It is set to zero in the processing.

TABLE 12-7
Timing Offset and the Pulses in the Scan

No.	Pulses	PRF	Duration(s)	Length (km)
1	247	1,692.0	0.146	0.993
2	356	2,369.7	0.150	1.02
3	274	1,715.2	0.160	1.09
4	355	2,159.8	0.164	1.12
5	327	1,915.7	0.171	1.16
Total			0.791	5.413

Note: PALSAR has five patterns: 3 Scan, 4 Scan, 5 Scan, and short burst and long burst. This table shows the short burst and 5 Scan cases only. PRF in the table is an example.

12.11.3.2   Azimuth Interpolation and Scaling

Azimuth interpolation of range-compressed data at a reference PRF is key. Master and slave orbits (and SAR) were not intensively operated to synchronize the burst locations and the spacing in general, while ALOS-2 was operated to equalize the burst start timing only. As a result, PALSAR could not adjust the two parameters and ALOS-2 could not adjust the spacing parameters. In order to obtain the same azimuth spacing, the slave reference PRF should satisfy the condition:

$$f_s = \frac{v_s T_s}{v_m T_m} f_m \tag{12.101}$$

Here, f_s is the modified PRF for the slave image, f_m is the PRF for the master image, v is the ground speed, T is the period of one scan, and the suffixes "m" and "s" are for "master" and "slave."

12.11.3.3   Beam Synchronization

The beam synchronization is implemented in such a way that only coexisted bursts in time between two images are extracted and the rest is zero-filled at the azimuth correlation. ALOS-2/PALSAR-2 operation has already considered the beam synchronization and may not have much of an impact, but PALSAR could improve the coherence.

12.11.3.4   Co-Registration

ScanSAR has a wide swath of several hundred kilometers, and B_{perp} varies greatly in azimuth and range directions. The more the two orbits are separated, the more the slant range and azimuth distance differences to the target behave more nonlinearly. These nonlinearities cannot be handled by an empirical model such as that within which the normal co-registration frame works (piling the tie point, finding the effective tie points, and modeling the co-registration frame by using the pseudo affine model or by the inclusion of higher-order terms), but two-step modeling (correcting the first-order model based on the theoretical co-registration model using the master-slave orbits and target and modeling the remained offsets in range and azimuth by pseudo affine) is more reliable and robust.

Thus, the co-registration frame can be designed as follows:

$$X = ax + by + cxy + d + \Delta X(x)$$
$$Y = a'x + b'y + c'xy + d' + \Delta Y(x). \tag{12.102}$$

Here, x (y) are the slant range (azimuth) pixel address at the master image; X and Y correspond at the slave image; and $\Delta X(x)$ ($\Delta Y(x)$) is the predetermined slant range (azimuth distance) difference, which is given by the real orbit data and the position determination model. Thus, the co-registration frame are the functions of x; and a, b, c, d, a', b', c', and d' are the pseudo affine coefficients.

12.11.4   Results

12.11.4.1   List of the Scenes

Three PALSAR Scan-InSAR are evaluated. The evaluated areas are Tanzania (as the case with no deformation) and Haiti and Wunshen/China (as the deformation cases). Table 12-8 shows the summary of the targets with some results.

12.11.4.2   Interferometric Results

The Tanzania images are displayed in Figure 12-22: (a) the flat Earth-corrected fringes, (b) the terrain-corrected image, and the qualified fringes with high coherence, two coherences: (c) by the proposed method and (d) by the standard method.

TABLE 12-8
Evaluated Images and Results

Area	Co-Registration	Master Date	Slave Date	B_{perp}(m)
Tanzania	0.10 (x) 0.44 (y)	20080417	20080302	158
Haiti M 7.0	0.30 (x) 0.31 (y)	20100211	20090926	241
Wunshen M. 7.8	–	20080520	20080103	781

Note: – means unavailability of measured data.

Deformation examples for (a) the Haiti Earthquake and (b) the Wunshen Earthquake of May 12, 2008, which was an M8.9 earthquake, are shown in Figure 12-23.

12.11.5 EVALUATION AND DISCUSSION

12.11.5.1 Excess Range and Co-Registration Residual

Using the orbital data, we can prepare the range-dependent ΔX and ΔY database before the co-registration processing. Depending on the baseline, both ΔX and DY should vary nonlinearly around several to several tens of pixels within a 200-km range distance. The Tanzania case varied 7 pixels

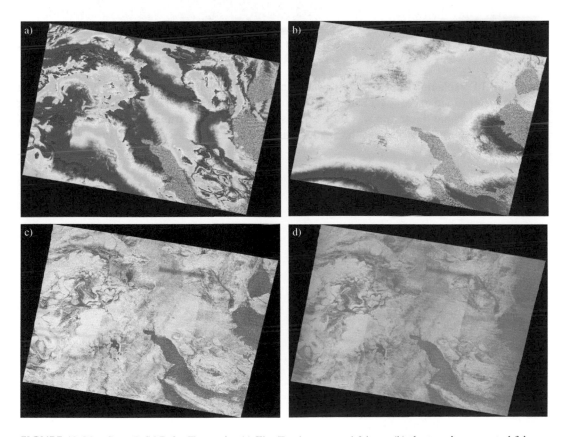

FIGURE 12-22 Scan-InSAR for Tanzania: (a) Flat Earth-corrected fringe; (b) the terrain-corrected fringes in (b); (c) coherence with the proposed method, and (d) coherence with the normal co-registration and without beam synchronization.

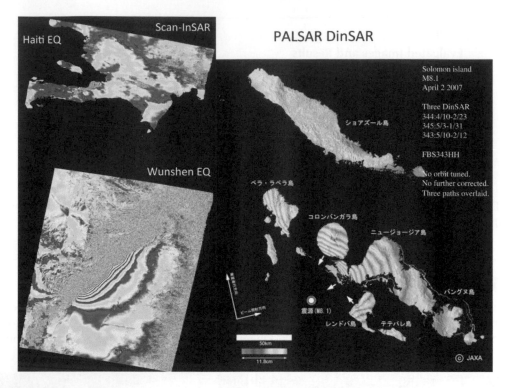

FIGURE 12-23 PALSAR DInSAR results (a) Haiti Earthquake of 20100112, (b) Wunshen Earthquake of 20080512 observed by Scan-InSAR, and (c) PALSAR DInSAR with the normal three strips combined.

in ΔX and 10 pixels in ΔY during a full swath range that was nonlinear for both. After substituting these predetermined excess data before co-registration, the co-registration frame for the remaining components became linear and was fully expressed by the pseudo affine. Figure 12-24a shows ΔX and ΔY, and Figure 12-24b shows the residual displacement that varies linearly in range direction (Tanzania case).

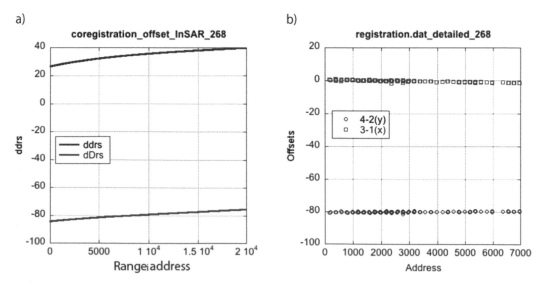

FIGURE 12-24 (a) Range dependence of ΔX and ΔY for Tanzania case. (b) Distribution of the residuals after subtraction of ΔX and ΔY, where the x-axis is the range pixel and the y-axis is the pixel number.

12.11.5.2 Coherence Improvement

All the representative cases (Tanzania, Haiti, and Wunshen) were processed using the proposed method, and their orbit-corrected fringe, further DEM-corrected fringe, and the interferometric coherence are shown in Figure 12-22a–c, respectively. Additionally, the coherence for Tanzania, which used the standard co-registration method and is without beam synchronization, is shown in Figure 12-22d as a comparison. It confirms that the new co-registration method as well as the beam synchronization significantly improves the coherence by 20% and improves the phase quality. Large-scale gradual phase variation in Figure 12-22b might be the ionospheric variation in the descending orbit. The other cases, Haiti and Wunshen, are shown in Figure 12-23, and the interferometric phases are clearly visible. However, the Wunshen case still looks noisy in the mountainous area (the upper half of the image). This could be due to the temporal baseline being more than 92 days and that the surface condition differs significantly between the snow coverage and the vegetation cover in the early summer (Shimada 2012).

12.12 TIME SERIES InSAR STACKING

12.12.1 INTRODUCTION

Time series InSAR analysis has become an important technique for monitoring regional land subsidence and speed. The persistent scatterer InSAR (PSInSAR) and SBAS approaches (Bernardino et al. 2002; Ferretti et al. 2001) are representative. DInSAR based on an image pair provides a real sign of the subsidence, while the external noises are convolved, making correction difficult. We introduce a time-series DInSAR stacking method because its basic idea is easy to understand and its application is simple. The key ideas of this method are calibration of each pair of images using GCPs and selection of an image pair that would not be affected by ionospheric disturbance.

12.12.2 STACKING DInSAR WITH A MOVING WINDOW

DInSAR detects the distance change (Δr) in the line-of-sight at any pixel between two acquisitions after correcting the reference terrain height and other offsets:

$$\phi = -\frac{4\pi}{\lambda}\left(\frac{B_{perp} \cdot z}{R \cdot \sin\theta} + B_{para} + \Delta r\right) - \Delta\phi_{atm} - \phi_{offset}. \tag{12.103}$$

A nonuniform refractive index in the signal propagation media decreases the accuracy of measurement, as does orbital determination error.

Assuming subsidence is the vertical motion, the subsidence speed (V) can be obtained by temporal differentiation of the distance change by

$$V = \frac{\Delta r}{T\cos\theta}. \tag{12.104}$$

Correcting the orbital errors and the atmosphere, then unwrapping the phase, we have

$$\phi_{US} = -\frac{4\pi}{\lambda}(V \cdot \cos\theta \cdot T) - \phi_{offset}. \tag{12.105}$$

The spatio-temporal weighted average of the phase stacking estimates the subsidence speed and its distribution, Equation (12.106), and the standard deviation, Equation (12.107):

$$\overline{V}_I = \frac{1}{G_I} \sum_i \frac{\lambda}{4\pi} \left(\phi_{US,i} + \phi_{offset,i} \right) \frac{1}{\cos\theta \cdot T_{,i}} \gamma_i$$

$$G_I = \sum_i \gamma_i \tag{12.106}$$

$$\sigma_{V_I} = \sqrt{\frac{1}{G_{I2}} \sum_i \left\{ \frac{\lambda}{4\pi} \left(\phi_{US,i} + \phi_{offset,i} \right) \frac{1}{\cos\theta \cdot T_{,i}} - \overline{V}_I \right\}^2 \gamma_i^2}$$

$$G_{I2} = \sum_i \gamma_i^2 \tag{12.107}$$

Here, G_I is the total summation of coherence, and G_{I2} is the squared summation of γ; $\overline{V}$ is the averaged subsidence speed; and γ is the coherence. The processing flowchart is shown in Figure 12-25.

In this case, the following need to be considered:

1. Are all the associated phases co-registered accurately?
2. Are the time series phases well screened (i.e., properly corrected or excluded)?
3. Is the weighting function (i.e., time span in a window) well selected?

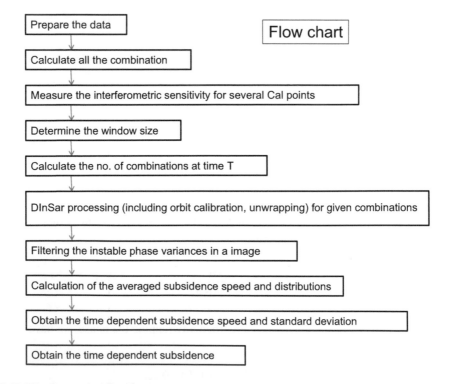

FIGURE 12-25   Processing flowchart.

TABLE 12-9

Statistics of the PALSAR Orbits of RSP422

Duration	January 6, 2007–March 4, 2011
Number of orbits	26
Combinations	378
Node of B_{perp}	−0.109 km
Standard deviation	0.632 km
Averaged temporal baseline	534 days

12.12.3 Data Analysis

Central Kalimantan, in Indonesia, which consists of peatland, peatland forest, and agricultural areas, is reported as being subsided due to global warming. This area was observed a total of 26 times from January 6, 2007, to March 4, 2011, by PALSAR with a 200-km strip. The orbits were well maintained, as listed in Table 12-9 (Shimada et al. 2010) and as represented by the averaged perpendicular baseline (B_{perp}) of −0.109 km (node) with a standard deviation of 0.632 km.

12.12.3.1 Coherence Analysis

This method depends on the temporal variation of the coherence; one example from Figure 12-26 shows that the time dependence of coherence at three typical sites decreases to about 600 days for the high coherence case and to 400 days even in the worst case and that this temporal decay occurs in a linear fashion. From this, we set the maximum extension of the time, B_t, as 365 days in which an image pair can be selected and be usable for this time series analysis. Coherence can be adopted as a weighting parameter in Equation (12.106) and Equation (12.107), and the averaging process on the interferometric phase can be applied after neglecting the outliers.

12.12.3.2 Preprocessing

Based on the previous investigation on B_t and B_{perp}, the DInSAR process matrix is prepared for the selection of the imaging pair. Under the selection scheme, the image was calibrated using calibration

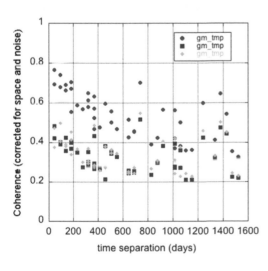

FIGURE 12-26   (a) Temporal coherence at three typical sites in Kalimantan. (b) Central Kalimantan sample images for amplitude (top) and coherence (bottom).

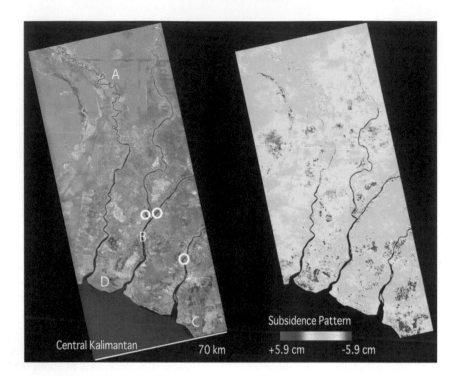

FIGURE 12-27 PALSAR image (left) and a representative deformation image (right) for the test site.

points selected from artificial targets—large bridges over rivers. Orbital estimation was conducted by a flattening process.

12.12.3.3 Results

Figure 12-27a shows an amplitude image of the target area where the calibration points were set at the bridges, and Figure 12-27b shows the corresponding subsidence. The number of the averages, as defined by the window function, varies with time such that a possible data combination is selected within 365 days (the maximum extension). A temporal history of subsidence of the target areas, A, B, C, and D, is finally obtained in Figure 12-28. As shown by figure, the area is linear in subsidence. The average subsidence speed is −2.02 cm/yr.

12.12.4 SUMMARY

This section described the estimation of a spatio-temporal subsidence distribution using time-series stacking of DInSAR. Using PALSAR data for Central Kalimantan, Indonesia, we conducted an experimental analysis of a 70 × 200 km area, including the Palangkaraya, and found that the estimated three-year average subsidence speed was 2.09 cm/year (Shimada et al. 2013).

12.13 AIRBORNE SAR INTERFEROMETRY

12.13.1 INTRODUCTION

While airborne SARs are associated with nonlinear flight trajectory supported by INS measurement, deriving the repeat-pass InSAR, providing the interferometric phase and the geophysical parameters, is challenging.

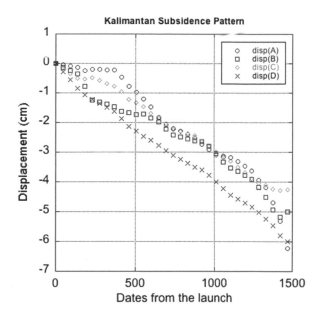

FIGURE 12-28   Subsidence history of four points in the image: A, B, C, and D of Figure 12-26.

12.13.2   PHASE DIFFERENCE

Because an airborne SAR has a shorter slant range than a spaceborne SAR, Taylor expansion of the master-slave difference is no longer valid. Thus, the interferometric phase should be given by

$$\phi = \frac{2\pi}{\lambda} r \left(1 - \sqrt{1 + \frac{2\mathbf{b} \cdot (\mathbf{r}_t - \mathbf{r}_m)}{r} + \frac{|\mathbf{b}|^2}{r^2}} \right) \tag{12.108}$$

$$\mathbf{b} = \mathbf{r}_m - \mathbf{r}_s.$$

Here, ϕ is the interferometric phase; r is the slant range between the master and the target; $\mathbf{r}_t$, $\mathbf{r}_m$, and $\mathbf{r}_s$ are the position vectors of the target, master orbit, and slave orbit, respectively; $\mathbf{b}$ represents the differential position vectors of $\mathbf{r}_m$ and $\mathbf{r}_s$; and $R(\mathbf{r}_t)$ is the Earth's radius at the target position.

From this,

$$2\mathbf{b} \cdot (\mathbf{r}_t - \mathbf{r}_m) = r \left\{ \left(1 - \frac{\lambda}{2\pi r} \phi \right)^2 - 1 - \frac{|\mathbf{b}|^2}{r^2} \right\} \tag{12.109}$$

can be solved to obtain the target position $\mathbf{r}_t$, and it provides the height (z):

$$z = |\mathbf{r}_t| - R(\mathbf{r}_t). \tag{12.110}$$

The surface deformation can be obtained similarly to the standard procedure as follows:

$$\phi = \frac{2\pi}{\lambda} r \left(1 - \sqrt{1 + \frac{2\mathbf{b} \cdot (\mathbf{r}_t - \mathbf{r}_m)}{r} + \frac{|\mathbf{b}|^2}{r^2}} \right) + \frac{2\pi}{\lambda} \Delta r \tag{12.111}$$

Tomakomai InSAR
HH Polarization

May 9 2013

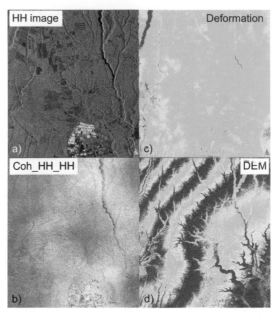

FIGURE 12-29 Pi-SAR-L2 interferometry: a) The amplitude image of the Hokkaido Tomakomai area, (b) interferometric coherence map, (c) the surface deformation map, and (d) the DEM.

Then, the deformation in the line of sight is obtained by

$$\Delta r = \frac{\lambda}{2\pi}\phi - r\left(1 - \sqrt{1 + \frac{2\mathbf{b}\cdot(\mathbf{r}_t - \mathbf{r}_m)}{r} + \frac{|\mathbf{b}|^2}{r^2}}\right). \qquad (12.112)$$

Figure 12-29 shows InSAR processing of Pi-SAR-L2 in Hokkaido: (a) HH amplitude image, (b) deformation, (c) HH InSAR coherence, and (d) orbit-corrected fringe (corresponding to the wrapped DEM).

12.14 ANALYSIS RESULTS

In this section, we will show several examples derived from interferometry at the three satellites:

1. JERS-1 SAR DInSAR: Two images were introduced for the Hanshinn Awaji Earthquake of January 16, 1995, and the related surface deformations: (a) The first results obtained at the very early phase in 1995, combining the images of February 6, 1995, and September 8, 1992. (b) The re-analyzed data, where the orbit correction, DEM correction, averaging for enhancing the phase, and the unwrapping the phase discontinuity over the isolated islands were performed. (c) The DEM generated from the JERS-1 SAR interferometry at Mt. Fuji (Figure 12-30).
2. PALSAR DInSAR: (a) The ScanSAR-InSAR for the 2010 Haiti Earthquake is shown, depicting the large areal deformation using the modified DInSAR method. (b) The Wunshen earthquake and related surface deformation observed by the two ScanSAR data in descending mode; this reduced the ionospheric disturbance much less than the ascending DInSAR. (c) This shows the three-pass mosaic of the strip interferometry covering the large areal deformation that occurred in the island areas (Figure 12-23).
3. PALSAR-2 DInSAR: These results are from the 2016 Kumamoto Earthquake. The Kumamoto area was densely observed by PALSAR-2 from the various observation directions, some of which decomposed the three-dimensional displacement. (a) This shows the deformation of the Kumamoto area using the 20160307-20160418 DInSAR. Square area was zoomed up to (b) where the amplitude image and deformation in slant rang was overlaid. (Figure 12-31).

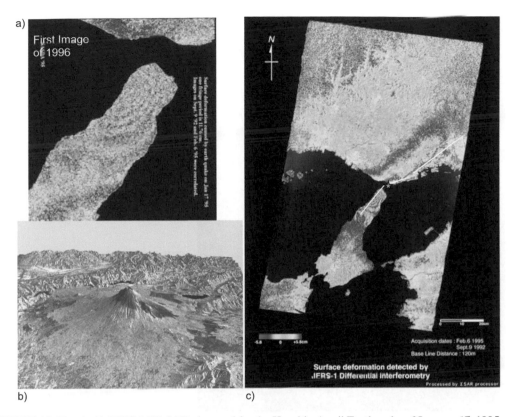

FIGURE 12-30 (a, b) JERS-1 DInSAR observed for the Hanshin Awaji Earthquake of January 17, 1995, and (c) three-dimensional view of the JERS-1 SAR ortho image corrected by the JERS-1 InSAR DEM.

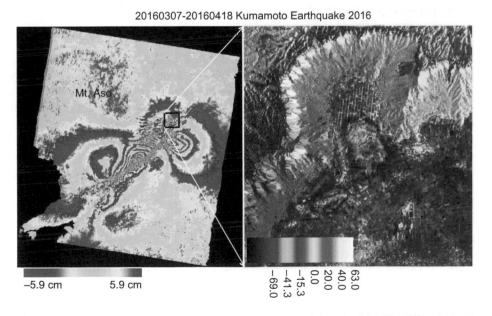

FIGURE 12-31 PALSAR-2 DInSAR deformation measurement using the 84-MHz HH polarization data acquired on March 7, 2016, and April 18, 2016, in the descending mode. (a) Deformation image in line of sight component covering 50 km and (b) zoom in of the Uchinomaki area, showing the heterogeneous deformation at several small regions.

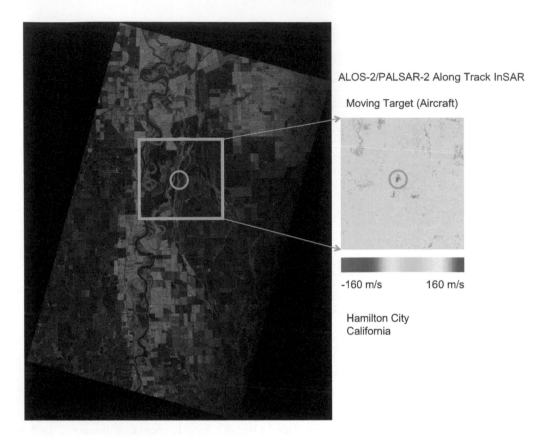

FIGURE 12-32 PALSAR-2 ATI image of Hamilton City, California. Left image is the amplitude image from ALOS-2 in descending node, and the right image is the InSAR phase showing the phase change possibly caused by the large-scale target—aircraft.

4. PALSAR-2 ATI results are shown in Figure 12-32. PALSAR-2 has experimentally prepared the ATI mode in several bandwidths. We used the data acquired for the city of Hamilton, California, using the UBS mode. The InSAR phase in the center of the right side of this picture shows the phase change of 93 degrees. Using 24 cm for λ, 5 m for baseline, and 6,700 m/s for ground speed, we detect the line of sight speed is 83 m/s. Using the incidence angle of 35 degrees, the target speed is 144 m/s or 520 km/hr. Thus, the target could be a large-scale aircraft, and the image dislocation caused by the Doppler shift could be 7700 m ($\Delta f_D / f_{DD} * vg = 7,700$ m where Δf_D is estimated as 691 Hz from a model, f_{DD} is -600 Hz/s, and vg is 6,700 m).

12.15   SUMMARY

This chapter reviewed SAR interferometry and differential interferometry, which accurately measure the topography and the surface deformation. We first discussed the measurement principles and derived a general equation relating the SAR measured phase, the geodetic quantities (DEM and surface deformation), and the error sources. Next, we evaluated those error sources quantitatively and discussed the quantitative relationship between each error source and the accuracy of the geodetic quantity. Finally, we evaluated the geodetic quantities derived from JERS-1 SAR, ALOS/PALSAR, ALOS-2/PALSAR-2, and Pi-SAR-L2.

APPENDIX 12A-1:   ATMOSPHERIC EXCESS PATH DELAY

We assume that the atmosphere between target A and satellite S is composed of N sub-layers, each one is parallel to the Earth's surface and has a constant humidity, temperature, and pressure (Figure 12A-1). For the two neighboring layers, we have a geometric relationship:

$$n_{i-1}r_{i-1}\sin\theta_{i-1} = n_i r_i \sin\theta_i, \text{ and} \tag{12A-1.1}$$

$$\frac{r_i}{\sin\theta_{i-1}} = \frac{r_{i-1}}{\sin\theta_i}, \tag{12A-1.2}$$

where the incidence angle θ_i and the refraction angle θ'_{i-1} through one border intersection are determined so that the fixed points A and S can be connected under the aforementioned condition. Since these two angles do not change much in a scene, we have

$$\theta_i = \theta_{i-1} + \frac{1}{\cos\theta_{i-1}}\left(\frac{n_{i-1}r_{i-1}}{n_i r_i} - 1\right)$$

$$\theta'_{i-1} = \theta_{i-1} + \frac{1}{\cos\theta_{i-1}}\left(\frac{n_{i-1}}{n_i} - 1\right). \tag{12A-1.3}$$

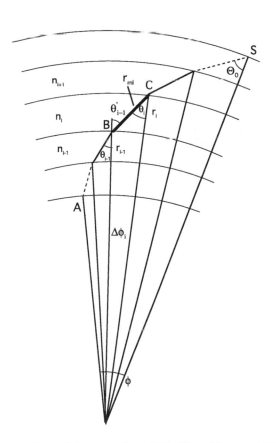

FIGURE 12A-1   Signal propagation path between the satellite (S) and the target (A). While the calculation in this paper was done over 3,000 10m-thick layers for one atmosphere, the atmosphere is approximated by five layers for ease of illustration.

The propagation length between two layers (r_{mi}) is given by

$$r_{mi} = r_{i-1} \frac{\sin \Delta\phi_i}{\sin \theta_i} = \frac{1}{\cos \theta_{i-1}} \frac{n_{i-1}}{n_i} \left(1 - \frac{r_{i-1}}{r_i} \right) \frac{r_{i-1}}{\sin \theta_i}. \tag{12A-1.4}$$

The corresponding propagation time is

$$t_{mi} = \frac{1}{\cos \theta_{i-1}} \frac{n_{i-1}}{n_i} \left(1 - \frac{r_{i-1}}{r_i} \right) \frac{r_{i-1}}{\sin \theta_i} \frac{n_i}{c} \cong \frac{2 r_{i-1} n_{i-1}}{c \sin \theta_{i-1}} \left(1 - \frac{r_{i-1}}{r_i} \right). \tag{12A-1.5}$$

Finally, we have the time difference between two observations:

$$\phi_3 = -4\pi \frac{1}{\lambda_0 \cos \Theta_0} \sum_i \left(n_{m,i} - n_{s,i} \right) \Delta r_i. \tag{12A-1.6}$$

REFERENCES

Bamler, R., Davidson, G. W., and Adam, N., 1996, "On the Nature of Noise in 2-D Phase Unwrapping, in: Microwave Sensing and Synthetic Aperture Radar," SPIE vol. 2958, pp. 216–225.

Bamler, R., Geudtner, D., Schattler, B., Vachon, P. W., Steinbrecher, U., Holzner, J., Mittermayer, J., Breit, H., and Moreira, A., 1999, "Radarsat ScanSAR Interferometry," Proc. IGARSS 1999, Hamburg, Germany, June 28–July 2, 1999, http://dx.doi.org/10.1109/IGARSS.1999.772005

Bamler, R. and Just, D., 1993, "Phase Statistics and Decorrelation in SAR Interferograms," Proc. IGARSS 1993, Tokyo, Japan, August 18–21, 1993, pp. 980–984.

Bernardino, P., Fornaro, G., Lanari, R., and Sansosti, E., 2002, "A New Algorithm for Surface Deformation Monitoring Based on Small Baseline Differential Interferograms," IEEE T. Geosci. Remote, Vol. 40, No. 11, pp. 2375–2383.

Cumming I. G. and Wong, F. G., 2005, Digital Processing of Synthetic Aperture Radar Data, Artech House, Norwood, MA, pp. 369–423.

Delacourt, C., P. Briole and J. Achache (1998): Tropospheric correction of SAR interferograms with strong topography: Application to Etna, Geophys. Res. Lett., 25, 2849–2852.

Ferretti, A., Prati, C., and Rocca, F., 2001, "Permanent Scatterers in SAR Interferometry," IEEE T. Geosci. Remote, Vol. 39, no. 1, pp. 8–20.

Gatelli F., Monti Guamieri, A., Parizzi, F., Pasquali, P., Prati, C., and Rocca, F., 1994, "The Wavenumber Shift in SAR Interferometry," IEEE T. Geosci. Remote, Vol. 32, No. 4, pp. 855–865.

Ghiglia, D. C. and Pritt, M. D., 1998, Two-Dimensional Phase Unwrapping: Theory, Algorithms, and Software, Wiley, Hoboken, NJ.

Goldstein, R. M., Zebker, H. A., and Werner, C. L., 1988, "Satellite Radar Interferometry: Two-Dimensional Phase Unwrapping," Radio Sci., Vol. 23, No. 4, pp. 713–720.

Gray, A. L. and Farris-Manning, P. J., 1993, "Repeat Pass Interferometry with Airborne Synthetic Aperture Radar," IEEE T. Geosci. Remote, Vol. 31, No. 1, pp. 180–191.

Hunt, B. R., 1979, "Matrix Formation of the Reconstruction of Phase Values from Phase Difference, J. Opt. Soc. Am., Vol. 69, pp. 393–399.

JMA 1997, Issued the Japan Meteorological Agency (1997): High Rise meteorological annual report, CD-ROM.

Kerr, D. E., Ed., 1951, Propagation of Short Radio Waves, McGraw-Hill, New York.

Li, F. and Goldstein, R. M., 1990, "Studies of Multibaseline Spaceborne Interferometric Synthetic Aperture Radars," IEEE T. Geosci. Remote, Vol. 28, No. 1, pp. 88–97.

Madsen, S. N., Martin, J. A., and Zebker, H. A., 1995, "Analysis and Evaluation of the NASA/JPL TOPSAR Across-Track Interferometric SAR System," IEEE T. Geosci. Remote, Vol. 33, No. 2, pp. 383–391.

Massonnet, D. and Rabaute, T., 1993, "Radar Interferometry: Limits and Potential," IEEE T. Geosci. Remote, Vol. 31, No. 2, pp. 455–464.

Massonet, D., Rossi, M., Carmona, C., Adragna, F., Peltzer, G., Feigl, K., and Rabaute, T., 1993, "The Displacement Field of the Landers Earthquake Mapped by Radar Interferometry," Nature, Vol. 364, No. 6433, pp. 138–142.

Moccia, A. and Vetrella, S., 1992, "A Tethered Interferometric Synthetic Aperture Radar (SAR) for a Topographic Mission," IEEE T. Geosci. Remote, Vol. 30, No. 1, pp. 103–109.

Monti-Guarnieri, A., Parizzi, F., Pasquali, P., Prati, C., and Rocca, F., 1993, "SAR Interferometry Experiments with ERS-1," *Proc. IGARSS 1993*, Tokyo, Japan, August 18–21, 1993, pp. 991–993.

Mushiake, Y., 1961, *Antenna and Wave Propagation*, Corona Co, Tokyo, pp. 138–141.

Prati, C., Rocca, F., and Guarnieri, A. M., 1993, "SAR Interferometry Experiments with ERS-1," *Proc. First ERS-1 Symposium–Space at the Service of Our Environment*, Cannes, France, November 4–6, 1992, pp. 211–218.

Prati, C., Rocca, F., Guarnieri, A. M., and Damonti, E., 1990, "Seismic Migration for SAR Focusing: Interferometrical Applications," *IEEE T. Geosci. Remote*, Vol. 28, No. 4, pp. 627–640.

Press, W. H., Flannery, B. P., Teukolsky, S. A., and Vetterling, W. T., 1989, *Numerical Recipes in C, the Art of Scientific Computing*, Cambridge University Press, Cambridge and New York, pp. 487–490; 517–547.

Rodriguez, E. and Martin, J. M., 1992, "Theory and Design of Interferometric Synthetic Aperture Radars," *Proc. IEEE*, Vol. 139, No. 2, pp. 147–159.

Shimada, M., 1999, "Correction of the Satellite's State Vector and the Atmospheric Excess Path Delay in the SAR Interferometry–An Application to Surface Deformation Detection (in Japanese)," *J. Geodet. Soc. Japan*, Vol. 45, No. 4, pp. 327–346.

Shimada, M., 2000, "Correction of the Satellite's State Vector and the Atmospheric Excess Path Delay in SAR Interferometry–Application to Surface Deformation Detection," *Proc. IGARSS 2000*, Honolulu, HI, July 24–28, 2000.

Shimada, M., 2008, "PALSAR SCANSAR SCANSAR Interferometry," *Proc. IGARSS 2008*, Boston, MA, July 7–11, 2008.

Shimada, M., 2009, "A New Method for Correcting SCANSAR Scalloping Using Forest and Inter SCAN Banding Employing Dynamic Filtering," *IEEE T. Geosci. Remote*, Vol. 47, No. 12, pp. 3933–3942.

Shimada, M., 2012, "PALSAR ScanSAR Interferometry Using the Modified Full Aperture Processing," *Proc. IGARSS 2012*, Munich, Germany, July 22–27, 2012.

Shimada, M., Tadono, T., and Rosenqvist, A., 2010, "Advanced Land Observing Satellite (ALOS) and Monitoring Global Environmental Change," *Proc. IEEE*, Vol. 98, No. 5, pp. 780–799.

Shimada, M., Watanabe, M., and Motohka, T., 2013, "Subsidence Estimation of the Peatland Forest in the Central Kalimantan Using the PALSAR Time Series Differential Interferometry," *Proc. IGARSS 2013*, Melbourne, Australia, July 21–25, 2013.

Sourifu 1997, Prime Minister's Office Headquarters for Earthquake Research Promotion Earthquake Research Committee ed. (1997): Japan's seismic activity - by region of the features seen from the affected areas, 193–195.

Spagnolini, U., 1995, "2-D Phase Unwrapping and Instantaneous Frequency Estimation," *IEEE T. Geosci. Remote*, Vol. 33, No. 3, pp. 579–589.

Stiefvater Consultants, 2006, "Along Track SAR Techniques for Ground Moving Target Detection," AFRL-SN-RS-TR-2005-410, Final Technical Report, Stiefvater Consultants, Marcy, NY.

Takajo, H. and Takahashi, T., 1988, "Least-Squares Phase Estimation from the Phase Difference," *J. Opt. Soc. Am. A*, Vol. 5, No. 3, pp. 416–425.

Ulaby F., Moore, R., and Fung, A., 1982, *Microwave Remote Sensing, Active and Passive, Vol. 1, Radar Remote Sensing and Surface Scattering and Emission Theory*, Addison-Wesley, Boston, MA, pp. 257–266.

Zebker, H. A. and Goldstein, R. M., 1986, "Topographic Mapping from Interferometric Synthetic Aperture Radar Observation," *JGR*, Vol. 91, No. B5, pp. 4993–4999.

Zebker, H., Madsen, S. N., Martin, J., Wheeler, K. B., Miller, T., Lou, Y., Alberti, G., Vetrella, S., and Cucci, A., 1992, "The TOPSAR Interferometric Radar Topographic Mapping Instrument," *IEEE T. Geosci. Remote*, Vol. 30, No. 5, pp. 933–940.

Zebker, H. A. and Villasenor, J., 1992, "Decorrelation in Interferometric Radar Echoes," *IEEE T. Geosci. Remote*, Vol. 30, No. 5, pp. 950–959.

Zebker, H. A., Werner, C. L., Rosen, P. A., and Hensley, S., 1994, "Accuracy of Topographic Maps Derived from ERS-1 Interferometric Radar," *IEEE T. Geosci. Remote*, Vol. 32, No. 4, pp. 823–836.

13 Irregularities (RFI and Ionosphere)

13.1 INTRODUCTION

Strange noise patterns or artificial types of noise frequently appear that are definitively apart from SAR system noise in the L-band and/or even the C-band SAR images and also in the InSAR phase. The types of patterns, shapes, sizes, areal densities, and frequency of appearance vary in time and space. However, the latter gradually increased since the 1990s, as the SARs were better prepared experimentally, and after 2000s, when the various SARs were operational in different frequencies. The main reason these strange noise patterns or artificial types of noise is that the SAR signal is interfered with by the ionosphere/troposphere and outer signals (Figure 13-1). There are four representative phenomena. The first is the radio frequency interference (RFI), an interaction with outside radio signals having different purposes that increase in level and frequency bandwidth because of the expansion in electric usability. The second is made up of imaging artifacts resulting from the ionosphere's non-uniformity, which causes streaks in SAR images and phase variation in interferometry. The third phenomenon is polarization rotation caused by the Faraday rotation. Last is the tropospheric phase delay, which often appears in InSAR analysis around water vapor mountain interaction areas as a ghost-like phase variation. Typical examples of all four are shown in Figure 13-2. We on focus on the first three types of interference in this chapter because tropospheric interference has been already discussed in Chapter 12.

13.2 RADIO FREQUENCY INTERFERENCE (RFI) AND NOTCH FILTERING

Any SAR must be assessed and approved for its frequency and bandwidth allocation by the International Frequency Registration Board. Since microwave remote sensing from space has a lower priority than microwaves utilized for infrastructural monitoring on the ground—such as aviation radio-controlling, ground communication, and GPS—SARs are easily affected by stronger signals. Recently, RFI, which is defined as the status of the preferred and un-preferred signal mixture, has become one of the major noise sources. SAR imaging is a correlation process between the signals from the target and the reference. Thus, the imaging process gain for the target signal is supposed to be much larger than that for the uncorrelated signal (i.e., noises from the outside). When the outer signal source has an extremely high power level, an SAR image will be interrupted easily and the outer signal source will become a problem (white noise).

13.2.1 INEQUALITY

In this subsection, we consider a condition in which RFI becomes a problem in an SAR image. We will start with a consideration of the raw data.

SAR and RFI are not correlative at all because their signal sources are totally different. Thus, the appearance of RFI in an SAR image is just based on the inequality of the raw signal power.

Let p_s be the raw signal power, p_{EX} the external noise, and p_{IN} the internal noise; the SAR image power (intensity) is expressed by

$$p = G^2 \cdot p_s + G \cdot (p_{EX} + p_{IN}), \tag{13.1}$$

where p is the SAR intensity and G the imaging correlation gain.

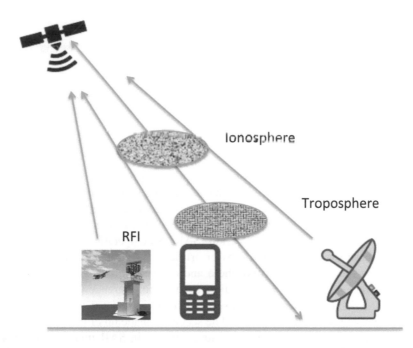

FIGURE 13-1   RFI and ionospheric irregularity in an SAR image.

When the external noise level increases more than the SAR signal, such the inequality shown in Equation (13.2) is valid, and the image would be covered by white noise:

$$G \cdot p_s < \left(p_{EX} + p_{IN} \right). \tag{13.2}$$

Here, G depends on the SAR performance—resolutions such as $G = 1,000,000$ at the range and azimuth compression has 1,000 each in the JERS-1 SAR case, and $G = 100,000,000$ at the range and

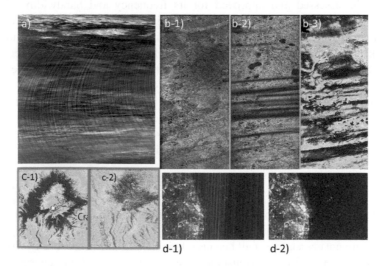

FIGURE 13-2   Typical examples of irregularities: (a) streaks observed by PALSAR over the Amazon; (b) PALSAR DinSAR observed for Siberia: (b-1) amplitude image, (b-2) coherence containing the azimuth streaks, and (b-3) interferometric phase; (c) atmospheric disturbance over the Mt. Ontake summit observed by ALOS-2/PALSAR-2: (c-1) DinSAR for the area containing the phase disturbance and (c-2) DinSAR after the atmospheric phase delay; and (d) RFI contamination at Japan's Akashi Channel by JERS-1 SAR: (d-1) before the correction and (d-2) after the RFI removal.

azimuth compression has 10,000 each in the ALOS-2/PALSAR 2 case. At that point, the external noise should be at quite a high level.

13.2.2 Power Spectrum and RFI Suppress Algorithm

If we consider the behavior of the raw data in the frequency domain, the power spectrum of the received signal is given by

$$\overline{f}(\omega) = F(f(t)),$$ (13.3)

where $F(\cdot)$ is the Fourier transform, ω is the angular frequency, and t is the time.

Figure 13-3 shows examples for three satellites: JERS-1, PALSAR, and PALSAR-2, where only the HH polarization is shown. In general, the received signal (frequency spectrum) looks similar to the transmission signal when RFI is small (such as in Brazil). The received signal is affected more as RFI becomes larger. These examples show the busy and calm RFI conditions for each generation of satellite. From the 1990s to the 2010s, the RFI has increased more severely, and the spectrum tends to be disturbed. The RFI spectrum widens from a spike type (narrow band width) to wide band-type noises. Notch filtering, which replaces the affected frequency bins by zero, can be one solution for suppressing the RFI noise (white line type noise) and is implemented as a part of the range compression:

$$F^{-1}\left\{F(S_r) \cdot A(\omega) \cdot F(f_r^*)\right\},$$ (13.4)

where $A(\omega)$ is an adaptive filter.

It should be noted, however, that notch filtering is only effective when the RFI bandwidth is narrower than 3 to 5 MHz; a wider band RFI is hard to correct.

The schematic process flow is shown in Figure 13-4, where the filter consists of the following four steps:

1. Measure the frequency spectrum of the transmission signal (from a replica).
2. Measure the spectrum of the received signal every 512 pulses.
3. Create the notch filter from Equation (13.5) if condition (4) is valid:

$$A(\omega) = \begin{cases} 0 & \Delta \geq 2dB \\ 1 & else \end{cases}$$ (13.5)

$$\Delta = |F(S_r) - F(f_r)|$$

4. Measure the total bandwidth RFI affected:

$$r_p = \frac{B_{N,p}}{B_w}$$ (13.6)

$$B_{N,p} = \Delta B \sum_{i \in I} I_i.$$ (13.7)

Here, $A(\omega)$ is the notch filter in the frequency domain; Δ is the difference between the ideal and measured spectra; $B_{N,P}$ is the total sum of the RFI contaminated and zero-filled bands; and r_p is the ratio of $B_{N,P}$ and total bandwidth.

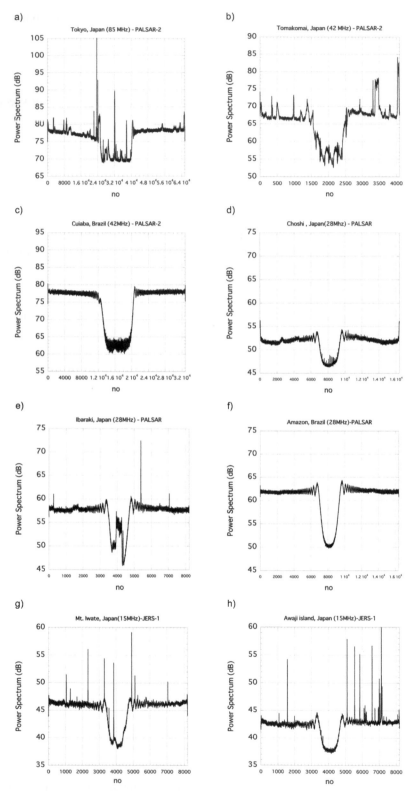

FIGURE 13-3 Frequency spectra of the SAR data, (a)–(c) PALSAR-2, (d)–(f) ALOS PALSAR, and (g)-(i) JERS-1 SAR.

i)

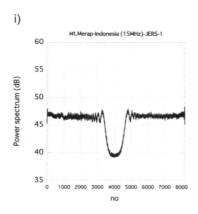

FIGURE 13-3 *(Continued)*

When $A(\omega)$ is largely dominated by zero values regardless of the distribution in a band, the resolution becomes seriously worse. Monitoring r_p, selecting whether the notch filtering should be applied or not will be interactively judged for the further process, we could generate an RFI-suppressed SAR image if the resolution were to worsen (Figure 13-5 and 13-6).

Recently, RFI status has become worse due to the wide use of mobile phones and television communications. The aforementioned parameter r_p indicates how greatly the SAR signal is contaminated by the external signals. At SAR operation processing, all the data are routinely browse-processed, and r_p is routinely monitored. Using r_p, we can find out the global distribution of the RFI contamination. Figure 13-7 shows the global distribution of the RFI where the percentile of

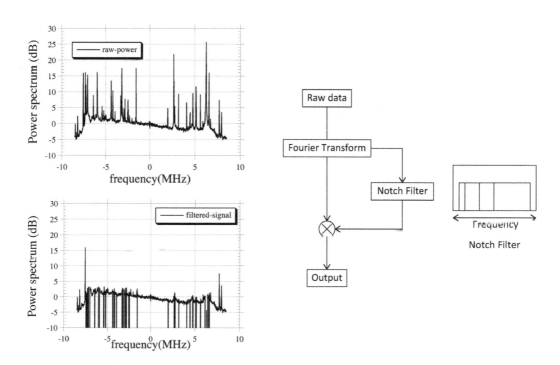

FIGURE 13-4 Thematic process flow of notching the RFI signal contaminated in the SAR image.

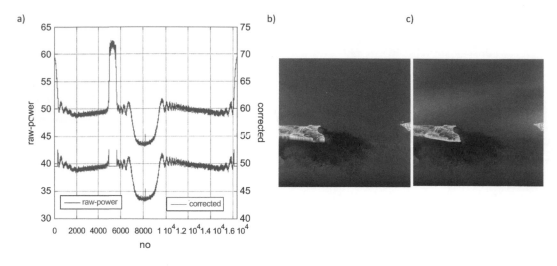

FIGURE 13-5   Examples of the correction process: (a) two spectra (before and after), (b) after notch correction, and (c) before notch correction.

the zero padded in a band is color-coded. The JERS-1 SAR of 15 MHz bandwidth represents RFI in the 1990s; the ALOS/PALSAR bandwidth of 14 MHz represents RFI in 2010; and the ALOS-2/PALSAR-2 bandwidth of 28 MHz represents a global map of RFI in 2015. It can be seen that the main RFI-contaminated areas in the 1990s (Japan, Korea, and the Yellow Sea), were widely expanded to include more global regions (the Persian Gulf, United States, and Central Europe). L-band SAR primarily is interrupted by a variety of outer signals and the bandwidths are expanding (Shimada et al. 2010; Shimada et al. 2014).

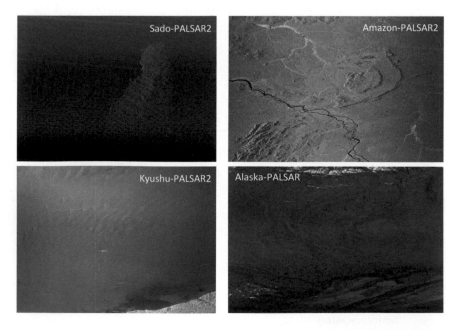

FIGURE 13-6   RFI recovered SAR images, each of which corresponds to the frequency spectra data shown in Figure 13-3.

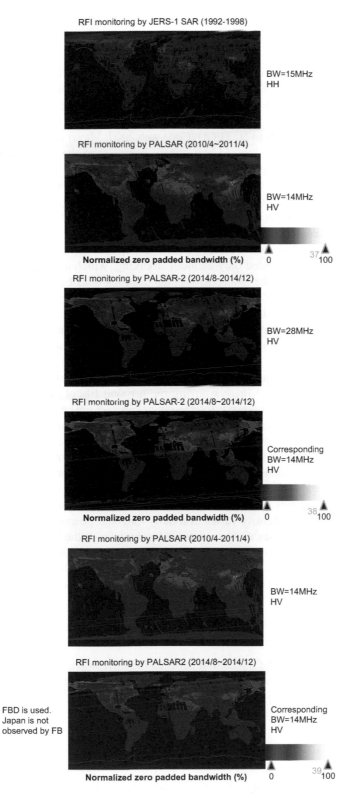

FIGURE 13-7   Global distribution of the RFI contamination that appeared in the SAR images of JERS-1 SAR (1992–1998), PALSAR (2010), and PALSAR-2(2014–2015).

13.3 IONOSPHERIC IRREGULARITY

13.3.1 INTRODUCTION

The ionosphere (called the F-layer), which is located at 300 km above sea level, changes its plasma density from 5.1×10^{11}/m^3 to 8.9×10^{11}/m^3 from day to night in times of low solar activity; but during high solar activity, this value advances one order higher than the aforementioned range. These variations occur daily and in eleven-year cycles. They can be represented by total electron content unit (TECU), ranging from 5 TECU to 60 TECU (Code web 2007; Meyer et al. 2006; Kelley 1989).

After dark, neutralization of the ionized atmosphere begins over the land's surface, with its nonuniform diffusion toward the higher altitudes sometimes observable in the Amazon region of South America (Knight et al. 2000). GPS measurements of ionospheric properties often suffer from scintillation phenomenon, the beating together of two path signals propagating in the same neighborhood but with different ionospheric properties. More serious GPS is locked out (Matsunaga et al. 2004; Aarons et al. 1996; Basu et al. 1978; Beach and Kintner 1999).

With regard to the relation between L-band SAR and ionospheric properties, research using an L-band SAR onboard JERS-1 revealed only azimuth streaking, which appeared as a phase shift in the azimuthal direction due to the ionospheric heterogeneous distribution (Gray et al. 2000). The interaction between the L-band SAR and the other ionospheric properties when the emitted radio wave passes the ionospheric disturbance region are topics to discuss at another time (Otsuka et al. 2002). In this paper, we will provide evidence that differences in the path of the radar signal through the turbulent ionosphere may have produced the stripes that were observed by the PALSAR onboard ALOS.

Although the ionosphere was thought to distort SAR images both at amplitude and phase, it has not provided purely evidential images during the mission life of JERS-1 SAR, although two phenomena might have been observed: interferometric phase variation in the Amazon and amplitude variation in Russia. In early 2006, when ALOS/PALSAR was launched and thrown into the polar orbit, primarily night observations might have been made; since then, the situation has changed drastically. There were many night images that captured irregular patches in SAR images—all of which were estimated as having resulted from ionospheric disturbances caused by scintillation, a plasma bubble, and traveling ionospheric disturbances (TIDs) as introduced in Figure 13-2 (http://www.geocities.jp/hiroyuki0620785/intercomp/wireless/ionosphere.htm). These occur because the local electron density varies as a daily diagonal phenomenon that increases in the morning until noon and decreases after dawn to midnight through the Rayleigh Taylor instable variation of the TEC. All of the ionospheric disturbances are caused by these instable decays of the TEC that occur in night time. In this section, we will describe the basic interactions between SAR imaging and ionospheric irregularities. Also, we will introduce some phenomena related to these irregularities.

13.3.2 REFRACTIVE INDEX

This irregularity only depends on the interaction between the microwave propagation in the media with a refractive index $n = c / c_0$, which is defined as a ratio of light speed (c) to light speed in free space (c_0). The ionosphere has

$$n = \sqrt{1 - \frac{Ne^2}{\varepsilon_0 4\pi^2 f^2 m}}, \tag{13.8}$$

where N is the total number density of the electron (number/m^3); e is the electric charge (1.602 e^{-19} Coulomb); ε_0 is the dielectric constant of the space (8.854 e^{-12}Fm^{-1}); m is the mass of the electron (9.109 e^{-31}kg), f is the carrier frequency, and c_0 is 299,792,458 m^{-1}s^{-1} (Kelley 1989).

13.3.3 Ionospheric Path Delay

Ionospheric phase delay occurs as a phase difference in the propagation length with and without the ionosphere during a given propagation time. This can be derived by differentiating the total phase experienced through the round-trip time for no ionosphere and subtracting that from the ionosphere:

$$\phi_{IONO} = \frac{4\pi f}{c_0}(r_0 - r_{iono}) = \frac{4\pi f}{c_0}\int_0^{t/2}\frac{c_0}{n}dt' = \frac{4\pi f}{c_0}\int_0^{t/2}c_0\frac{Ne^2}{2\varepsilon_0\omega^2 m}dt'$$

$$= \frac{4\pi}{c_0 f}\frac{e^2}{\varepsilon_0 8\pi^2 m}\int_0^{t/2}c_0 N\,dt' = \frac{4\pi k}{c_0 f}\cdot TEC \tag{13.9}$$

$$k = \frac{1}{2}\frac{e^2}{\varepsilon_0(2\pi)^2 m} = 40.28 \tag{13.10}$$

where t is the round-trip time, n the refractive index depending on the TEC density, r_0 is the distance between the SAR and the target at no ionosphere condition, and r_{iono} is that for a non-zero ionosphere condition. Equation (13.9) is the TEC line-of-sight through which the radio-wave propagates. Another description of the phase delay is that the ionospheric path delay increases the group delay of the wave, but the dispersive ionospheric propagation advances the wave phase at the same magnitude as the group delay.

13.3.4 Range Shift Due to TEC Variation

Another phenomenon is the image shift in range direction. The SAR–target distance is expressed by

$$r = \int_0^{t/2} c_0 \cdot n^{-1} dt'. \tag{13.11}$$

The differentiation becomes

$$\Delta r = -\int_0^{t/2} c_0 \cdot \Delta n \cdot n^{-2}\,dt'$$

$$\cong \frac{1}{2}\frac{e^2}{\varepsilon_0(2\pi)^2 f^2 m}\int_0^{t/2}c_0\Delta n\,dt' \tag{13.12}$$

$$= \frac{k}{f^2}\Delta TEC.$$

Figure 13-8 shows the dislocation of the SAR image measured by CR deployed in the Amazon; each shows components in the East-West (circle) and the North-South (square), and total length (plus). E-W distribution exceeds N-S distribution. All the data were acquired in the night. The negative distribution of the E-W seems to be less TEC distribution at the time of the measurement. Three case studies: 500,000 TEC for 12.4 m in positive value (10^9 * 500,000 * 40.28/1.27e9^2 = 12.4 m); 100,000 TEC for 2.48 m; and −500,000 TEC for −12.4 m, could explain that why the CR deviate in range direction on the order 10 m is that the TEC is reduced in range. Here, PALSAR geometry is calibrated for the global CRs using the ascending/descending paths as no shift in range. Thus, the previous explanation is the qualitative one. One big shift at the date 850 depends on the large TEC that remains even at the night and causes the azimuth and range shift.

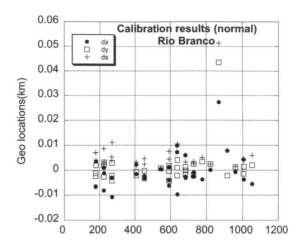

FIGURE 13-8 Location shift measured by the CR deployed in the Amazon and ascending ALOS PALSAR.

The range shift decreases the co-registration accuracy of the InSAR pair and the InSAR coherence.

13.3.5 AZIMUTH SHIFT DUE TO DOPPLER FLUCTUATION

The Doppler frequency parameters described in Chapter 3 were from a situation where no ionosphere existed. The Doppler frequency of the electromagnetic wave that propagates through the ionosphere can be derived as follows: The SAR received signal is expressed by $f(-2r/c)$, where $f()$ is the transmission signal, and the r varies with the slow time, T. Then, its time derivative becomes

$$\frac{df}{dT} = j\omega f\left(-\frac{2}{c_0/n}\frac{dr}{dT} + \frac{2r}{(c_0/n)^2}n^2\frac{dn}{dT}\right). \tag{13.13}$$

Here, the first term is the Doppler frequency of the signal through a propagation media with a constant thickness; the second term is the Doppler frequency variation due to the media whose thickness varies with the satellite passing time, T. This term causes the shift of the ground target in the azimuth direction.

Using Equation (13.8), Equation (13.13) can become

$$\begin{aligned}
\frac{df}{dT} &= \frac{df}{d(-2r/c)}\frac{d(-2r/c)}{dT} \\
&= \dot{f}\cdot\left(-\frac{2\dot{r}}{c} + \frac{2r}{c^2}\frac{dc}{dT}\right) \\
&= j\omega f\cdot\left(-\frac{2\dot{r}}{nc_0} + \frac{2r}{n^2 c}\frac{dn}{dT}\right) \\
&= j\omega f\cdot\left(-\frac{2\dot{r}}{nc_0} + \frac{2r}{n^2 c}\frac{-e^2}{2\varepsilon^0\omega^2 m}\frac{dn}{dT}\right)
\end{aligned} \tag{13.14}$$

Then, the Doppler frequency change due to the ionospheric variation is given by

$$f_{de} = f_0\cdot\left(\frac{2r}{n^2 c}\frac{-e^2}{2\varepsilon^0\omega^2 m}\frac{dN}{dT}\right). \tag{13.15}$$

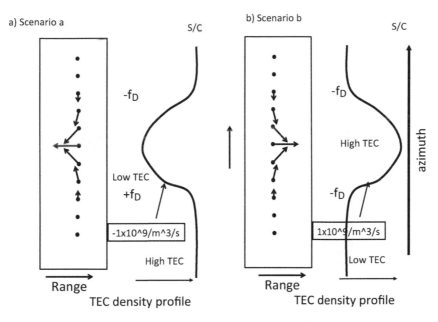

FIGURE 13-9 Two scenarios for explaining the CR shift in range and azimuth at the Amazon nighttime observations.

The azimuth location shift is given by

$$\Delta y = \frac{\Delta f_D}{-f_{DD}} v_g.$$ (13.16)

Here, f_{DD} is the Doppler chirp rate, and v_g is the ground speed. Typical f_{DD} for PALSAR is -500 Hz/s and v_g is 7.0 km/s.

In Figure 13-8, two signal patterns are visible: a negative shift in dx and a small positive in dy and one big positive shift in dx and dy at one point, where $+dx$ is defined as the east shift (slant range expansion) and $+dy$ is defined as the north shift (azimuth shift). These shifts could be explained by the scenarios on the distribution of the TEC in locations and heights (Figure 13-9). If we assume that $N = 500,000$ e9/m^3 and dN/dT is approximately 1.0 e9/m^{3}s^{-1}, such as the foothills of an ionospheric valley or mountain, we have f_{de} is approximately -0.2 Hz at the positive slope and $+0.2$Hz at the negative slope. The corresponding azimuth shift is estimated as ± 2.6 m (0.2 Hz). If we further assume $N = 500,000$ e9/m^3, then the range shift could be 12 m in range. The sign of that shift depends on the relative position of the CR, either before or after the peak of a mountain or in a valley.

One big dislocation of $+30$ m to approximately 40 m shift in azimuth and $+30$ m in range probably is due to the large deviation in TEC, such as at the scenario (b) before the peak.

13.3.6 SCINTILLATION

Figure 13-10a is a ScanSAR image acquired over the Amazon on April 12, 2006, UTC03: 24:29, centered at -9.338 degrees latitude and -71.556 degrees longitude projected on a Mercator map and sized at 350 km in range width and 800 km in azimuthal length. ALOS moves from south to north observing off the right-hand side with an incidence angle of from 18 to 42 degrees. In this figure, the gray area denotes the fully forested region and the dark line represents the river, while a white zone indicates the slightly flooded region around the river. The double bounce of the radar signal between the flooded surface and the forest makes the radar backscattering brighter. The image has

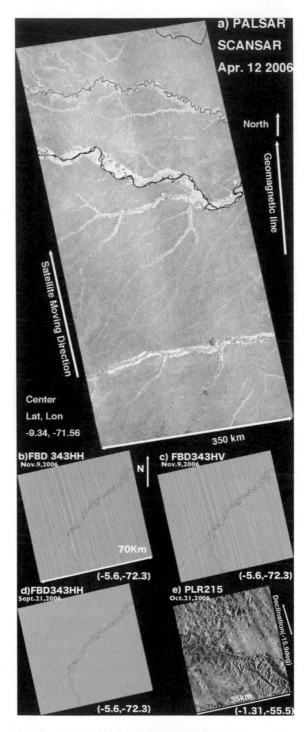

FIGURE 13-10   Samples of the streaks that appeared in the PALSAR images: (a) ScanSAR image of the central part of the Amazon obtained on April 12, 2006, during the night. The imaged area is 350 km wide and 800 km long. Stripes are visible along the satellite track. The gray area shows the forest, the dark line the river, and the white area the flooded region. (b) HH of FBD image on November 9, 2006. (c) HV of FBD on November 9, 2006. (d) HH of FBD on September 24, 2006, as a stripe-free case. (e) PLR on October 21, 2006, as a low-frequency case where HH+VV is assigned to red, HH-VV to green, 2HV to blue. Latitude and longitude of the scene center are shown in the bracket.

been enhanced to emphasize the stripes. It has been confirmed that the stripes are slightly curved, exist nearly along the satellite track with a range spacing of 3 km, and run nearly parallel to the geomagnetic lines—to the declination line.

Other examples of observed stripes are seen in Figure 13-10. Figure 13-10b and c shows fine beam dual (FBD) images from 03:43:49 UTC on November 9, 2006, for both horizontal (H) and vertical (V) polarizations in reception and for H polarization in transmission. These exhibit similar interference from along-track stripes, with stripe width measuring approximately 600 m to 900 m in the range direction. This 180-km-long image was extracted from the mother image, where the stripe extended more than 1,200 km. Figure 13-10d depicts a normal (non-scintillated) image acquired at the same location as Figure 13-10b but on a different date. The images behind the stripes are the rainforest and rivers in the Amazon. Figure 13-10e is an example of low frequency stripes observed by polarimetry. The stripes may express the variation in the Faraday rotation angle and thus represent the TEC variation across the track.

The radar has sensitivity to both the scattering property of the ground surface and the ionospheric irregularity. The streaks are generated due to refraction of the SAR signals both ways (in transmission and reception) through the eddy of the ionospheric cubes located in the *F*-layer (Figure 13-11).

A complete understanding of image irregularity (in amplitude and phase) interacting with the ionospheric variation layer (especially with the nighttime variations in the equatorial region, the so-called plasma bubble) is quite difficult. But, we can introduce the basic concept of how the streaks appear in the image.

Here, we consider a night observation scenario where the electron density gradually decreases through the generation and adaptation process of the plasma bubbles, which have less electron density than the outside. An *r*-radius, long, thin tube positioned at the height *H* is filled with the electron

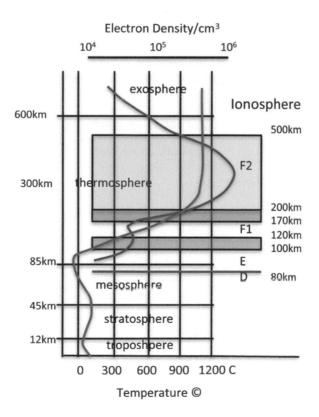

FIGURE 13-11 Electron density distribution at various heights.

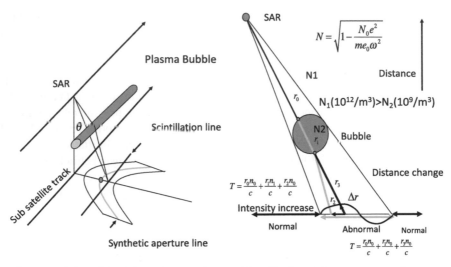

a) General View of the SAR imaging over the plasma bubble

b) Side view of the wave propagation through the plasma bubble.

FIGURE 13-12 (a) General view of the SAR imaging over the plasma bubble and (b) side view of wave propagation through the plasma bubble.

density of N_1, which is the constant or gradual distribution on r, while the outer electron density is N_2. The SAR flies along the tube at a much higher altitude so that the time series IFOV over the Earth, which appears as the black line-covered area, is produced by the cascaded pulses; the red line curve (the range compressed line) also is produced; and a focused point is created by the SAR imaging (Figure 13-12a). A signal emitted from the SAR propagates through three medias—Media-a, Media-b, and Media-c, before approaching the Earth's surface as the side view of the signal refraction, which is indicated in Figure 13-12b. We consider one signal trajectory (S-A-B-C), where the swell's conditions are valid on A and B, respectively, as:

$$n_{i-1}r_{i-1}\sin\theta_{i-1} = n_i r_i \sin\theta_i \tag{13.17}$$

$$n_{i-1}r_{i-1}\sin\theta_{i-1} = n_i r_i \sin\theta_i. \tag{13.18}$$

Then, the total time delay becomes

$$T_1 = \frac{r_0 n_0}{c} + \frac{r_1 n_1}{c} + \frac{r_2 n_0}{c}. \tag{13.19}$$

On the contrary, a normal pass with no plasma bubble ($n_1 = n_0$) has

$$T_2 = \frac{r_0 n_0}{c} + \frac{r_1 n_0}{c} + \frac{r_3 n_0}{c}. \tag{13.20}$$

Although $n_1 > n_0$, condition $T_1 = T_2$ requires

$$r_2 < r_3 \tag{13.21}$$

$$\Delta r = \frac{2(r_3 - r_2)}{\lambda}. \tag{13.22}$$

Here, Δr is the shortening distance passing through the bubble in the unit of cycle.

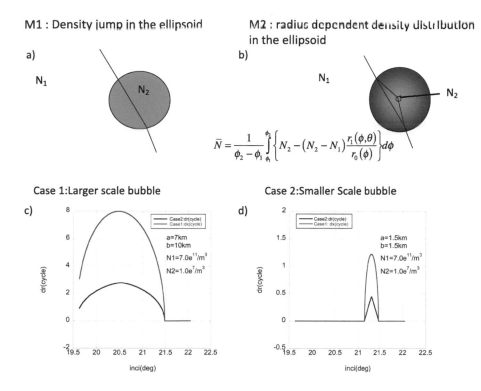

FIGURE 13-13 Simulation conditions for the smaller and larger scale bubbles and the case studies on the shortage in traveling distance. (a) Density distribution of M1. (b) Density distribution of M2. (c) Case 1 simulation for large-scale bubble. (d) Case 2 simulation for small-scale bubble.

Thus, the path length through a tube is always shorter than the normal path with no plasma bubble and the image is focused at a nearer range. This process is similar to the foreshortening effect in SAR imaging. The foreshortening depends on the TEC conditions: electron density distribution, height of tube, and number of tubes. Two models were examined on the refraction paths through two cylindrical models: the circular bubble with the constant density inside and discontinued at outer space (M1) and the radially ramp distribution connected continuously with outer space (M2). Two simulations were performed: 7-km to 10-km scale bubbles for two densities (Case 1) and 1.5-km scale bubbles (Case 2), both of which represent the typical TEC nighttime density (Figure 13-13). The blue line shows the phase delay as the shortening distance of Equation (13.22) at M1 and the red shows M2. M2 is always less effective than M1 because the large density gradient of TEC across the border gives the large variation of the intensity and phase variation as seen from M1. M1 has a larger irregular area than M2 because of the larger bubble.

A PALSAR image of September 20 2006 over the Amazon was severely interfered with by streaks, as shown in Figure 13-14, and its interferometric phase difference associated with November 5 2006 of the ionospheric calm condition showed similar phase streaks. The simulation study assuming the five (one large and four small) tubes located at the 100-km altitude could explain the cause of this phenomena, where the streaks were visible at $N2 \sim 10^9$ and $N1 \sim 10^{11}$ or more.

13.3.7 SCINTILLATION FREQUENCY

All the stripes interfered with by scintillation were manually detected and used for spatio-temporal evaluation. Figure 13-15 plots the geographical distribution of all the streaks that appeared

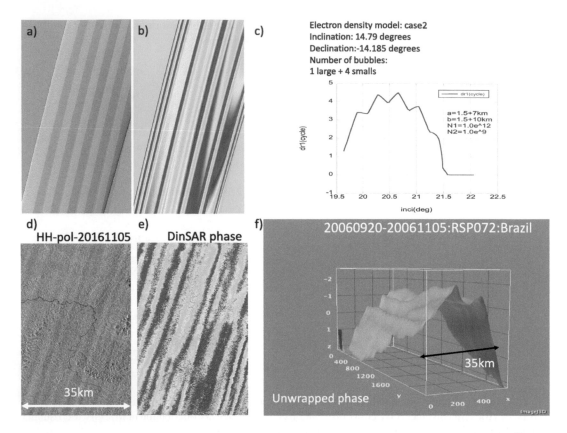

FIGURE 13-14 (a) Simulated amplitude with one large and four smaller bubbles and a TEC of 10^{12} outer space and 10^9 inside the bubbles; (b) simulated phase; (c) cross section of the InSAR phase under the five bubbles; (d) PALSAR image, November 5, 2006, of Brazil; (e) InSAR phase from September 20 to November 5, 2006; and (f) three-dimensional view of the unwrapped InSAR phase.

between April 2006 and April 2009. Figure 13-16a and Table 13-1 rearranged the data for calendar time dependency, and Figure 13-16b for the annual dependency. Figure 13-17 arranges the phenomena as season-longitude dependence and shows the very good correlation with the Defense Meteorological Satellite Program (DMSP) measurement in 0.47 μm to approximately 0.95 μm (Gentile et al. 2006).

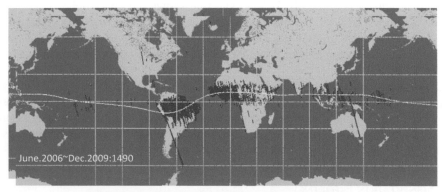

Total number of appearance : 1490 : June 2006~Dec. 2009

FIGURE 13-15 Distribution of 1,490 streaks from June 2006 through December 2009.

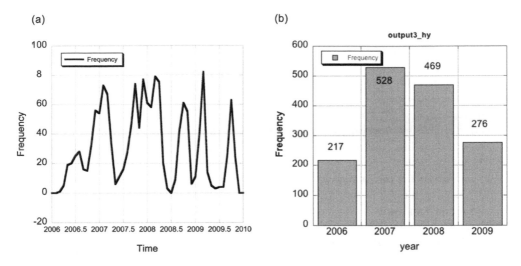

FIGURE 13-16   (a) Seasonal dependence of the streaks. (b) Year dependence of the streaks.

TABLE 13-1

Annual Frequency of PALSAR Scintillation Events

Year	Numbers	Duration
2006	217	June 2006–Dec. 2006
2007	528	Jan. 2007–Dec. 2007
2008	469	Jan. 2008–Dec. 2008
2009	276	Jan. 2009–Dec.2009
Total	1,490 (64,500:2.5%)	June 2006–Dec. 2009

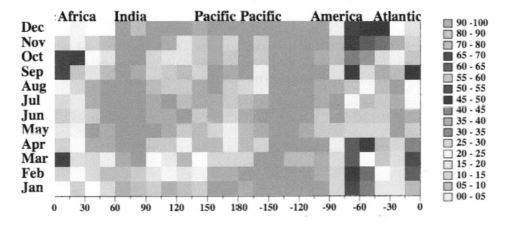

FIGURE 13-17   Season-longitude dependence of the ionospheric irregularity events measured by ALOS/PALSAR (2006–2009).

From these, we observe the following:

1. In 1,490 path observations out of total 64,500 paths, the stripes were seen during four years. The total number of events was 2.5% of all the images.
2. Most of the anomalies lie around the geomagnetic Equator in the Amazon region while other examples have been obtained in West Africa, Southeast Asia, east of New Guinea, and Hawaii. A small number of events over Africa and Southeast Asia just reflect an observation effect; the observation time over Africa and Southeast Asia was limited during the wintertime in local areas.
3. The events appear to be season-dependent as they are maximized in winter and minimized in summer, while the small peaks are visible in 2009.
4. The events decrease in time and are strongly correlated with the solar cycle of 11 years (number of sun spots).
5. There is a very good agreement between the optical sensor events measured by the DMSP and PALSAR. The DMSP (optical sensor) is verified to be sensitive to the light emission on the scintillation. This close relationship shows that PALSAR detected the scintillation events.

13.3.8 FARADAY ROTATION

When the electromagnetic wave propagates through the geomagnetic fields, its polarization plane rotates around the radar line of sight by Faraday rotation. This rotation angle is theoretically expressed by

$$\Omega = \frac{K}{f^2} \overline{B \cdot \cos \psi \cdot \sec \theta_0 \cdot TEC} \tag{13.23}$$

where $K = 2.365 \times 10^4$ in SI units, f is the transmission frequency (Hz), TEC is the total electron contents (electrons/m^2), B is the geomagnetic flux density (Tesla), ψ is the angle between the geomagnetic field vector and the radar line of sight (radians), θ_0 is the incidence angle, and the over-bar indicates averaging. One TECU corresponds to 1×10^{16} electrons/m^2.

Thus, the Faraday rotation angle is proportional to the TEC and $\cos \psi$. The inclination angle of the ALOS orbit plane is 98.16 degrees and $\cos \psi$ in ascending and descending at the test site in the Amazon region is more than 0.96. Here, the inclination of the geomagnetic line in Rio Branco (S9.76, W68.07) is −7.188 degrees (Shimada et al. 2010), and the inclination of the satellite is −8.19 degrees in ascending and $\cos \psi$ in ascending is approximately 1.0. The angular difference between the geomagnetic line and the direction of the satellite motion is around 15 degrees and $\cos \psi$ is approximately 0.96. Thus, Ω mainly depends on TEC when discussing the possible change in ascending and descending.

From the calibrated polarimetric data, we can estimate Ω by one proposed method and the reference method, both give here. When we place a condition of two cross-polarized powers for a distributed target (i.e., the Amazon Rainforest) they should be the same,

$$\left\langle \mathbf{S}_{hv} \cdot \mathbf{S}_{hv}^* \right\rangle = \left\langle \mathbf{S}_{vh} \cdot \mathbf{S}_{vh}^* \right\rangle, \tag{13.24}$$

and we have

$$\alpha \cdot \tan \Omega \left(1 + \tan^2 \Omega\right) - \beta \left(1 - \tan^4 \Omega\right) = 0, \tag{13.25}$$

where

$$\alpha = \left\langle \left(\mathbf{Z}_{hv} + \mathbf{Z}_{vh}\right) \cdot \left(\mathbf{Z}_{hh} + \mathbf{Z}_{vv}\right)^* \right\rangle + \left\langle \left(\mathbf{Z}_{hv} + \mathbf{Z}_{vh}\right)^* \cdot \left(\mathbf{Z}_{hh} + \mathbf{Z}_{vv}\right) \right\rangle$$

$$\beta = \left\langle \mathbf{Z}_{hv} \cdot \mathbf{Z}_{hv}^* - \mathbf{Z}_{vh} \cdot \mathbf{Z}_{vh}^* \right\rangle. \tag{13.26}$$

Here, $\mathbf{Z}$ is the polarimetrically calibrated SAR data.

The second iteration, Ω_1, can be solved as follows:

$$\Omega_1 = \tan^{-1}\left\{\frac{\beta}{\alpha}\left(1 - \tan^4\Omega_0\right) - \tan^3\Omega_0\right\}$$

$$\Omega_0 = \tan^{-1}\left(\frac{\beta}{\alpha}\right)$$

(13.27)

One other reference is simply from the cross correlation of the LR and RL (Bickel and Bates 1965):

$$\Omega = \frac{1}{4}Arg\left(Z_{LR} \cdot Z_{RL}^*\right)$$

(13.28)

We show these two Faraday rotations in Figure 13-18 with GPS-TEC values given from the Website of Switzerland's University of Bern.

These measurements at the test site imply (1) that GPS-TEC in ascending (ASD) orbit has 10 TECU, while in descending (DSD) orbit it has 20 TECU constantly; (2) that the Faraday rotation angles measured are almost always independent of ASD and DSD; and (3) that the two methods, Bickel and Bates (BB) (1965) and Equation (13.27) (S), show the same values of 0.25 degrees for ASD and DSD with the standard deviation of 0.099 degrees. From these measurements, we can say that geomagnetic flux density in the radar line of sight does not change significantly at the Rio Branco test site both in ASD and DSD (over 3 years from 2006 to 2008). When we select the average value of the Faraday rotation angle of 0.177 degrees, cos (0.177) approximately 0.99999, and sin (0.177) approximately 0.00309, Faraday rotation is not the issue with the polarimetric calibration. Since the solar activity declined from a peak in 2002 toward an estimated minimum near the end of 2009, we consider the test site a Faraday rotation-free target. However, increased solar activity was observed at the end of 2009 that could affect future results.

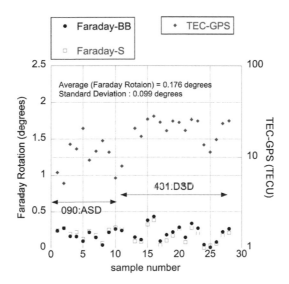

FIGURE 13-18 Faraday angles of the data points. Black circles indicate Faraday rotation angle measured using the Bickel and Bates (1965) method. The open squares show the proposed method of Equation (13.27). The filled blue squares display GPS-measured TEC values.

13.4   ESTIMATION OF THE TECU USING THE InSAR AND POLARIMETRY

The phenomenon was investigated by two different data taken by PALSAR InSAR and polarimetry. Both measurements gave the same profiles for space irregularity.

   Since all of the aforementioned phenomena are associated with the fact that the signal from the SAR is modified when propagating through a varying ionosphere, we can estimate the TEC variations and the absolute value from the SAR data. We propose the following two methods: One is to measure the Faraday rotation from Equation (13.28) and convert to the TEC using Equation, (13.23), and the other is to measure the slant range change using the SAR interferometry and convert to the TEC using Equation, (13.12), referred to as InSAR hereafter. The former method requires the calibration of four scattering components at two polarizations for both transmission and reception.

   Rewriting them as follows

$$TEC = \Omega \frac{f^2}{K \cdot B \cdot \cos \psi \cdot \sec \theta_0} \tag{13.29}$$

$$\Delta TEC = \frac{\lambda}{2} \frac{f^2}{k} \phi_{UW} \tag{13.30}$$

where ϕ_{UW} is the unwrapped InSAR phase.

   We demonstrate the aforementioned method using an Amazon night image acquired by PALSAR PLR on October 21, 2006, at 02: 21:09 UTC (Figure 13-10e). The area is located 500 km east of Manaus, and the image has 35 km by 35 km coverage centered at −1.31 degrees latitude and −55.52 degrees longitude. The color-coded image means that the flat green area is the forest and the red area is the water. Dark stripes are just visible in the 330 degrees azimuthal direction while the declination of the geomagnetic line is 345 degrees.

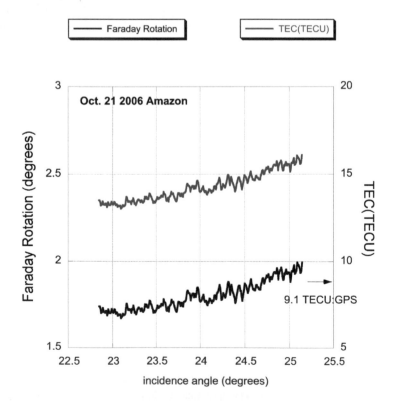

FIGURE 13-19   Faraday rotation angle measured by PALSAR polarimetry in red and by TECU in blue.

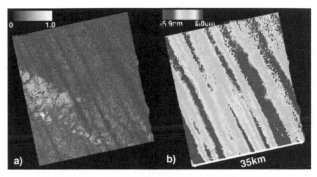

FIGURE 13-20 InSAR analysis of the Amazon striping area is made using two images from October 21, 2006, and September 5, 2006, separated by 46 days in time and 20 m in space. (a) The left side figure represents the coherence and (b) the right side figure shows the phase difference due to the ionospheric disturbance. The phase values, corrected for orbit and topography, reveal a spatial distortion that is possibly due to ionospheric nonuniformity. (c) Spatial variation in the delta TEC for November 9, and September 24, 2006. TEC varies both in range and azimuth direction, and its maximum change within 100 km is 3 TECU.

Figure 13-19 indicates the incidence angle dependence of the averaged Faraday angle at around 2 degrees. Equation (13.27) is a small-medium value compared with previous research (Wright et al. 2003). The TEC variation was around 15 TECU. A TEC value from the Global Ionosphere Map produced by CODE was 9.1 TECU (Code web).

Applying InSAR to images from October 21, 2006, and September 5, 2006, yields maps of the coherence, comprising their electromagnetic similarity, and the phase, their change in distance, as seen in Figure 13-20a and b, respectively. Both images exhibit stripes in the 330-degree azimuthal direction, the same as in the amplitude image (Figure 13-10e). Figure 13-20a contains a mixture of a white area of high co-registration and a dark area of low co-registration. This means that spatial TEC variations cause nonuniform stretching of the satellite–ground distance in an image. Figure 13-20b depicts the phase difference between two images, where a color cycle represents half of the wavelength. There are two color patterns across the track, one changing in green-red-purple, and the other in green-red-green. The former is a simple increase or decrease of phase and the latter is the phase variation. The differential TEC map of Figure 13-20c was made using Equation (13.9).

We derived a differential TEC map, seen in Figure 13-20c, of a part of the Amazon that was affected by major stripes on November 9, 2006 (Figure 13-10b), with reference to the stripe-free data of September 24, 2006 (Figure 13-10d). This indicates that the TEC increases as we move to the north with 3 TECU and varies across the track. A browse image of Figure 13-20c says that the stripes gradually increase from the south, reaching the maximum in the middle of the observation path near the Equator. This TEC pattern also depicts a gradual increase in the TEC, coinciding with the browser image. InSAR provides the structure of the TEC perturbation.

REFERENCES

Aarons, J., Mendillo, M., Yantosca, R., and Kudeki, E., 1996, "GPS Phase Fluctuations in the Equatorial Region During the MISETA 1994 Campaign," *J. Geophys. Res.*, Vol. 101, No.A12, pp. 26851–26862.

Basu, S., Basu, S., Aarons, J., McClure, J. P., and Cousins, M. D., 1978, "On the Coexistence of Kilometer- and Meter-Scale Irregularities in the Nighttime Equatorial *F* Region," *J. Geophys. Res.*, Vol. 83, No. A9, pp. 4219–4226.

Beach, T. L. and Kintner, P. M., 1999, "Simultaneous Global Positioning System Observations of Equatorial Scintillations and Total Electron Content Fluctuations," *J. Geophys. Res.*, Vol. 104, No. A10, pp. 22553–22565.

Bickel, S. H. and Bates, R. H. T., 1965, "Effects of Magneto-Ionic Propagation on the Polarization Scattering Matrix," *Proc. IEEE*, Vol. 53, No. 8, pp. 1089–1091.

Code web: http://aiuws.unibe.ch/ionosphere/

Gentile, L. C., Burke, W. J., and Rich, F. J., 2006, "A Global Climatology for Equatorial Plasma Bubbles in the Topside Ionosphere," *Ann. Geophys.*, Vol. 24, No. 1, pp. 163–172.

Gray, L. A., Mattar, K., and Sofko, G., 2000, "Influence of Ionospheric Electron Density Fluctuations on Satellite Radar Interferometry," *Geophys. Res. Lett.*, Vol. 27, No. 10, pp. 1451–1454.

Kelley, M. C., 1989, *The Earth's Ionosphere, Plasma Physics and Electrodynamics*, Academic, San Diego, CA.

Knight, K., Cervera, M., and Finn, A., 2000, "A Comparison of Measured GPS Performance with Model Based Prediction in an Equatorial Scintillation Environment," *Proc. 56th Annual Meeting of the U.S. Institute of Navigation*, San Diego, CA, June 26–28, 2000, pp. 588–601.

Matsunaga, K., Hoshino, K., and Igarashi, K., 2004, "Observation and Analysis of Ionospheric Scintillation on GPS Signals in Japan," *ENRI Papers*, Issue No. 111.

Meyer, F., Bamler, R., Jakowski, N., and Fritz, T., 2006, "The Potential of Low-Frequency SAR Systems for Mapping Ionospheric TEC Distribution," *IEEE T. Geosci. Remote S.*, Vol. 3, No. 4, pp. 560–565.

Otsuka, Y., Shiokawa, K., Ogawa, T., and Wilkinson, P., 2002, "Geomagnetic Conjugate Observations of Equatorial Airglow Depletions," *Geophys. Res. Lett.*, Vol. 29, No. 15, pp. 43-1–43-4, http://dx.doi.org/10.1029/2002GL015347

Shimada, M., Tadono, T., and Rosenqvist, A., 2010, "Advanced Land Observing Satellite (ALOS) and Monitoring Global Environmental Change," *Proc. IEEE*, Vol. 98, No. 5, pp. 780–799.

Shimada, M., Watanabe, M., Motooka, T., and Ohki, M., 2014, "Calibration and Validation of Pi-SAR-L2 and Cross-Calibration with ALOS-2/PALSAR-2," *Proc. GARSS-2014*, Quebec, Canada, July 13–17, 2014.

Wright, P. A., Quegan, S., Wheadon, N. S., and Hall, C. D., 2003, "Faraday Rotation Effects on L-Band Spaceborne SAR Data," *IEEE T. Geosci. Remote*, Vol. 41, No. 12, pp. 2735–2744.

14 Applications

14.1 INTRODUCTION

Since the JERS-1 SAR era, sensitivity studies on L-band SAR have examined and investigated how to utilize SAR data in the scientific, research, and practical fields. There is a great deal of literature on the use of L-band SAR that has introduced various trials and success/failure stories (Rosenqvist et al. 2000; Rosenqvist et al. 2007; Lucas et al. 2010; Shimada et al. 2010; Shimada 2010; Buono et al. 2014; and many more).

All of these studies sought successful results using the time-series L-band data at amplitude and phase while highlighting sensitivity from the interaction between the longer wavelength L-band SAR and targets on Earth. L-band SAR has a relatively higher signal penetration through vegetation. Because the land surface is mainly covered by vegetation, the higher signal penetration and interaction with that vegetation is advantageous. On the other hand, the ionosphere and the troposphere also interact with the signal making corrections necessary.

There are two major fields of application to which the L-band SAR contributes: deformation detection, using SAR interferometry, and forest monitoring, using time series backscatter data. While some details regarding deformation have been already introduced in Chapter 12, this chapter will summarize general applications, including examples of methodology. This chapter also will describe forest-non-forest (FNF) mapping using rigorous application of calibrated gamma-zero (which will be detailed further in Chapter 15).

Recent progress in space technology allows the maintenance of orbital corridors within several hundred meters. Sensor sensitivity prepares the time series SAR data in SLC (as well as the amplitude data) under the same observation conditions much better than before. Although a single event could be detected by radiometric change as differentiation, more data could increase the detection probability for the sake of latency. Thus, recently, the use of multiple SAR data is becoming more common than dual data use, which is simple but less accurate. Also examples in this book are limited they could be considered as representative. Table 14-1 summarizes the data application and introduces relevant studies.

14.2 PARAMETERS FOR APPLICATIONS

In use of the SAR data, there are several data combinations: the SAR image of gamma-zero, interferometry, or polarimetry; differential analysis; time series analysis with more than two pieces of data; and conversion to the geophysical parameter through the model. Although there are many parameters, representative examples for application analysis are listed in Table 14-2. Since single or dual polarization SAR data are the first available, with polarimetric observation being second or less (due to the operational constraints), most application examples appear with interferometric phase detection and gamma-zero change as the main drivers. In the following sections, we limit our examples to forests and natural disasters.

14.3 DEFORMATION

There are many deformation examples produced from the variety of SARs. Those discussed here are selected from the L-band SAR of JERS-1 to ALOS-2. From JERS-1, we have the very first monumental DinSAR image for the M 7.0 Hanshinn-Awaji Earthquake of January 17, 1995, which was obtained from the JERS-1 SAR images taken in September 5, 1992, and February 6, 1995 (Figure 12-30).

TABLE 14-1
SAR Application List

Category	Data or Numbers	Main Parameters
Landslide—surface collapse	1 or 2-Polarimetry	Incoherence decomposition
Detection of the landslide		HH-VV coherence, alpha-entropy
		HH/HV ratio
Landslide—slow movement or subsidence speed	≥ 2	DInSAR-phase and temporal analysis
Earthquake deformation and decomposition	≥ 2 at the same orientation	(InSAR-phase or and speckle tracking) and decomposition
	≥ 3 pairs at different orientations	
Volcanic eruption	≥ 2 acquisitions	InSAR-phase and coherence change
Surface uplift/subside		Time series amplitude
Lava flow channel		
Growing volcanic island		
River flooding	≥ 2 acquisitions	Amplitude change or coloring
Rice paddy estimation	≥ 2 or more acquisitions	Model-based classification
Oil spill	1 acquisitions	Supervised classification
Forest monitoring		
1) Quick deforestation (Illegal logging monitor)	≥ 2	Amplitude change
	1	Supervised classification
2) Forest classification	1	Supervised classification
3) FNF	1	
4) Biomass estimation	1 or 2	Model-based estimation
5) Wetland + mangrove		Supervised classification
6) MRV system + REDD+		
5) Fire scar		Supervised classification
Sea ice monitoring		
-Sea ice density	≥ 1	Amplitude analysis
-Moving vector	≥ 2	Speckle tracking
-Glacier movement	≥ 2	InSAR or speckle tracking
Coastal erosion	≥ 2 or all the year data	Amplitude change detection
Ocean wind speed	≥ 1 and wind direction given from the wind field.	Model-based calculation under the given wind direction
DEM (DSM)	1 or 2	Single-pass InSAR or repeat-pass InSAR corrected for tropospheric delay and ionospheric forward modeling
Ionospheric disturbance	1 or 2	Interferometry: coherence and phase
		Polarimetry: Faraday rotation
		Amplitude analysis
RFI	As many as possible	Amplitude: spatial distribution
Map with ortho-rectification	1	Amplitude
Ship detection	1	Amplitude and spatial deviation

Note: MRV = measurement, reporting, and verification; REDD = reducing emissions from deforestation and forest degradation; DSM = digital surface model; RFI = radio frequency interference; DInSAR = differential interferometric synthetic aperture radar.

These SAR images as well as other deformation analysis (Massonnet et al. 1993, and many more publications) opened a door for utilizing spaceborne SAR interferometry to monitor the land surface deformation caused by movement of the Earth's surface and interior. Even though the JERS-1 was not designed especially for interferometric Earth observation, especially with regard to maintenance of the orbital corridor, the resulting images clearly depicted the surface deformation. The quality

TABLE 14-2

Parameters to Be Used for the Applications

Parameters	Contents	Application
Gamma-zero	Values and the spatial variation	
InSAR or DinSAR or amplitude detection	Interferometric phase and/or interferometric coherence, correlation (speckle tracking)	Landslide and subsidence speed
		Surface deformation by volcanic eruption and earthquake
Amplitude change	Amplitude change	
Polarimetric analysis	Polarimetric coherence (HH-VV) and eigenvalue analysis	Discriminate the surface and volume scatterings

of the JERS-1 SAR images as well as the orbital accuracy are not as good as those of present-day SARs, but necessary correction of the data successfully depicts the deformation pattern.

14.4 FOREST

There are three major applications in forest observation: (1) Deforestation detection, (2) forest classification, and (3) biomass estimation. Other than these three, more complex applications are under the investigations, such as forest height estimation and scattering density observations. The latter two are very important in measuring forest biomass more quantitatively, and they are achieved using polarimetric SAR interferometry and tomography (Cloude and Papathanassiou 1998). We will describe the principles of the first three applications in the following subsections.

14.4.1 DEFORESTATION DETECTION

Representative forest losses are the result of deforestation and degradation. The former is represented by the large aerial clear cut and the spot-like thinning of the forest. The latter losses range from natural decay of the forest to the thinning of trees. The capability of detecting these phenomena depends on their radar response and various possibilities exist, such as changes in radar backscatter, polarimetric parameters, and interferometric parameters. Theoretically, deforestation can be explained as a change in the radar backscattering mechanism from volume scattering to surface scattering. Thus, the following parameters can be used for deforestation monitoring:

1. Polarimetric decomposition change
2. Interferometric coherence change
3. Gamma-zero change

Although all of these can be implemented only by time-series polarimetry, polarimetric observation is still experimental because of resource limitations. The second parameter utilizes interferometric synthetic aperture radar (InSAR) operation and gamma-zero change. Figure 14-1 shows the HH image (a) as an HV image, (b) as an HH-HH InSAR coherence image, (c) in an HH-HH InSAR phase, and (d) its location in Indonesia's Riau province Indonesia, which depicts a mixture of an acacia plantation, a peat forest, and a clear-cut area. This image shows the representative features of the forest; the clear cut shows much lower backscatters in HV and only slightly higher coherence in HH than in the other area. The acacia shows the darker radar backscatter in HH and HV and higher InSAR coherence area; the peat forest has stable radar backscatter in HH and HV and lower coherence; and interferometric

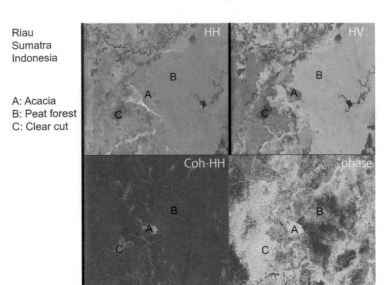

FIGURE 14-1   Comparison of the SAR image for three forest stands: (upper left) the HH image; (upper right) HV image; (lower left) HH-HH InSAR coherence; and (lower right) HH-HH InSAR phase, where the image is located in Indonesia's Riau province and is a mixture of acacia plantation, peat forest, and clear-cut area.

phase shows the height of the scattering points. From this data, it could be said that the coherence is more sensitive than the gamma-zero. However, SAR processing for the amplitude is easier than InSAR's, mainly because of the efficiency, and gamma-zero-based deforestation monitoring is being performed. Chapter 15 discusses JAXA's forest and non-forest (FNF) measurements.

Figure 14-2 shows the footrest of 15-year FNF gamma-zero variations, and classification of the data from the Amazon where the left columns are JERS-1 SAR HH data and the right columns are from ALOS/PALSAR in HV. In the left image, deforestation runs vertically and is represented as the gray value, which changes from bright to dark. The temporal process of deforestation can be described as a series of epochs that the natural forest undergoes—the clear-cutting of the trees, the mixed conditions during the brief period of time when the lumber is distributed on the ground, the burning or removal of timber from the field, and the clearing of the surface for secondary use. Depending on the time section, the radar backscatter varies as it is modified by the processes.

14.4.2 Classification

Classifiers have been developed based upon the many published examples (Longepe et al. 2011; Preesan et al. 2010; Hoekman 2007). Classification reduces the accuracy as the number of classes increases. Otherwise, FNF classification is one of the representative applications to which the annual global PALSAR mosaic can be applied.

14.4.3 Early Warning Detection

ScanSAR features wide coverage within a brief period of time. PALSAR2/ScanSAR is used operationally to find deforestation areas quickly based on two-time differentiation or multi-temporal differentiation of the slope-corrected ortho-rectified gamma zero (Watanabe et al. 2018; JICA). Japan International Cooperation Agency–Japan Aerospace Exploration Agency (JICA-JAXA) has developed a quick detection system to locate areas of deforestation. The system uses a rule for long-term observation that HH and HV are decreased by the deforestation process. The system also recognizes that HH quickly increases shortly after deforestation occurs when the timbers remain on the ground. Figure 14-3 shows one detection example from Mato Grosso in Brazil during the rainy season.

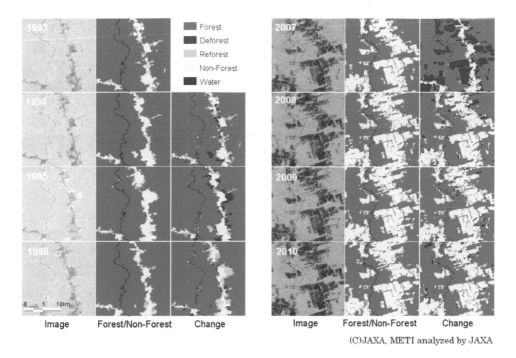

FIGURE 14-2 Time series SAR images of the deforestation area in Brazil, where the left column is from the JERS-1 SAR HH in the 1990s and the right column is from the ALOS/PALSAR HV in the 2000s.

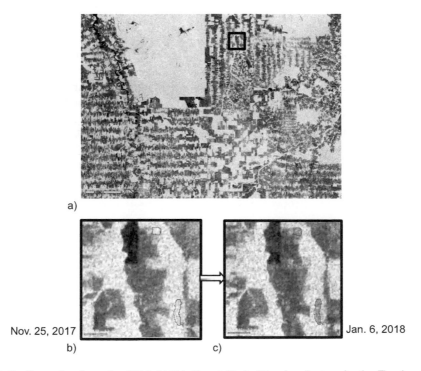

FIGURE 14-3 Examples from the JICA-JAXA Forest Early Warning System in the Tropics (JJ-FAST). (a) The PALSAR-2 ScanSAR HV image overlaid by the dark polygon representing the deforestation area detected using the following two images: (b) the "before event" image from November 25, 2017, and (c) the "after event" image from January 6, 2018. Courtesy: M. Watanabe.

14.5   LANDSLIDE MONITORING

14.5.1   INTRODUCTION AND TOTSUKAWA-MURA DISASTER

In September 2011, Japan was hit by Talas (Wikipedia 2011) and Typhoon No. 12, and severe countrywide devastation resulted in 82 causalities and 16 missing people ((http://www.fdma.go.jp/bn/ Disaster status by Typhoon no. 12_台風12 号による被害状況について (Report No. 20、20巻).pdf). The storm passed slowly over central Japan, delivering more than 1,800 mm of precipitation (http://www.jma.go.jp/jma/menu/h23t12-portal.html). Landslides occurred over a vast mountainous area forested with 20-m to 30-m trees, leaving behind bare soil (Landslide.dpri.kyoto-u.ac.jp/ Typhoon-12-RCL-Report-1.pdf). A natural dam was created by the build-up of the landmass by the landslide movement that could have potentially caused a secondary disaster.

There are two types of landslides: corruptive large-scale landslides (such as the previous example) and small-scale, slow-moving landslides. In this chapter, we will discuss the first type. Such a landslide is in a transition process from the forest area (or volume scattering area) to the rough surface area (surface scattering area) and can be discriminated by a radiometric differentiation of the co-registered polarimetric channel. This type of feature classification can only take place using the images after the event.

14.5.2   POLARIMETRIC PARAMETERS

Possible parameters for detecting a mountainous landslide from the forests are listed in Table 14-3.

14.5.2.1   Color Composite

Two polarimetric color assignments are available—an HV-basis that assigns HH-HV-VV for red-green-blue (R-G-B), and a PAULI-basis that assigns HH+VV, HH-VV, and 2HV to R-G-B, respectively. Figure 14-4 shows the Totsukawa test site in Wakayama, Japan, sized in 20 km × 20 km azimuth and range directions, with HV-basis color assignments in Figure 14-4a, a Google Earth image in Figure 14-4f, as well as an aerial photo of the Nagatono landslide area (aeroasahi.co.jp) in e). Although HH and VV are about 6 dB higher than HV, the image is adjusted because all band intensities are similar: HH~VV~HV. Thus, the image in Figure 14-4a looks mostly green at the HV exceeding area of forest and purple at the limited areas. The purple-colored areas represent bare surfaces where HH and VV have almost the same high backscatter and HV is low. Thus, the purple indicates a landslide area.

TABLE 14-3
Comparative Parameters

Parameters	Contents
Color composites	HH-HV-VV for R-G-B and visual interpretation
Correlation	Correlation of HH-VV, HH-HV, HV-VH
Incoherent decomposition	Decompose into the surface, double, and volume scatterings
Entropy Alpha	Randomness and the phase of the scattering target
Power ratio	HH/HV, HH/VV, VV/HV

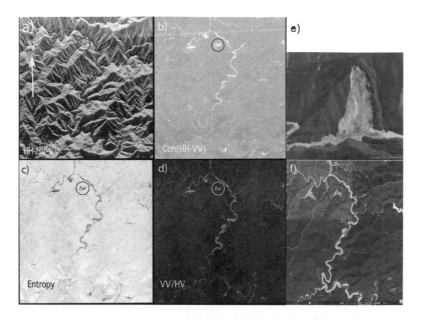

FIGURE 14-4 Totsukawa-mura landslide images. a) Pi-SAR-L2 HV-basis of Totsukawa with color assigned as follows: HH (red), HV(green), and VV(blue); b) HH-VV coherence; c) polarimetric entropy; d) HH/HV ratio; e) aerial photo; and f) Google Earth image with a red circle covering the large Nagatano landslide. In a) several red-to-light purple areas are distributed within the 20 × 20 km location and appear to represent areas of landslide acquired on June 18, 2012. In f) sevela muddy colored areas are confirmed as the land slide areas.

14.5.2.2 Correlation of Two Polarizations

Correlation of two like components can be calculated by

$$\gamma = \frac{\langle \mathbf{a} \cdot \mathbf{b}^* \rangle}{\sqrt{\langle \mathbf{a} \cdot \mathbf{a}^* \rangle \langle \mathbf{b} \cdot \mathbf{b}^* \rangle}}$$

$$c = |\gamma|$$

$$\varphi = \arg(\gamma)$$

(14.1)

where γ is the complex coherence; c is amplitude of the coherence; φ is the phase, $\mathbf{a}$ and $\mathbf{b}$ are the complex backscattering coefficients; * is the complex conjugate notation; and <> is the ensemble average. For the sensitivity analysis, we calculated three cases: $\mathbf{a} = $ HH; $\mathbf{b} = $ VV; $\mathbf{a} = $ HV and $\mathbf{b} = $ VH, $\mathbf{a} = $ HH and $\mathbf{b} = $ HV; c shows a high value at landslides and smooth surfaces and a low value at the forest. The other parameters have low or zero values. Thus, HH-VV coherence is an effective indicator of landslide areas within a mountainous forest.

14.5.2.3 Entropy

Polarimetric entropy, as a measure of randomness, is given by

$$H = \sum_{i=1}^{3} -P_i \log_n P_i, P_i = \frac{\lambda_i}{\sum_{j=1}^{n} \lambda_j},$$

(14.2)

where n is 3, and λ_i is the Eigen value of the covariance matrix. Polarimetric entropy (Cloude and Pottier 1996) measures the spectrum of the radar backscattering mechanism.

14.5.2.4 Ratio of Two Polarizations

The next parameter is the power ratio of HH/HV or VV/VH. Either component, HH or VV, shows a larger backscatter from the forest than from the landside area, and the cross component, HV or VH, shows a similar feature but with smaller magnitude. However, the cross component shows a larger difference than the like component between the forest and land surface. Thus, the ratio HV/HH or VV/VH shows a larger magnitude in the landslide area than in the forested area. In this evaluation, we selected VV/VH:

$$R = P_{pp} / P_{pq} \tag{14.3}$$

where R is the ratio of two backscatters, and p and q are the polarizations (i.e., horizontal or vertical).

Figure 14-4 shows that all the parameters (a, b with HH-VV coherence, c, and d) effectively discriminate the landslide areas from the forested mountain. To evaluate the sensitivity of these parameters, a magnification of the Nagatono area, circled in red in Figure 14-4f, is selected.

14.6 POLARIMETRIC DATA

We have prepared several data observing the landslide area.

14.6.1 Pi-SAR-L2 Data

In order to perform this evaluation, Pi-SAR-L2 images were acquired from two flight course types—the West (W) flight course observing in the southern direction, which hits the landslide surface in an almost perpendicular direction, and the North-North-East (NNE) course observing the target area from an almost western direction and resembling the course of a spaceborne SAR on June 18, 2012. The azimuth direction of the second flight course is 11.1 degrees.

14.6.2 PALSAR-2

PALSAR-2 polarimetry was activated to observe the Totsukawa-mura area on August 26, 2014, three years after the disaster, to investigate how a spaceborne L-band SAR could detect the landslide areas under similar observation geometry.

14.6.3 X-Band SAR

The X-band SARs (XSARs), TerraSAR-X and COSMO-SkyMed, have resolutions that are superior to the L-band SARs because of high bandwidth allocations. They are not fully polarimetrically operational but do operate with dual-like polarization or cross polarization.

14.7 COMPARATIVE STUDY

Figure 14-4a–d shows a comparison of the four parameters using the Pi-SAR-L2 and observing the landslide area from perpendicular directions. We observe that the entropy and HH-VV coherence are highly sensitive but complementary in value and that HH/HV is secondary.

Figure 14-5a–d shows polarimetric parameters observed from the W course, and Figure 15-5e–h shows NNW course observations. The latter four do not provide a clearer signal for the landslide than former four but still provide sensitive information. This means that the volume and the surface

TABLE 14-4
SARs Used for Nagatono-Landslide Area, Wakayama, Japan

	2011	2012	2014
Pi-SAR-L (airborne)	9/30 (QP: 30.5): EW 9/30(QP: 45): AL		
Pi-SAR-L2 (airborne)		6/18 (QP: 30.5: EW) 6/18 (QP: 45:AL)	
TerraSAR-X (spaceborne)		7/31(HH+VV: 39.34:AR)	
COSMO-SkyMed (spaceborne)	9/28 (HH: 30.85:AR) 9/25(HH: 29.55:AR)	6/16(HH+VV: 25.8:AL) 6/18 (HH+HV: 30.85:AR)	
PALSAR-2 (spaceborne)			8/26(QP:DL)-39.00

Note: QP (quad polarization): HH-HV-VH-VV; EW: Flight from east to west; AR: Right observation from ascending flight; AL: Left observation from ascending flight. Here, HV stands for the transmission polarization (H) and the received polarization (V).

scatterings can be discriminated effectively using the L-band full polarimetry no matter what the beam illumination angle is.

Figure 14-6 shows similar sensitivity using the spaceborne SARs: PALSAR-2 (Figure 14-6a–d), TSX (Figure 14-6e–f), and CSK (Figure 14-6g–h). The radar line-of-sight of PALSAR-2 hits the surface from the EES direction from the descending left observations. From these figures, the landslides at Nagatono (and Akatani) are clearly detected by the polarimetric analysis.

XSAR was not acquired for the full polarimetry. A similar evaluation was conducted for the coherence between two channels and only the HH-VV correlation was evaluated.

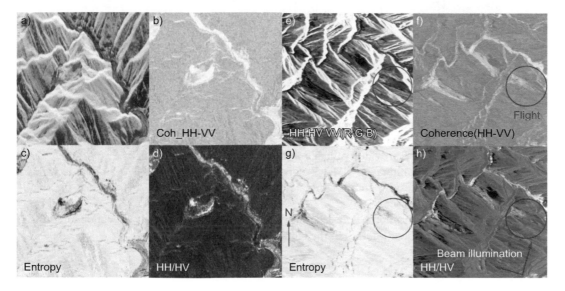

FIGURE 14-5   Close up of Nagatono landslide area observed by Pi-SAR-L2. Upper and lower left side images are from W flight viewing S, showing (a) color composite, (b) HH-VV coherence, (c) entropy, d) HH/HV ratio. Upper and lower right side images are from NNE flight viewing WNW, showing (e) color composite, (f) HH-VV coherence, (g) entropy, and (h) HH/HV power ratio. Acquisition date is June 18, 2012.

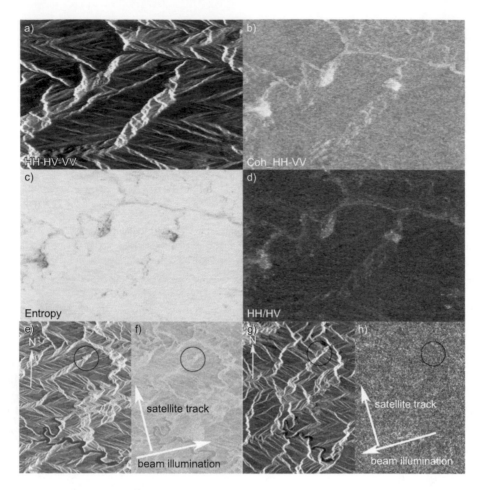

FIGURE 14-6 Spaceborne SAR examples: (a–d) PALSAR-2 L-band polarimetry; TSX-X band SAR: (e) HH, (f) HH-VV coherence; CSK-X-band SAR: (g) VV, (h) HH-VV coherence. The area covered by the black circle corresponds to Nagatono. The acquisition dates for PALSAR-2, TSX, and CSK are August 24, 2014; July 31, 2012; and June 16, 2012; respectively.

From these observations, the following became apparent:

1. L-band polarimetry provides valuable information for discriminating the landslide areas through several parameters: HH-VV correlation, entropy, and HH/HV amplitude ratio. However, it is necessary to discriminate between the smooth, flat surfaces and the flat, steeply inclined surfaces.
2. While X-band polarimetry is less sensitive to the landslide, its high resolution supports manual interpretation (finding) of the landslide area.

14.8   EVALUATION AND DISCUSSION

14.8.1   EVALUATION

We evaluated the performance of the proposed parameters using the following formula:

$$R = \frac{|\mu_1 - \mu_2|}{\sqrt{\sigma_1^2 + \sigma_2^2}} \tag{14.4}$$

TABLE 14-5

Sensitivity Evaluations

Parameter	Forest	Landslide	R
Entropy (PSL)	190.3(8.2)	86.9(22)	3.4
Coh (HH-VV): PSL	102.0(13.9)	195.6(22)	2.6
Coh (HV-VH): PSL	102(13.5)	102(13.5)	0.0
Ratio (HH/HV): PSL	63.0(13.6)	189.1(38.8)	2.4
Decomposition (Volume): PSL	91 (9)	44(14.5)	2.0
Entropy (PS2)	190.98(8.8)	104.1(28.9)	2.3
Coh (HH-VV): PS2	106.5(16.5)	193.9(28.8)	2.0
Ratio (HH/HV): PS2	80.3(33.2)	33.9(16.3)	0.93
Coh (HH-VV): TSX	142 (18)	174(20)	0.84
Coh (HH-VV): CSK	81 (37)	81.8(37)	0.01

Note: Coh = coherence; PSL = Pi-SAR-L2; PS2 = PALSAR-2.

Here, R is the separation parameter, μ_1 is the average of the proposed parameter within the forest area, μ_2 is the average for the landslide area, and σ_1 and σ_2 are the standard deviations of the forest and landslide areas, respectively.

The results are summarized in Table 14-5. The parameter, R, has a maximum value at entropy, and a minimum value at HH-VV coherence for XSARs. Overall, the L-band SAR exceeds the X-band SAR for landslide discrimination. Because of the wavelength difference, this result could be obvious.

14.8.2 DISCUSSION

14.8.2.1 Frequency Dependence of the Correlations

Results show that the L-band is more sensitive than the X-band for discriminating the landslide area within the forest region. Based on the incoherent decomposition method (Freeman and Durden 1998), we have the following model expression for the HH-VV correlation:

$$C = a \cdot f_v + b \cdot f_d + c \cdot f_s \tag{14.5}$$

Here, f_v, f_d, and f_s are the volume scattering, double bounce, and surface scattering components, respectively; a, b, and c are the coefficients of the respective components. Since the L-band can penetrate through a forest canopy, the forest areas have the parameters $a \sim 0.3$, $b \sim 0.1$, and $c \sim 0.0$, and the surfaces have the parameters $a \sim 0$, $b \sim 0$, and $c \sim 1$ (Shimada 2011). Thus, the landslide area has a correlation that is three times brighter than the forest area. In the case of X-band SAR data, the coefficients are $a \sim 0$, $b \sim 0$, and $c \sim 1$ for the landslide and forest areas. Thus, the X-band data show no clear difference between the coefficients for the landslide and forest areas. In that regard, the L-band SAR is more appropriate for detecting landslide areas.

14.8.2.2 Incidence Angle dependence

Sensitivity depends on observation geometry: local incidence angle, which is defined as the normal of the local surface and the line-of-sight vector of the radar. Figure 14-5a–d shows clearer information than Figure 14-5e–h. This is because Figure 14-5a–d has a small local incidence angle (approximately 20–30 degrees) and Figure 14-5e–h has a large angle (probably close to 80–90 degrees). The dependence of the local incidence angle, and possibly beam direction, on detection sensitivity (R in Equation [14.4]) have not been well investigated. Future case studies need to be undertaken for this task.

14.8.2.3 Satellite versus Aircraft

From Table 14-5, we see that the Pi-SAR-L2 result exceeds the PALSAR-2 result at R. This obviously is due to the signal-to-noise ratio issue. As introduced in Chapter 9 (on calibration), the noise equivalent sigma-zero of these two results differ and the Pi-SAR-L2 reports much better than PALSAR-2. This could differentiate the results.

14.9   DETECTION IMPROVEMENT

To suppress false estimation, surface slope data are necessary. The algorithm consists of three steps: First, the landslide candidates are screened using HH-VV correlation in either threshold screening or probability calculation, where the "true" data may be contaminated by the "false data" (i.e., flat land). Second, the surface slope angle is calculated using the digital elevation model (DEM) and the geometric information provided by the SAR line of sight information. Third, the areas with slopes exceeding 5° are selected as possible landslide areas. Last, the screened areas are red-colored and overlaid on the HV-basis color-composite SAR image as shown in Figure 14-7.

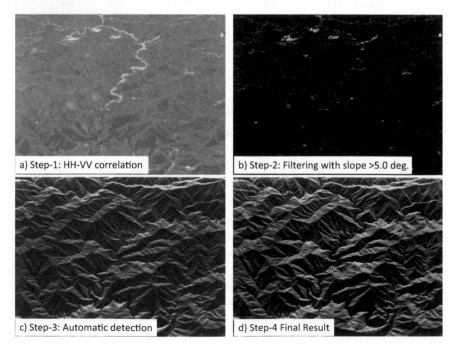

FIGURE 14-7   Improved detection of landslide areas using the HH-VV coherence and slope information derived from the GSI's 10-m DEM. Red-colored areas in the bottom-right image correspond to possible target areas. Acquisition date is June 18, 2012.

14.10 SUMMARY

We learned the effectiveness of the L-band polarimetric SAR parameters for detecting a mountainous landslide. We found that all three polarimetric parameters—HH-VV coherence, polarimetric entropy, and the power ratio of HH/HV in the L-band—work effectively because of various scattering mechanisms in the L-band forest. A comparative study using the X-band SAR did not show a similar result for the landslide possibly because the scattering mechanism has less variety compared to that of the L-band. More robust landslide area detection is combined with surface slope information from the DEM. Other examples can be easily found in journals or on the Internet.

REFERENCES

Buono, A., Lemos Paes, R., Nunziata, F., and Migliaccio, M., 2015, "Synthetic Aperture Radar for Oil Spill Monitoring: A Brief Review," presented at *Anais XVII Simpósio Brasileiro de Sensoriamento Remoto— SBSR*, João Pessoa-PB, Brazil, April 25–29, 2015.

Cloude, S. R. and Papathanassiou, K.P., 1998, "Polarimetric SAR interferometry," *IEEE Trans. Geosci. Remote Sensing*, vol. 36, pp. 1551–1565, Sept.

Cloude, S. R. and Pottier, E., 1996, "A Review of Target Decomposition Theorems in Radar Polarimetry," *IEEE T. Geosci. Remote*, Vol. 34, No. 2, pp. 498–518.

Freeman, A. and Durden, S. L., 1998, "A Three-Component Scattering Model for Polarimetric SAR Data," *IEEE T. Geosci. Remote*, Vol. 36, No. 3, pp. 963–973.

Hoekman, D., 2007, "Radar Backscattering of Forest Stands," *Int. J. Remote Sens.*, Vol. 6, No. 2, pp. 325–343.

Japan International Cooperation Agency (JICA), "Watching on Deforestation," *JICA-JAXA Forest Early Warning System in the Tropics (JJ-FAST)*, http://www.eorc.jaxa.jp/jjfast (October 31, 2017).

Longepe N., Preesan, R., Isoguchi, O., Shimada, M., Uryu, Y., and Yulianto, K., 2011, "Assessment of ALOS PALSAR 50m Orthorectified FBD Data for Regional Land Cover Classification by Using Support Vector Machines," *IEEE T. Geosci. Remote*, Vol. 49, No. 6, pp. 2135–2150.

Lucas, R., Armston, J., Fairfax, R., Fensham, R., Accad, A., Carreiras, J., Kelly, J., et al., 2010 "An Evaluation of the ALOS PALSAR L-Band Backscatter—Above Ground Biomass Relationship, Queensland, Australia: Impacts of Surface Moisture Condition and Vegetation Structure," *IEEE J-STARS Special Issue on Kyoto and Carbon Initiative*, Vol. 3, No. 4, pp. 576–593.

Massonnet, D., Rossi, M., Carmona, C., Adragna, F., Peltzer, G., Feigl, K., and Rabaute, T., 1993, "The Displacement Field of the Landers Earthquake Mapped by Radar Interferometry," *Nature,* Vol. 364, pp. 138–142.

Preesan, R., Longepe, N., Isoguchi, O., and Shimada, M., 2010, "Mapping Tropical Forest Using ALOS PALSAR 50m Resolution Data with Multiscale GLCM Analysis," *Proc. IGARSS 2010*, Honolulu, HI, July 25–30, 2010, pp. 1234–1237.

Rosenqvist, A., Shimada, M., Chapman, B., Freeman, A., De Grandi, G., Sacchi, S., and Rauste, Y., 2000, "The Global Rainforest Mapping Project—A Review," *Int. J. Remote Sensing*, Vol. 21, Nos. 6 &, pp. 1375–1387.

Rosenqvist, A., Shimada, M., Itoh, N., and Watanabe, M., 2007, "ALOS PALSAR: A Pathfinder Mission for Global-Scale Monitoring of Environment," *IEEE T. Geosci. Remote*, Vol. 45, No. 11, pp. 3307–3316.

Shimada, M., 2010, "On the ALOS/PALSAR Operational and Interferometric Aspects (in Japanese)," *J. Geodet. Soc. Japan*, Vol. 56, No. 1, pp. 13–39.

Shimada, M., 2011, "Model-Based Polarimetric SAR Calibration Method Using Forest and Surface Scattering Targets," *IEEE T. Geosci. Remote*, Vol. 49, No. 5, pp. 1712–1733.

Shimada, M., Tadono, T., and Rosenqvist, A., 2010, "Advanced Land Observing Satellite (ALOS) and Monitoring Global Environmental Change," *Proc. IEEE*, Vol. 98, No. 5, pp. 780–799.

Watanabe, M., Koyama, C., Hayashi, M., Nagatani, I., and Shimada, M., 2018, "Early-Stage Deforestation Detection in the Tropics with L-Band SAR," *IEEE J. Sel. Top. Appl.*, Vol. 11, No. 6, pp. 2127–2133.

Wikipedia, "Tropical Storm Talas," 2011, http://en.wikipedia.org/wiki/Tropical_Storm_Talas.

http://www.jma.go.jp/jma/menu/h23t12-portal.html

Landslide.dpri.kyoto-u.ac.jp/Typhoon-12-RCL-Report-1.pdf

(http://www.fdma.go.jp/bn/Disaster status by Typhoon no. 12_台風12号による被害状況について (Report No. 20、20巻).Pdf

15 Forest Map Generation

15.1 INTRODUCTION

15.1.1 THE IMPORTANCE OF FORESTS

Over the past 8,000 years, more than 50% of the world's forests have been lost, primarily because of the demand for food, products, and energy, either from the forests themselves, such as timber, or due to the land use that replaces these forests—agriculture, aquaculture, water reservoirs, better communications, roads, and the need for living spaces by an ever-increasing population (Bryant et al. 1997). Until recently, these forests have provided resources to these populations (e.g., timber, water), maintained the biological interactions (i.e., biodiversity) required to sustain healthy functioning of ecosystems, and moderated the concentration of atmospheric greenhouse gases (e.g., CO_2, CH_4) and hence climate. However, loss and degradation of these forests has led to a disruption of these ecosystem services (Solomon et al. 2007) as manifested in, for example, permanent loss of plant and animal habitats, accelerated soil erosion, and a reduction in forest carbon stocks. For this reason, there is an ever-increasing need to assess the extent and state of the forest resource, and how this is changing, and likely to change, in the future.

15.1.2 REQUIREMENTS FOR FOREST/NON-FOREST MAPPING

At a global level, maps showing areas of forest and non-forest (FNF) typically have been generated at coarse (250 m to 1 km) spatial resolution using data from the National Oceanic and Atmospheric Administration (NOAA) Advanced Very High-Resolution Radiometer (AVHRR; Loveland et al. 2000; Hansen et al. 2000; Hansen et al. 2004), SPOT VEGETATION (Bartholomé and Belward 2005), Terra-1 Moderate Resolution Imaging Spectroradiometer (MODIS) (Friedl et al. 2002; Hansen et al. 2003), or from Envisat MERIS (GlobCover; 300 m; Arino et al. 2007). Other mapping efforts have focused on particular forest types (e.g., mangroves; Spalding et al. 2010; Giri et al. 2010) or structures (Lefsky 2010), latitudes (e.g., tropics; Achard et al. 2004), continents (e.g., South America; Eva et al. 2004), or countries (e.g., Brazil; Shimabukuro et al. 2013), either at coarse or finer (typically ~30 m) spatial resolution depending on the sensor(s) used (Hansen et al. 2008). In all global mapping efforts and the majority of the others, data from optical sensors have been used and maps have been generated for a single year or period—largely because of the necessity for multiple acquisitions to obtain cloud-free images. Repeat mapping typically has occurred at the country level (e.g., Brazil, Australia) to support national and international reporting requirements for forest monitoring and forest carbon accounting. The exception is the Food and Agriculture Organization of the United Nations (UN FAO), which has provided estimates of global forest cover at five- to ten-year intervals since 1946 based largely on national reporting but increasingly by using satellite sensor data (FAO 2012a, 2012b). Global mosaics of Landsat sensor data have also been generated for 1975, 1990, 2000, 2005, and 2010 (Hansen et al. 2009), and, in 2013, the first forest cover change maps from 2000 to 2012 were produced using time series of cloud-free Landsat data with the annual forest gain and loss shown for all countries (Hansen et al. 2013).

Based on these mapping efforts, estimates of the global extent of forests range from 3.5 billion hectares (ha) to just over 4 billion ha (31% of the land area). However, the mapped area differs depending on the definition, which varies among countries and institutions, and how the remote sensing data are interpreted. For example, the FAO defines all contiguous areas less than 0.5 ha where the crown cover is greater than 10% as forest (FAO 2000). Other countries also include a minimum height in the definition, with this typically ranging from 2 m to 5 m. Although definitions based on cover alone favor the use of optical sensors, separating woody from herbaceous vegetation can be problematic; hence, information on three-dimensional structure (including height) is desirable, with this either inferred from optical data or requiring spaceborne Southeast (Lefsky 2010) and/or low frequency (e.g., L-band) SAR (Cloude and Papathanassiou 2003). The use of high frequency X- and C-band SARs (e.g., Envisat ASAR, RADARSAT, COSMO-SkyMed and TerraSAR/TanDEM-X) for FNF mapping is more limited, again because of the difficulty in distinguishing woody and herbaceous vegetation.

15.1.3   FOREST MAPPING AND CHARACTERIZATION USING L-BAND SAR

The first cloud-free L-band observations of forests from space were provided by Seasat and the Shuttle Imaging Radar (SIR-C) SAR in 1978 and 1994, respectively. These missions were short-lived, although they provided important proof of concept for later missions (Way and Smith 1992). JERS-1 SAR (from 1992 to 1998) provided the first systematic global spaceborne observations, and significant advances in understanding the use of L-band SAR for forest characterization, mapping, and monitoring were made (Rosenqvist et al. 2000; Shimada and Isoguchi 2002). These and other studies established that non-forest areas typically exhibited a lower L-band backscatter compared to forests, facilitating mapping of deforested areas in some regions. However, observations by the successor mission, ALOS PALSAR (from 2006 to 2011), highlighted the fact that the backscatter (particularly at HH polarization) often increased above values typical for forests following an initial clearance event because of enhanced scattering from cut stumps and woody debris and, particularly, following rainfall events (Dos Santos et al. 2007). However, thereafter values declined such that the backscatter of the deforested area became similar to that of the original forest for a short period and then declined further because of the transition from double bounce to surface scattering associated with removal of woody debris and active use of the land for agriculture. At this point, greatest separation of forest and non-forest generally was achieved. However, discrimination of tree crops, forests regenerating subsequently on abandoned land, or degraded forests from those that were intact was compromised because of similarities in the backscatter at both L-band HH and HV, with both relating to the structural development and accumulation of aboveground biomass (AGB) by the forests. Using JERS-1 SAR, Luckman et al. (1998) defined this limit as approximately 60 Mg ha^{-1}, with saturation occurring thereafter. The sensitivity to AGB up to the saturation level was explained by the ability of L-band microwaves to penetrate through the canopy and interact with the larger woody branches and trunks. Beyond this, the signal was attenuated by the higher biomass in the canopy and hence the backscatter remained similar. Other studies on biomass retrieval were conducted (Watanabe et al. 2006; Lucas et al. 2010; Mitchard et al. 2011) with greater focus on the use of the JAXA's airborne L-band SAR HV or ALOS PALSAR L-band HV data because of the greater sensitivity of this channel to AGB levels. However, although of similar form, the consistency of the relationship within and between regions was compromised by differences in vegetation structure and also surface moisture conditions. For example, Lucas et al. (2010) established that the saturation level could be raised to more than 200 Mg ha^{-1} in Australia but only when data were acquired under relatively dry conditions. Similarly, the use of the retrieved AGB thresholds for separating forest and non-forest and forest growth stages was compromised. These and other studies, however, paved the way for the advancement of methods and algorithms for retrieving AGB from ALOS PALSAR data, either singularly or in combination with other data (e.g., optical, LiDAR), with potential provision of input to programs such as REDD+ (Asner et al. 2010; Englhart et al. 2011; Saatchi et al. 2011).

15.1.4   THE ALOS PALSAR: GLOBAL OBSERVATIONS AND CALIBRATION

The ALOS mission was launched on January 24, 2006, and operated until April 22, 2011. Three sensors were carried, including the PALSAR, which acquired 2.1 million scenes (70 km × 70 km in dimension) in more than five years of observations, with this corresponding to an average of 16 for each scene location (Shimada 2010a; Shimada et al. 2010). Observations were acquired systematically through what was termed the basic observation scenario (BOS) (Rosenqvist et al. 2007), with two observations over the period June to September and one from November to January each year. A major driver for the BOS was the acquisition of data that were consistent in time and space and could be used for forest and also deformation monitoring (Rosenqvist et al. 2007). Throughout, robust calibration of the data was undertaken and radiometric and geometric accuracies of 0.6 dB and 7.8 m dB, respectively, were maintained (Shimada et al. 2009).

15.1.5 POTENTIAL FOR FNF MAPPING

Observations using the JERS-1 SAR and also the European Remote-Sensing Satellite (ERS-1)/RADARSAT-1 SAR (Hawkins et al. 2000; Shimada 2005) confirmed that the radar backscatter from dense forests generally was very stable. On this basis, the Committee on Earth Observation (CEOS) SAR working group made the recommendation to use the Amazon Rainforest to calibrate these SAR data but also for most of those missions that followed (Shimada et al. 2009; Shimada and Freeman 1995; Zink and Rosich 2002; Lukowski et al. 2003). However, investigations into the regional and temporal variations in the backscattering coefficient were not undertaken because SAR data had not been acquired globally or for that many years. Nevertheless, several studies (Hoekman and Quiriones 2000; Longepe et al. 2011; Rakwatin et al. 2012) started to use these data for classifying the extent of different land cover types, with many suggesting that accuracies of up to 90% could be achieved in the discrimination of forest and non-forest (Shiraishi et al. 2014; Motohka et al. 2013; Thapa et al. 2013). However, there has been no concerted effort to generate FNF maps from these data.

Given the availability of the global archives of ALOS PALSAR data from 2006 to 2011, this study sought to establish the consistency of backscatter thresholds for separating forest from non-forest within and between regions. Once established, the primary objective was then to generate annual mosaics of these data and, from these, derive FNF maps and forest change maps. Although the intention of the study was not to retrieve AGB, the sensitivity of the L-band backscatter to lower levels provided the basis for the use of thresholds for mapping. The study also sought to evaluate changes in L-band backscatter over the periods of observation in relation to forest change.

The paper is structured as follows: In Section 2, the methods used to generate separate global mosaics of L-band HH and HV data for each of four years and the variation in the L-band backscattering coefficient (γ^0) in space (globally and within regions) and time are described. The approaches to deriving and validating the FNF maps are outlined. The maps of FNF are illustrated in Section 3 as are FNF change maps for selected regions. Estimates of classification accuracy are conveyed. The discussion (Section 4) provides a critical overview of the mosaics generated and the classifications derived, highlighting the benefits (e.g., availability of cloud-free observations) but also the limitations of the approaches used. Differences in the areas of forest and forest change obtained in this study and that of others (FAO 2000; FAO 2012a; FAO 2012b; Hansen et al. 2013) are also discussed. The study is concluded in Section 5.

15.2 MOSAIC GENERATION AND THE FNF CLASSIFICATION ALGORITHM

15.2.1 AVAILABLE DATA, DATA PRE-PROCESSING, AND MOSAIC GENERATION

ALOS PALSAR HH and HV data acquired between June and October during 2007, 2008, 2009, and 2010, were used for the generation of 25-m global mosaics (Shimada and Ohtaki 2010; Shimada 2010b). The processing steps were as follows:

a. The raw data were prepared such that they covered individual processing units 500 km × 500 km in dimension, with consideration given to the orbit inclination and pass overlaps. For each year and location, the strip data were selected through visual inspection of the browse mosaics available over the period, with those showing minimum response to surface moisture preferentially used. In cases where the availability was limited (e.g., because of the requirement for observations during specific emergencies), data were necessarily selected from the year before or after, including from 2006.

b. Each raw image was calibrated using published coefficients (Shimada and Ohtaki 2010) and output with 16-looks to achieve speckle reduction. All data were ortho-rectified using the 90-m SRTM DEM and slope-corrected using these same data to account for the variation in the backscattering coefficient with topography. A destriping process (Shimada and Isoguchi 2002; Shimada and Ohtaki 2010) was applied to equalize the intensity differences between neighboring strips, which were attributed largely to seasonal and daily differences in surface moisture conditions, and each was projected to a geographic (latitude and longitude) coordinate system. In addition to the mosaic of HH and HV polarization data, masks identifying areas of layover, shadowing and oceans were generated to limit the area of the image considered to be valid for FNF mapping. Information on the local incidence angle and days since launch was also included.

Orbit information for the ALOS mission is stated to be accurate to within 30 cm; hence, a high geometric accuracy of ±8 m was therefore obtained for the PALSAR data. However, the geometric accuracy of the ortho-product was degraded because of the relatively coarse spatial resolution of the SRTM DEM and was estimated at 12 m (Shimada and Ohtaki 2010). As the ortho-rectification was undertaken using the SRTM DEM, a close correspondence with this elevation data set was obtained; hence, no post-processing adjustments were necessary. As with previous SAR missions, the PALSAR L-band data were considered very stable across densely forested areas (e.g., the Amazon) (Shimada 2005), with the variation in the wet and dry season backscatter being less than 0.2 dB in both polarizations; hence, no inter-calibration of the signal based on geometric ground control points (GCPs) representing dense forest was necessary.

The mosaic data were expressed in the form of the normalized radar cross section with gamma-naught (γ^0) rather than sigma-naught (σ^0) or beta-naught (β^0) adopted as the expression unit (Small 2011) since the backscatter is normalized by the realistic illumination area under an assumption of scattering uniformity (i.e., lambertian) where:

$$\gamma^0 = \frac{\sigma^0}{\cos \theta_{local}} \frac{\cos \psi}{\sin \theta_{inci}}. \tag{15.1}$$

Here, γ^0 is the gamma-naught, σ^0 is the sigma-naught, θ_{local} is the local incidence angle, θ_{inci} is the incidence angle at the GRS80, and Ψ is the angle between the local normal vector and a tangential vector to the radar line of site at the target point.

The mosaic was then tiled into 1 degree × 1 degree areas (4,500 × 4,500 pixels), with these referenced by the integer latitude and longitude coordinates of the northwest corner (Table 15-1).

The BOS ensured that acquisitions in each year were achieved at least once during the period June to September. A summary of the ALOS PALSAR data used to generate the global mosaics, including the number of data acquisitions in each year used to construct the mosaics for each year, is given in Table 15-2.

The global mosaics for 2007, 2008, 2009, and 2010 are illustrated in Figure 15-1a–d, with L-band HH and HV and the HH: HV ratio displayed in RGB, respectively. Areas of forest and non-forest are broadly indicated in green and purple, respectively. In each year, darker stripes are observed in Siberia and Alaska, with these attributed to differences in freeze/thaw conditions. Snow-covered surfaces typically are associated with a lower backscatter because of signal attenuation. However, across the majority of the globe, the uniformity of the areas was largely preserved for both forested (e.g., Amazonia, central Africa, and islands in Southeast Asia) and non-forested regions

TABLE 15-1

Characteristics of the Tiles Used to Generate the Global ALOS PALSAR HH and HV γ^0 Mosaics

Characteristic	Description
Reference location	Latitude and longitude of northwest corner
Coordinate system	Latitude-longitude coordinate
Spacing	0.8 arcsec unit providing spacing of 25 m
Resolution of SAR image	36 m (azimuth) × 20 m (range)
Number of pixels	4500 columns × 4500 rows
Data volume	40.5 MB (per tile)
Contents	Normalized radar cross-section, gamma-naught in HH and HV, mask information (ocean flag, effective area, void area, layover, shadowing), local incidence angle, total dates from the ALOS launch (1:30: January 24, 2006: UTC).
No. of tiles per year	2007: 27,062; 2008: 27,163; 2009: 27,703; 2010: 27,923

(e.g., southern Argentina, Mongolia, Pakistan, central Australia, and the Saharan, Takla Makan and Nefud Deserts). Notable differences in the backscatter were observed in Greenland, where the coastal areas exhibited a yellow coloration (indicating a strong like- and cross-polarized response and a small HH/HV ratio), whereas the central ice sheets were darker.

15.2.2 METHODS LEADING TO FNF CLASSIFICATION

In general, the variability of γ^0 from forests will be a function of the observing configuration of the radar (e.g., incidence angle) and surface characteristics, namely the moisture content (dielectric constant) of the vegetation and underlying surface and size and geometric arrangement of structural

TABLE 15-2

The Number and Proportion of Tiles from Each Year Used to Generate the Final ALOS PALSAR Mosaics for 2007, 2008, 2009, and 2010

Product Year		Observation Year				Total Number
		2007	2008	2009	2010	
2007	% of tiles	91.04	6.26	2.603	0.098	100
	# paths	3918	434	165	100	4,617
	% of area	84.86	9.40	3.57	2.17	100
2008	% of tiles	4.545	93.06	2.219	0.179	100
	# paths	243	4159	157	88	4,647
	% of area	5.23	89.50	3.38	1.89	100
2009	% of tiles	0.148	6.153	91.19	2.511	100
	# paths	9	371	4044	303	4727
	% of area	0.19	7.85	85.55	1.89	100
2010	% of tiles	0.252	1.734	3.695	94.32	100
	# paths	15	116	153	4517	4,801
	% of area	0.31	2.42	3.19	94.08	100

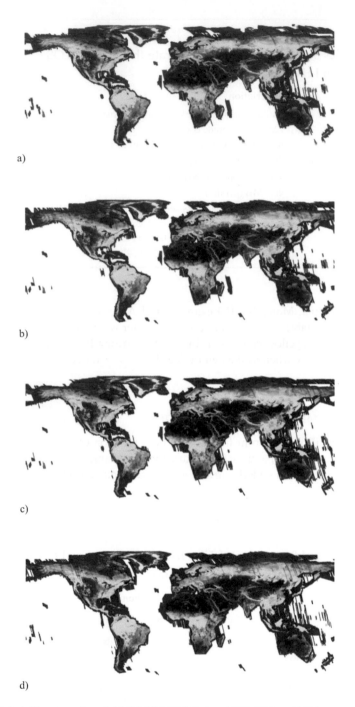

FIGURE 15-1   Global 25-m mosaics of ALOS PALSAR L-band HH, HV, and HH/HV displayed in RGB for (a) 2007, (b) 2008, (c) 2009, and (d) 2010.

elements. At the L-band, these are primarily the trunks and branches, the arrangement of which depends on tree architecture as a function of species and growth form. As the key parameter for the forest/non-forest estimation is γ^0, detailed investigation into the characteristics of both forest and non-forest surfaces was necessary at global and local geographical scales in order to develop the algorithm for FNF mapping.

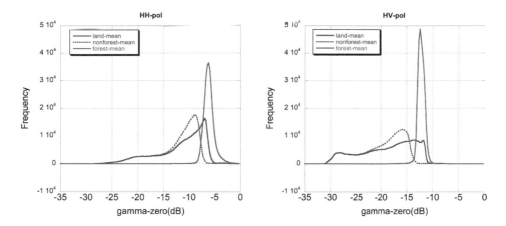

FIGURE 15-2 Radar backscatter distribution for forest and non-forest at HH and HV polarizations derived from the previous FNF (Shimada et al. 2011).

15.2.2.1 Introductive Property of the γ^0 and Forest and Non-Forest (FNF)

In preparation for the generation of FNF maps, a multi-resolution segmentation of each of the ALOS PALSAR HH and HV mosaics was first performed using eCognition software, following the application of a 5 × 5 median filter. Objects known to be either forest or non-forest were then identified with reference to Google Earth imagery (GEI) and, from these, frequency distributions were generated for both the HH and HV polarizations. Typically, and as illustrated in Figure 15-2 from the early version of global FNF (Shimada et al. 2011), forests exhibited a normal distribution with γ^0 at HH always exceeding that at HV. The range of γ^0 was also relatively narrow for forests compared to the wide range observed for non-forest areas above a noise threshold of −34 dB. On this basis, the crossover point between forest and non-forest at both polarizations was considered an appropriate threshold between these broad land cover types. At a global level, these thresholds were determined to be −7.5 and −14.0 dB for HH and HV polarizations, respectively.

15.2.2.2 Spatio-Temporal Variability of γ^0

a. **PALSAR stability evaluation**

The ability to use a single threshold for FNF mapping depends upon the consistency of the HH and HV γ^0 spatially and over time on the assumption that variations are not attributed to sensor performance. The PALSAR consists of 80 transmit-receive (TR) modules, each of which produces 25 W of transmission power to generate a total of 2,000 W. For each module and for a period of 20 minutes every 46 days, the transmission power was monitored, with this relating to γ^0 as:

$$\gamma^0 \propto \frac{\hat{\gamma}^0 P_t}{P_t(t)} \tag{15.2}$$

where $\hat{\gamma}^0$ is the true γ^0, P_t is the nominal (averaged) transmission power, and $P_t(t)$ is the real transmission power as a function of time, t. SAR processing assumes that P_t is constant. For the ALOS PALSAR, the transmission power remained stable (0.065 dB) over the five-year period of observation. The true (averaged) transmission power was 33.527dBW (2,252.7 W), with this being larger than the specification of 2,000 W (Figure 15-3).

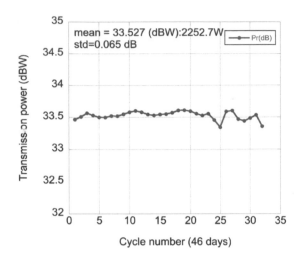

FIGURE 15-3   The time variation of the transmission power over the lifetime of the ALOS PALSAR.

b. **Definitions of geographical regions, region of interests (ROIs), and forest areas**

By knowing the transmission power was relatively stable over the PALSAR mission life, the spatio-temporal variability of γ^0 over forests and non-forests at three geographical scales and over four different years was investigated. For this purpose, three ROIs (i.e., ROIs A, B, and C) were defined (Figure 15-4).

  a. ROIs A were established over smaller areas that were distributed in 15 different regions: Indonesia (Sumatra), Papua New Guinea, Borneo, Malaysia, Philippines, East Asia, Japan, India, Europe-Russia, Australia, Amazon, Chile, Africa, North America, and Central America. These ROIs were used to investigate the local forest-type dependence of the scattering property and to determine thresholds for FNF mapping.

  b. ROIs B were established to investigate regional (rather than local) spatio-temporal scattering properties, within the same 15 regions selected.

  c. ROIs C were established to investigate the temporal variation of the scattering property at a near continental level.

Forests were defined as areas where the cover of woody vegetation exceeded 10%, as determined from the GEI and through reference to the Forest Resources Assessment (FRA) (FAO 2012a, 2012b). Examples of areas with sparse and dense forests are illustrated in Figure 15-5.

c. **Regional scale spatio-temporal variability of γ^0**

For each year of observation and for each of the 15 regions, the statistics of γ^0 for both forest and non-forest were extracted from ROIs B (Figure 15-6a and b, respectively) and summarized for all years (Table 15-3). The minimum area was 200,000 pixels (approximately 125 km^2; Figure 15-4a, Appendix 15A-1). The comparison revealed differences in HH and HV γ^0 among regions. For example, in Amazonia, HH and HV γ^0 were, on average, −6.84 and −11.85 dB, respectively (Table 15-3), but were lower in Indonesia (Sumatra) being −7.68 and −12.54 dB. Forests also exhibited a normal distribution compared to non-forested areas, with the standard deviation being approximately 30% smaller compared to non-forest. The mean differences in the HH and HV γ^0 for forest and non-forest were 3.97 and 6.42 dB.

Over time, γ^0 was relatively stable for all forest areas, with HH and HV values averaging −6.89 ± 0.95 dB and −12.07 ± 1.52 dB, respectively, over the four years (Table 15-3). A normal distribution was also followed, with the standard deviation being 2.13 and 2.04 dB,

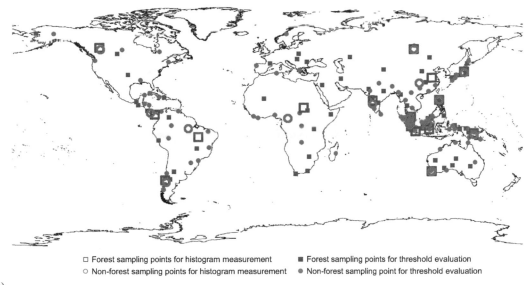

a)

b)

FIGURE 15-4 (a) Distribution of ROIs A (solid symbols) for determining thresholds for FNF discrimination and ROIs B (hollow symbols) for evaluating changes in HH and HV γ^0 at regional to global levels, (b) distribution of zones globally, at ±30° and ±15°, from extensive areas of forest in Indonesia, Amazonia, and Central Africa over which ROIs C were located.

respectively (Figure 15-7), within all years, but with annual averages of the standard deviation being as small as 0.21 ± 0.18 dB and 0.21 ± 0.19 dB when all four years were considered. Values of γ^0 were lower and more variable for non-forest areas (Figure 15-8), averaging -10.86 ± 4.78 dB and -18.49 ± 3.84 dB at HH and HV polarization, respectively (Table 15-3).

d. **Continent scale temporal variability of γ^0**

To establish whether there were differences in the temporal changes in HH and HV γ^0 as a function of latitudinal zones, γ^0 was extracted for forest and non-forest areas (ROIs C) from sub-tiles 900×900 pixels in dimension distributed globally, at ±30 degrees and

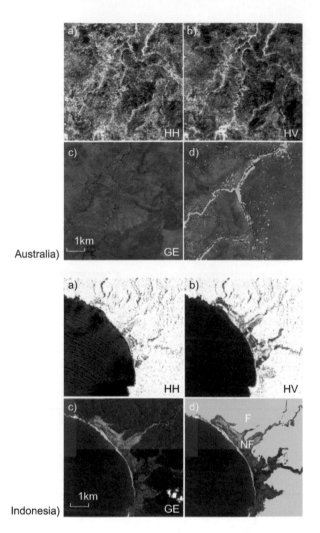

FIGURE 15-5 Examples of ALOS PALSAR data acquired over Australian (left) and Indonesian (right) forests; (a) HH and (b) HV γ^0, (c) GEI; and (d) the mapped forest area.

±15 degrees and from extensive areas of forests in Indonesia, Amazonia, and Central Africa (Figure 15-4b). It was expected that these averaged data should be distributed normally with 95% and 99% confidence levels as follows:

$$\Delta_{95\%} = 1.960 \frac{\sigma}{\sqrt{n}}$$

$$\Delta_{99\%} = 2.576 \frac{\sigma}{\sqrt{n}}.$$

(15.3)

On this basis, the standard deviation for forests at HH and HV polarizations was 2.25 dB and 2.15 dB, with this being 0.0049 dB and 0.0047 dB at the 95% confidence level and 0.0063 dB and 0.0061 dB at the 99% confidence level.

For the majority of areas and over the course of the observation period (2007 to 2010), an overall decrease in γ^0 was observed at both HH and HV polarization, with this being (at the global level) −0.040 and −0.028 dB yr^{-1}, respectively, for forest and non-forest combined, −0.106 and −0.031 for forest, and −0.032 and −0.016 for non-forest areas

a) HH

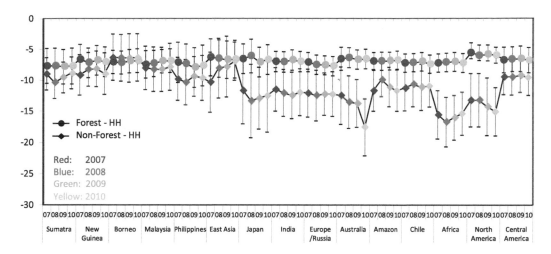

b) HV

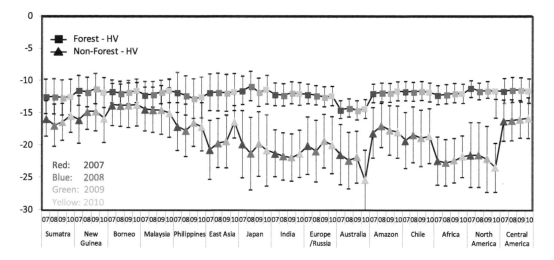

FIGURE 15-6 The γ^0 statistics for forested areas observed in 2007, 2008, 2009, and 2010 in HH and HV.

(Table 15-4 and 15-5). The decreases in HH and HV degrees for the forest areas in Southeast Asia, Amazonia, and Central Africa were $-0.004 \sim -0.045$ and $-0.007 \sim -0.038$ dB yr^{-1}. These differences are larger than the 99% confidence level (i.e., 0.0063 and 0.0061 dB at HH and HV polarizations, respectively). The overall decrease was also evident within the histograms of these two polarizations (Figure 15-9).

e. **Forest type dependency of γ^0**

In the previous sections, we have learned that the forest γ^0 was temporally stable at global and continental scales, but there was some regional dependency. Hence, region-specific thresholds were considered necessary for classification. However, in each region, forests are comprised of many different types; hence, statistics from a single ROI (i.e., B) were considered insufficient for representing this diversity (Figure 15-4). For this reason, statistics were extracted from a greater number of ROIs (A) within each of the 15 regions such that a region-dependent threshold for FNF classification could be better defined. In each

TABLE 15-3

Four-Year L-band HH and HV Backscatter Statistics Extracted from ROIs for 15 Representative Forest and Non-Forest Areas

No.	Regions	Forest HH (μ, δ1, δ2; dB)	Forest HV (μ, δ1, δ2; dB)	Non-Forest HH (μ, δ1, δ2; dB)	Non-Forest HV (μ, δ1, δ2; dB)
1	Sumatra	−7.68, 0.06, 2.84	−12.54, 0.08, 2.71	−9.41, 0.69, 2.56	−16.11, 0.66, 2.87
2	New Guinea	−6.86, 0.25, 2.41	−11.62, 0.28, 2.32	−8.61, 0.49, 3.06	−15.32, 0.68, 3.34
3	Borneo	−6.96, 0.13, 1.90	−11.77, 0.15, 1.80	−6.34, 0.06, 3.81	−13.74, 0.07, 3.34
4	Malaysia	−7.09, 0.28, 2.04	−11.96, 0.29, 1.94	−8.06, 0.20, 3.61	−14.59, 0.25, 3.60
5	Philippines	−7.47, 0.37, 3.25	−12.40, 0.42, 3.10	−9.76, 0.41, 3.48	−17.10, 0.47, 3.77
6	East Asia	−6.48, 0.18, 3.03	−11.83, 0.04, 2.88	−8.27, 1.43, 4.48	−19.00, 1.85, 3.80
7	Japan	−6.55, 0.41, 2.50	−11.41, 0.40, 2.23	−12.58, 0.63, 5.57	−20.38, 0.73, 5.32
8	India	−6.89, 0.13, 1.66	−12.09, 0.14, 1.74	−11.97, 0.40, 3.70	−21.50, 0.32, 3.63
9	Europe/Russia	−7.36, 0.24, 1.49	−12.34, 0.16, 1.52	−12.25, 0.16, 3.75	−20.05, 0.63, 4.42
10	Australia	−6.50, 0.13, 1.66	−14.45, 0.16, 1.43	−14.30, 2.22, 4.31	−22.73, 1.81, 4.40
11	Amazon	−6.84, 0.08, 1.46	−11.85, 0.13, 1.45	−11.07, 0.83, 3.16	−17.64, 0.51, 3.80
12	Chile	−7.16, 0.14, 1.49	−11.70, 0.09, 1.47	−10.97, 0.42, 3.62	−18.75, 0.38, 4.29
13	Africa	−7.11, 0.09, 1.52	−12.10, 0.12, 1.48	−15.90, 0.60, 3.71	−22.27, 0.39, 3.46
14	North America	−5.79, 0.19, 1.47	−11.51, 0.20, 1.58	−13.95, 0.91, 4.27	−22.12, 0.92, 4.73
15	Central America	−6.64, 0.09, 2.05	−11.54, 0.10, 1.91	−9.42, 0.13, 2.87	−16.01, 0.18, 3.10
	Mean (min. − max.)	−6.89(−5.79~−7.68),	−12.07(−11.41~−14.45)	−10.86(−6.34 ~−15.90)	−18.49(−14.59 ~−22.27)
	SD of 4-year means (range)	0.21(0.06 ~ 0.41)	0.21(0.04 ~ 0.42)	0.84(0.06 ~ 2.22)	0.83 (0.07 ~ 1.85)
	SD (range)	2.13(1.46 ~ 3.25)	2.04(1.43 ~ 3.10)	3.79(2.56 ~ 5.57)	3.90(2.87 ~ 5.32)

Note: μ = four-year average of γ^0 for 2007–2010; $\delta1$ = standard deviation of the four-year averaged γ^0; $\delta2$ = standard deviation of the γ^0 over four years, SD = standard deviation.

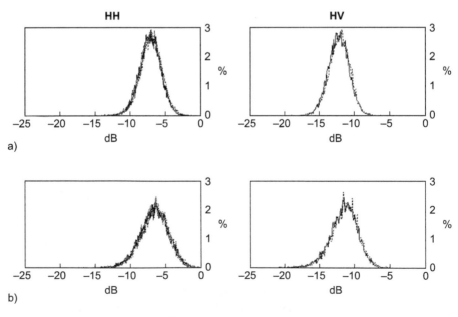

FIGURE 15-7   Histograms of L-band HV and HH data for 2007, 2008, 2009, and 2010 for forests in (a) Africa and (b) Central America.

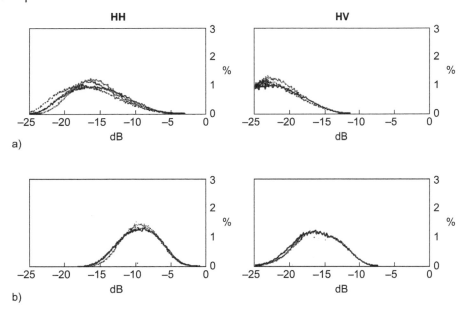

a)

b)

FIGURE 15-8 Histograms of L-band HV and HH data for 2007, 2008, 2009, and 2010 for non-forested areas in (a) Africa and (b) Central America.

region, three to ten ROIs (ROIs A) were selected with reference to GEI (Figure 15-5), with these representing a wide range of forests types as well as non-forest categories. In each case, the ROIs ranged in size from 65,000 to 80,000 pixels. As an example, in Riau Province in Indonesia (Figure 15-10), plantations dominated by *Cocos nucifera* (coconut) and *Elaeis guineensis* (oil palm) typically exhibited a lower HV γ^0 compared to natural (wet and dry) forests; plantations dominated by *Hevea brasiliensis* (rubber) and *Acacia* species were variable in their response but exhibited a response more similar to natural forests. Mangroves dominated by species (e.g., *Rhizophora stylosa*) of higher biomass and with large supporting root systems also exhibited a low HV γ^0. The data from all forest types were then combined to produce single histograms. For each region, similar histograms were generated from the ROIs A and associated with different forest types based on knowledge from the FRA (where available) and also local information. Data were also extracted from non-forest

TABLE 15-4

Annual γ^0 Change within Six Areas in HH (dB)

Region	Global Land 2007–2008	2008–2009	2009–2010	Mean	Forest 2007–2008	2008–2009	2009–2010	Mean	Non-Forest 2007–2008	2008–2009	2009–2010	Mean
World	−0.093	0.039	−0.066	−0.040	−0.106	−0.441	0.087	−0.106	−0.159	0.205	−0.143	−0.032
	0.828	0.848	0.917	0.867	0.847	1.082	0.993	1.008	0.875	1.021	1.057	1.002
<30°	−0.096	0.000	0.058	−0.012	0.003	0.025	−0.020	−0.088	−0.19	0.082	0.063	−0.016
	0.739	0.742	0.827	0.773	0.819	0.889	0.820	0.851	0.831	0.904	0.954	0.907
<15°	−0.066	−0.019	0.002	−0.028	0.018	−0.100	−0.087	−0.057	−0.217	0.026	0.102	−0.030
	0.709	0.740	0.706	0.719	0.590	0.645	0.608	0.617	0.894	0.938	0.886	0.917
SEA	−0.033	−0.235	0.324	0.019	−0.011	−0.297	0.297	−0.004	−0.108	0.062	0.089	0.014
	0.433	0.523	0.539	0.551	0.440	0.550	0.565	0.575	0.514	0.851	0.832	0.753
Brazil	0.139	0.010	−0.311	−0.053	0.085	−0.009	−0.212	−0.045	0.104	−0.055	−0.154	−0.035
	0.539	0.586	0.510	0.577	0.427	0.434	0.365	0.428	0.896	0.890	0.796	0.869
Africa	−0.137	0.315	0.574	−0.016	−0.050	−0.016	−0.032	−0.031	−0.459	−0.042	0.344	−0.052
	0.391	0.420	0.419	0.420	0.334	0.346	0.338	0.340	0.777	0.713	0.801	

TABLE 15-5

Annual γ^0 Change within Six Areas in HV (dB)

Region	Global Land				Forest				Non-forest			
	2007–2008	2008–2009	2009–2010	Mean	2007–2008	2008–2009	2009–2010	Mean	2007–2008	2008–2009	2009–2010	Mean
World	−0.104	0.061	−0.040	−0.028	0.027	−0.130	0.010	−0.031	−0.141	0.182	−0.087	−0.016
	0.789	0.823	0.906	0.844	0.512	0.611	0.602	0.581	0.823	0.937	0.980	0.927
<30°	−0.077	0.011	0.057	−0.003	0.017	−0.079	−0.016	−0.026	−0.123	0.069	0.040	−0.007
	0.595	0.582	0.657	0.614	0.457	0.545	0.488	0.500	0.696	0.741	0.703	0.746
<15°	−0.055	0.005	0.010	−0.014	−0.000	−0.022	−0.042	−0.021	−0.155	0.046	0.052	−0.018
	0.562	0.586	0.576	0.576	0.328	0.560	0.328	0.339	0.768	0.796	0.753	0.779
SEA	−0.049	−0.118	0.179	0.004	−0.016	−0.068	0.028	−0.019	−0.040	0.111	−0.093	−0.008
	0.389	0.399	0.468	0.439	0.248	0.256	0.269	0.261	0.445	0.696	0.732	0.643
Brazil	0.040	0.016	−0.192	−0.045	−0.055	0.019	−0.077	−0.038	0.044	−0.103	−0.013	−0.024
	0.393	0.420	0.392	0.415	0.219	0.228	0.197	0.219	0.814	0.769	0.650	0.750
Africa	−0.134	0.096	0.091	0.018	−0.062	0.034	0.067	−0.007	−0.328	0.053	0.025	−0.009
	0.297	0.384	0.377	0.371	0.194	0.236	0.232	0.225	0.660	0.703	0.734	0.740

Note: The mean and standard deviation (in dB) is reported for each column.

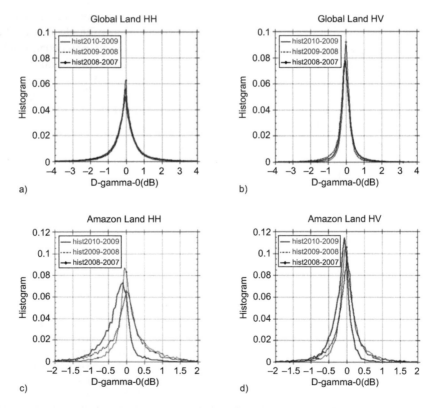

FIGURE 15-9 Comparative histograms of HH and HV γ^0 for 2007, 2008, 2009, and 2010 for (a–b) the global and (c–d) the Amazon land areas, (e–f) the global and (g–h) the Amazon forest areas, and (i–j) the global and (k–l) Amazon non-forested areas.

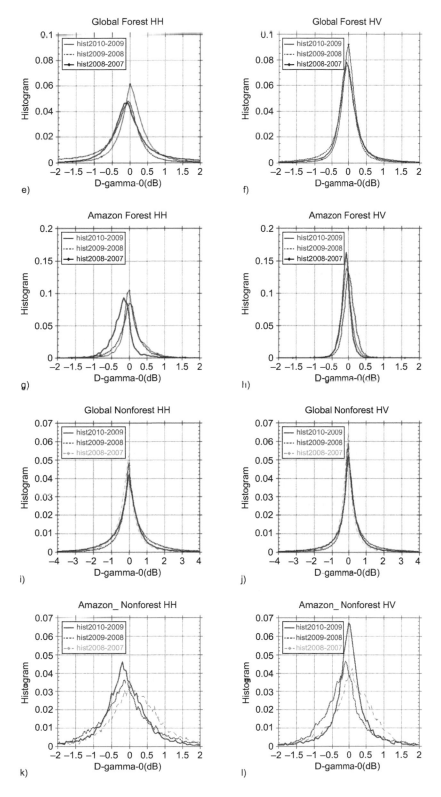

FIGURE 15-9 *(Continued)*

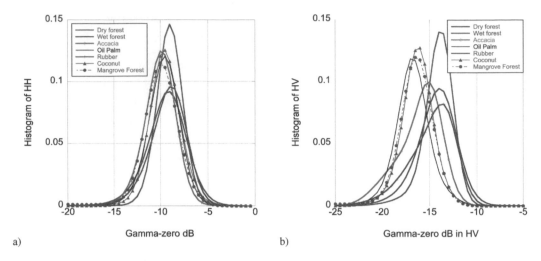

FIGURE 15-10   Examples of the frequency distribution of (a) HH and (b) HV γ^0 for a range of forest types, including plantations, in Riau Province, Indonesia.

areas within those including artificial (e.g., urban infrastructure) and natural (e.g., water, snow, ice, and deserts) non-vegetated as well as vegetated environments (e.g., tundra, grasslands, and agricultural land). In each case, the majority of non-forest areas exhibited a lower HV γ^0 compared to forests.

15.2.3   FNF CLASSIFICATION ALGORITHM AND VALIDATION

15.2.3.1   Extracting Forest and Non-Forest Statistics

As confusion between forest and non-forest areas was greater at HH polarizations, HV thresholds alone were used for FNF mapping. For the 15 regions, histograms of HV γ^0 for forest types and non-forest areas were generated using data from ROIs A along with cumulative distribution functions, from which the threshold separating forest and non-forest, notated by x, was determined as follows:

$$F_F(x) = 1 - F_{NF}(x)$$

$$F_F(x) \equiv \int_x^\infty f_F(x')dx' \qquad\qquad (15.4)$$

$$F_{NF}(x) \equiv \int_{-\infty}^x f_{NF}(x')dx'$$

where F_F is the accuracy in the estimate of forest area using the threshold of x, defined for each region, F_{NF} is the same but for non-forest, f_F and f_{NF} are the probability density functions of γ^0 for forest and non-forest, respectively, and x' is γ^0. To determine x, the value was increased in 0.1-dB increments until the solution giving the greatest accuracy in Equation (15.4) was obtained. The accuracy would be 50% when forest and non-forest regions had the same distributions but 100% when these were totally different. Hence, the thresholds varied depending upon the land cover occurring.

15.2.3.2   FNF Classification Algorithm

To assist in the classification of non-forest areas, a rule-based approach was applied within eCognition to first identify urban and water areas (Figure 15-11). To assist the classification, a geometric density (GD) function was calculated for objects, with this representing the number of pixels in the object

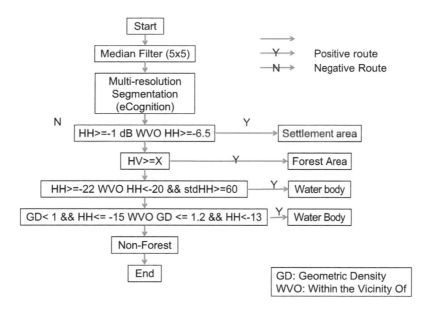

FIGURE 15-11 Rules used for the classification of forest and non-forest areas (including urban and water) from the ALOS PALSAR mosaic data.

divided by the radius of the objects and distinguishing between those that were compact (square or circular) or more elongated. Objects "within the vicinity" of others (WVO) were also defined. Urban areas (settlements) were then separated using HH and HV γ^0 thresholds of greater than −1 dB and −6.5 dB, respectively, with these being WVO each other (Figure 15-11). Water bodies were associated with objects where the HH was greater than or equal to −22 dB, the neighboring segment supported an HH less than −20 dB, and the standard HH was greater than or equal to 60 dB (Figure 15-11). The standard HH was defined as a ratio of the number of pixels within the segment to the standard deviation of the HH γ^0 (number of pixels/dB), which was found to be greater for more uniform targets. Water bodies were also defined where proximal objects supported a GD between 1 and 1.2 and where the HH backscatter was less than or equal to 15 dB and was less than or equal to 13 dB, respectively. Several images acquired in Alaska, Siberia, and Scandinavia supported HH and HV γ^0 values that were quite low (e.g., because of snow cover) and images from other years were therefore substituted. Forest and remaining non-forest areas were then separated using variable thresholds.

For each of the 15 regions, the histograms of HV γ^0 and the associated accumulative distribution functions for forest and non-forest are illustrated in Figure 15-12a and b), respectively. The associated statistics are given in Table 15-6 and indicate that, for forest, the average values of HV γ^0 varied from −11.34 to −16.75 dB, although for non-forest, these varied from −12.38 to −24.84 dB. For comparison, the average values of HH γ^0 varied from −6.35 to −9.21 dB and −6.50 to −11.85 dB for forest and non-forest, respectively. However, the actual thresholds defined for each of the regions (based on Equation 15.4) and used in the FNF mapping are listed in Table 15-7.

15.2.3.3 Validation

To provide validation of the FNF maps, data from GEI (http://www.google.co.jp/intl/ja/earth/), the Degree Confluence Project (DCP; http://confluence.org/), and the Global FRAs (FAO 2000, 2012a, 2012b) were used, with the location of points shown in Figure 15-13 and the number given in Table 15-8. The images within Google Earth were available from 2000 to 2012, with a date stamp given for each point. A total of 4,114 points (1,456 and 2,548 for forest and non-forest, respectively) were selected (Figure 15-13a) with the year of assignment provided in Table 15-8. Of these, 1,529 were for the period 2007 to 2010. The DCP data were collected by volunteers, with photographs

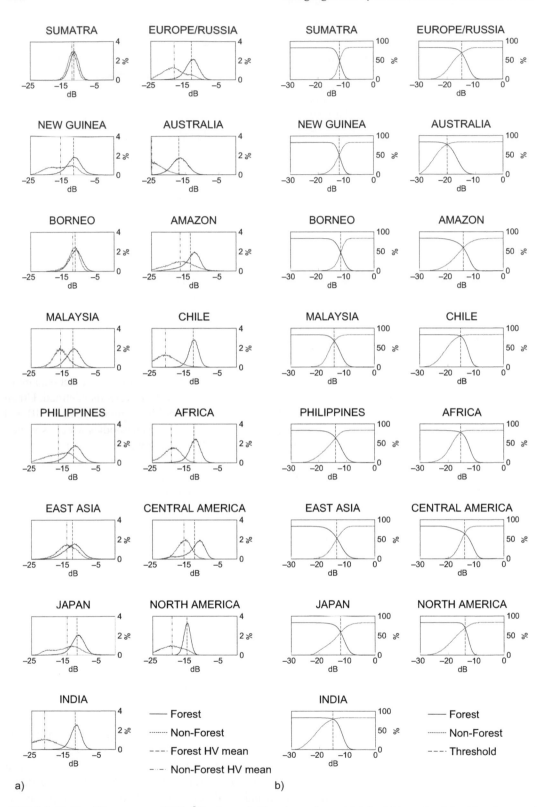

FIGURE 15-12 Histograms of HV γ^0 for forest (left) and non-forest (right) within each of the 15 regions, and cumulative histograms showing the cross-over point between forest (left) and non-forest (right).

TABLE 15-6

HH and HV γ^0 Statistics for Forest and Non-Forest Areas Located in Each of the 15 Regions Based on ROIs A

No.	Region	Forest HH μ	Forest HH δ	Forest HV μ	Forest HV δ	Forest ROIs No.	Non-Forest HH μ	Non-Forest HH δ	Non-Forest HV μ	Non-Forest HV δ	Non-Forest ROIs No.
1.	Sumatra	−7.19	1.46	−11.86	1.41	10	−7.35	1.36	−12.39	1.38	10
2.	New Guinea	−6.90	1.46	−11.98	1.45	6	−6.70	1.84	−13.15	2.22	5
3.	Borneo	−6.89	2.03	−11.69	1.92	8	−6.85	1.63	−12.38	1.92	7
4.	Malaysia	−7.67	2.29	−12.46	2.17	3	−8.80	2.30	−16.16	2.34	3
5.	Philippines	−7.36	2.74	−12.32	2.61	3	−9.44	3.58	−16.88	3.67	2
6.	East Asia	−7.22	3.01	−12.69	2.98	9	−7.64	2.26	−14.40	2.75	8
7.	Japan	−6.35	2.37	−11.34	2.27	5	−6.50	4.42	−14.33	4.44	5
8.	India	−6.77	1.78	−12.00	1.83	5	−11.85	3.70	−21.21	3.60	4
9.	Europe/Russia	−7.16	2.19	−13.07	2.09	4	−8.27	1.81	−18.10	3.30	4
10.	Australia	−9.21	2.08	−16.75	2.38	6	−10.47	1.74	−24.84	3.19	5
11.	Amazon	−7.40	1.50	−13.56	2.79	8	−8.21	2.91	−16.58	4.23	3
12.	Chile	−8.06	1.54	−12.85	1.58	4	−8.61	1.92	−21.00	3.19	2
13.	Africa	−7.30	1.60	−12.76	2.01	7	−11.13	2.34	−19.38	2.70	6
14.	North America	−6.36	1.67	−12.62	3.03	6	−7.53	1.47	−15.72	2.08	3
15.	Central America	−7.82	1.25	−12.92	1.26	6	−8.58	3.27	−18.18	3.95	5

TABLE 15-7

Thresholds of HV γ^0 Identified for the 15 Regions

No.	Region	Threshold (dB)	Accuracy (%) Forest	Accuracy (%) Non-Forest
1.	Sumatra	−12.1	57.92	57.61
2.	New Guinea	−12.3	59.51	61.31
3.	Borneo	−11.9	56.35	56.02
4.	Malaysia	−14.3	80.92	79.85
5.	Philippines	−13.9	76.01	76.24
6.	East Asia	−13.5	62.54	61.57
7.	Japan	−12.2	69.62	69.91
8.	India	−15.1	95.20	94.88
9.	Europe	−14.8	82.19	82.29
10.	Australia	−20.1	91.69	92.00
11.	Amazon	−14.4	70.28	70.46
12.	Chile	−15.5	95.20	95.32
13.	Africa	−15.6	92.80	93.23
14.	North America	−14.2	76.92	76.96
15.	Central America	−14.1	83.20	82.73

Note: The accuracy is calculated from Equation (15.4) using the ROIs A in Figure 15-4.

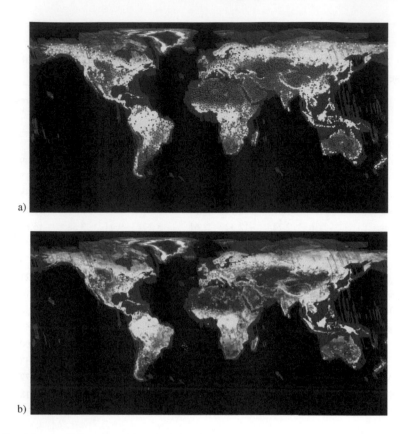

FIGURE 15-13   The spatial locations of (a) the 1,456 forest (green) and 2,548 non-forest (red) validation points generated from GEI and (b) points located in 2007 (red), 2008 (sky-blue), 2009 (blue), and 2010 (yellow) based on the DCP.

TABLE 15-8

The Number of GEI and DCP Points Used for the Validation of the Forest and Non-Forest Maps by Year. The Numbers in Brackets Indicate Those That Were Available for the Period 2007–2010

Acquisition Year	GEI	DCP
2000	29	
2001	58	
2002	148	
2003	335	
2004	354	
2005	394	
2006	440	
2007	500	940
2008	140	742
2009	304	690
2010	585	280
2011	621	
2012	236	
TOTAL	4,144 (1,529)	2,652 (2,652)

taken at each point (Iwao et al. 2006; Figure 15-13b). Each point was then associated with forest or non-forest class by Nagoya University. The Global FRA was conducted by the Food and Agricultural Organization (FAO), which involved all member countries and resulted in the production of a database of land use, land cover, and forestry (LULCF) categories (FAO 2000; Shimada et al. 2011). The FNF maps for 2007 and 2008 and also 2009 and 2010 were compared with the FRA2005 and FRA2010, respectively. Separate comparisons were conducted for Africa, Asia, Europe, North and South America, and South America as well as the total land area.

15.3 RESULTS

15.3.1 FOREST AND NON-FOREST MAPS

The maps of forest and non-forest for 2007, 2008, 2009, and 2010 are illustrated in Figure 15-14 with more detailed examples from Southeast Asia, Central Africa, and South America presented

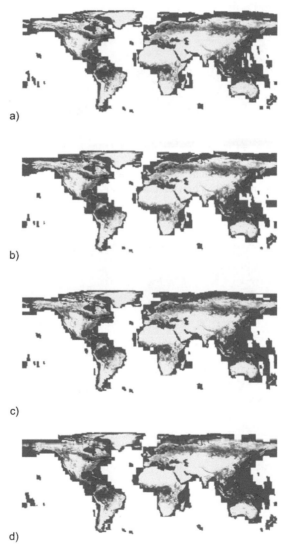

a)

b)

c)

d)

FIGURE 15-14 Global mosaics of forest and non-forest for (a) 2007, (b) 2008, (c) 2009, and (d) 2010 generated from ALOS PALSAR mosaics displayed in Figure 15-1.

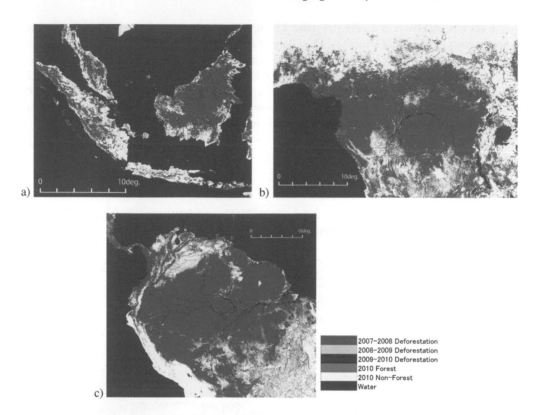

FIGURE 15-15   Deforestation in (a) Indonesia, (b) Central Africa, and (c) northern South America between 2007 and 2008 (pink), 2008 and 2009 (ochre), and 2009 and 2010 (red). Areas observed to remain in forest over the time-series are shown in green.

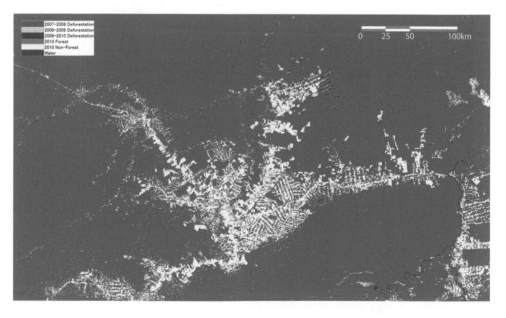

FIGURE 15-16   A subset of the mosaic for Acre, Brazil, showing deforestation between 2007 and 2008 (pink), 2008 and 2009 (ochre), and 2009 and 2010 (red). Areas observed to remain in forest over the time-series are shown in green.

TABLE 15-9

Comparison with FRA Forest Areas (× 1000 ha; 2005 and 2010)

Region	Landsat[a] (2000)	PALSAR (2007)	FRA (2005)	Diff (%)[b]	PALSAR (2008)	FRA (2005)	Diff (%)	PALSAR (2009)	FRA (2010)	Diff (%)	PALSAR (2010)	FRA (2010)	Diff (%)	PALSAR (2010– 2007)	FRA (2010– 2005)
Africa	664,834	635460	691369	−8.09	630185	691369	−8.85	641822	674318	−4.82	653447	674318	−3.10	17987	−17051
Asia	545,418	580807	584049	−0.56	581074	584049	−0.51	573354	592513	−3.23	594370	592513	0.31	13563	8464
Eurasia	992,909	945540	1009462	−6.33	935662	1009462	−7.31	926121	1013297	−8.60	933957	1013297	−7.83	−11583	3835
N/C America	778,456	697116	705183	−1.14	673179	705183	−4.54	686989	705281	−2.59	680659	705281	−3.49	−16457	98
Oceania	132,427	187538	196745	−4.68	177026	196745	−10.02	177597	191385	−7.20	179115	191385	−6.41	−8423	−5360
S. America	951,614	807790	874158	−7.59	825212	874158	−5.60	813203	856269	−5.03	811082	856269	−5.28	3292	−17889
Total	**4,065,657**	**3854250**	**4060966**	**−5.09**	**3822337**	**4060966**	**−5.88**	**3819087**	**4033063**	**−5.31**	**3852630**	**4033063**	**−4.47**	**−1620**	**−27903**

Note: [a]Derived from Hansen et al. (2013) for tree cover 26% to 100%.
[b]Derived from 1-PALSAR divided by FRA

in Figures 15-15 and 15-16. The total area estimated was 3,854.250; 3,822.337; 3,819.087; and 3,852.630 million ha for 2007, 2008, 2009, and 2010, respectively. The comparison with the FRA mapping (2007 and 2008 with FRA2005: 4,060.966 million ha; 2009 and 2010 with FRA2010: 4,033.063 million ha) suggested an underestimate of between 4.5% and 5.9% when using the ALOS PALSAR data (Table 15-9). The area was also smaller (by 5.20%) than that estimated by Hansen et al. (2013; 4,065.657 million ha) based on a tree cover of 26 to 100%. Despite the differences, there were broad similarities in the distribution of forests at the regional level when comparing the mapping generated by Hansen et al. (2013; based on 26–100% cover) and through this study (Table 15-10). The net forest decrease based on changes in the PALSAR FNF between 2007 and 2010 (Table 15-9) was estimated at 1.620 million ha (−0.042%) and that between FRA2005 and FRA2010 was 27.903 million ha (−0.687%). At a regional level, most decreases were observed in Eurasia, North and Central America, and Oceania (Table 15-11).

15.3.2 ACCURACY OF CLASSIFICATION

Validation against the GEI indicated an overall accuracy in the classification of forest and non-forest exceeding 91% (Table 15-12), although this was reduced to no less than 82% when compared with the DCP (Table 15-13). Average accuracies over the four years of mapping were 91.25% and 84.86% based on the GEI and DCP assessments, respectively. Accuracies were least for Oceania. Validation against the FRA indicated an overall accuracy of 94.81%. It should be noted that the areas used for training and the validation were prepared independently.

TABLE 15-10

Estimates of Forest Cover Obtained from ALOS PALSAR Data (2007) and Landsat Sensor Data (2000) (× 1,000 ha)

	ALOS PALSAR	Landsat	% Difference
Africa	635,460	664,834	95.6%
Asia	580,807	545,418	106.5%
Eurasia	945,540	992,909	95.2%
North/Central America	697,116	778,456	89.6%
South America	807,790	951,614	85%
Oceania	187,538	132,427	142.0%
TOTAL	3,854,250	4,065,657	94.8%

TABLE 15-11

Differences in the Forest Area (× 1,000 ha) Estimated from ALOS PALSAR Data

Region	2008–2007	2009–2008	2010–2009	2010–2007[a]
Africa	−5,276	11,637	11,626	17,987
	(−0.830)[b]	(1.847)	(1.811)	(2.83)
Asia	267	−7,719	21,016	13,563
	(0.046)	(−1.328)	(3.665)	(2.34)
Eurasia	−9,877	−9,541	7,835	−11,583
	(−1.045)	(−1.020)	(0.846)	(−1.23)
N/C America	−23,937	13,810	−6,331	−16,457
	(−3.434)	(2.051)	(−0.926)	(−2.36)
Oceania	−10,512	571	1,518	−8,423
	(−5.605)	(0.323)	(0.855)	(−4.49)
S. America	17,422	−12,009	−2,122	3,292
	(2.157)	(−1.455)	(−0.261)	(0.407)
Total	**−31,913**	**−3,250**	**33,543**	**−1,620**
	(−0.828)	**(−0.085)**	**(0.878)**	**(−0.042)**

Note: [a]Average; [b]ratio of areas, Year 2 to Year 1; number in bracket is percentile.

15.3.3 COMPARISON OF HH AND HV γ^0 OVER TIME

Over the four-year period, an overall decrease in HH and HV γ^0 was observed for both forest and non-forest areas (Table 15-14) with a 99% confidence that this was not attributed to changes in the system performance of the ALOS PALSAR.

15.4 DISCUSSION

15.4.1 APPROACH TO FNF CLASSIFICATION

The use of variable thresholds for the discrimination of forest and non-forest provided a more reliable estimation of the forest area compared to the previous version presented in Shimada et al. (2011), which focused on using a single threshold. The thresholds used were based on HV γ^0 from a range of forest types as determined from GEI, which allowed these to be encompassed in the forest mask. However, some forest types (e.g., regenerating forests, oil palm, coconut plantations, and high biomass mangroves) were classified as non-forest as these supported a lower HV γ^0 than the majority of forests. Others (e.g., rubber plantations) were unavoidably merged with natural forest. Nevertheless, the accuracies in the discrimination of known areas of forest and non-forest exceeded 82% and often 90%.

Discrepancies in the global extent of forests estimated by the FAO and also Hansen et al. (2013) were observed, with this attributed to the following.

a. The FRA2005 and FRA2010 were generated by combining 233 country reports with full or partial use of Landsat sensor data from 1990 to 2000 and 2005 to 2010. Hansen et al. (2013) used time-series of Landsat sensor data from 2000 to 2012, and our study used ALOS PALSAR data from 2007, 2008, 2009, and 2010. Hence, differences in forest extent would have resulted from the different dates of the imagery used. Changes in forest extent would also have occurred among the image acquisition dates, with Hansen et al. (2013) reporting losses and gains between 2000 and 2012 of 230 million ha and 80 million ha,

TABLE 15-12

Accuracy in the Classification of Forest and Non-Forest from ALOS PALSAR for Each Year Based on GEI

			PALSAR FNF 2007		
		Forest	Non-Forest	Total	PA (%)
GEI Points	Forest	1,300	274	1,574	**82.59**
	Non-Forest	64	2,332	2,396	**97.33**
	Total	1,364	2,606	3,970	-
	UA (%)	**95.31**	**89.49**	-	**91.49**
			PALSAR FNF 2008		
		Forest	Non-Forest	Total	PA (%)
GEI Points	Forest	1,277	297	1,574	**81.13**
	Non-Forest	77	2,319	2,396	**96.79**
	Total	1,354	2,616	3,970	-
	UA (%)	**94.31**	**88.65**	-	**90.58**
			PALSAR FNF 2009		
		Forest	Non-Forest	Total	PA (%)
GEI Points	Forest	1,283	291	1,574	**81.51**
	Non-Forest	67	2,329	2,396	**97.20**
	Total	1,350	2,620	3,970	-
	UA (%)	**95.04**	**88.89**	-	**90.98**
			PALSAR FNF 2010		
		Forest	Non-Forest	Total	PA (%)
GEI Points	Forest	1,285	289	1,574	**81.64**
	Non-Forest	63	2,333	2,396	**97.37**
	Total	1,348	2,622	3,970	-
	UA (%)	**95.33**	**88.98**	-	**91.13**

Note: UA: Users' accuracy; PA: producers' accuracy.

respectively, with 20 million ha of this area experiencing both a loss and a gain. Some areas of forest or non-forest may also not have been adequately detected with the Landsat sensor because of cloud cover.

b. The optical remote sensing data used for generating some of the FRA estimates and also those of Hansen et al. (2013) are more sensitive to the two-dimensional structure of vegetation, particularly the foliage cover. Hence the extent of woody vegetation may have been overestimated in some cases because of confusion with herbaceous vegetation. Conversely, some woody vegetation with deciduous or semi-deciduous characteristics may have been omitted using the Landsat sensor data, particularly in savannas or temperate areas.

c. The PALSAR resolution (with a 34-degree off-nadir viewing geometry) may be too coarse to allow sparse forests and woodlands to be captured in the classification (particularly toward the lower cover threshold of 10%), thereby leading to an underestimate in the total area. Some forest types, including lower biomass sparse forests and woodlands, some plantations, and also higher biomass mangroves, were also excluded because values of HV γ^0 were lower than the thresholds applied.

TABLE 15-13

Accuracy in the Classification of Forest and Non-Forest from ALOS PALSAR for Each Year Based on DCP

		PALSAR FNF 2007				
		Forest	Non-Forest	Other	Total	PA (%)
DCP 2007	Forest	172	59	19	250	**68.80**
	Non-Forest	27	616	32	675	**91.26**
	Total	199	675	51	925	-
	UA (%)	**86.43**	**91.26**	**0.00**	-	**85.19**
		PALSAR FNF 2008				
		Forest	Non-Forest	Other	Total	PA (%)
DCP 2008	Forest	127	39	19	185	**68.65**
	Non-Forest	26	497	28	551	**90.20**
	Total	153	536	47	736	–
	UA (%)	**83.01**	**92.72**	**0.00**	-	**84.78**
		PALSAR FNF 2009				
		Forest	Non-Forest	Other	Total	PA (%)
DCP 2009	Forest	89	34	22	145	**61.38**
	Non-Forest	30	434	26	490	**88.57**
	Total	119	468	48	635	-
	UA (%)	**74.79**	**92.74**	**0.00**	-	**82.36**
		PALSAR FNF 2010				
		Forest	Non-Forest	Other	Total	PA (%)
DCP 2010	Forest	44	14	8	66	**66.67**
	Non-Forest	5	200	9	214	**93.46**
	Total	49	214	17	280	-
	UA (%)	**89.80**	**93.46**	**0.00**	-	**87.14**

TABLE 15-14

Differences in Mean HH and HV γ^0 2007–2008, 2008–2009, and 2009–2010

Year		Global Land	Global Forest	Tropical Forest[a]	95% Confidence	99% Confidence
2007–2008	HH	−0.093	−0.106	0.008	0.0049	0.0064
	HV	−0.104	0.027	−0.044	0.0047	0.0061
2008–2009	HH	0.039	−0.441	−0.107	0.0049	0.0064
	HV	0.061	−0.130	−0.005	0.0047	0.0061
2009–2010	HH	−0.066	0.087	0.018	0.0049	0.0064
	HV	−0.040	0.010	0.006	0.0047	0.0061
Mean	HH	−0.040	−0.106	−0.027	0.0049	0.0064
	HV	−0.028	−0.031	−0.02	0.0047	0.0061

Note: [a]The values for tropical forest are averaged from Southeast Asia, Amazonia, and Central Africa. All the units are in decibels (dB).

Discrepancies also occurred because of the differences in the methods used for FNF mapping. In particular, the FAO used a sample-based approach and country reports, although a wall-to-wall mapping was used in this study and by Hansen et al. [2013]. The FAO, contributing countries, and the European Commission (EC) Joint Research Centre (JRC) merged country reports for the generation of the FRA2005 and FRA2010 but also referenced time series of Landsat data from 1990, 2000, 2005, and 2010. For each integer latitude and longitude intersect, 10 km × 10 km Landsat image samples were taken, giving a total of 15,779 image samples with each classified through multi-spectral comparison, supervised approaches, or manual delineation. The classifications were developed by contributing countries and through attendance at workshops, with four main land cover classes considered and more than 73 spectral libraries used (FAO 2012b). Forest areas were defined as supporting more than 10% canopy cover and being more than 0.5 ha in area. The FRA also defined forests in some countries (e.g., China) as both natural and managed (Christine et al. 2007). In this study, wall-to-wall mapping of FNF was achieved using ALOS PALSAR data acquired in a single year, with the FAO definition of forest cover also used. Four FNF maps were also generated at a global level, whereas the FRA produced single FNF maps based on three years of data. The spatial resolution of the FNF maps generated using the ALOS PALSAR data was also much finer (25-m spatial resolution) allowing greater detail to be resolved. The use of HV γ^0 only in the mapping allowed for better discrimination of forest from non-forest because at HH polarization, confusion with rough ground was more evident. Hansen et al. (2013) produced a new high-resolution map of global forest-cover change between 2000 and 2012 using the 30-m resolution Landsat 7 archives. In generating the maps, Hansen et al. (2013) converted all the cloud-free pixels to top of the atmosphere reflectance and calculated time-series of spectral metrics from the growing season during 2000 to 2012. A time-series decision algorithm was then applied. This used the MODIS/Landsat-based tree cover and was supported by manual interpretation of the target using high-resolution optical (e.g., GeoEye) data. On this basis, wall-to-wall maps of forest cover gain and loss were obtained for four different tree cover stratums (i.e., less than 25%, 26~50%, 51~75%, and 76~100%) at the levels of continents, sub-regions, and countries. To resolve the differences in the mapping, the integration of the various data sets (i.e., country reports and, particularly, Landsat and ALOS PALSAR data) should be considered particularly as the latter are complementary in the information that they can provide. The recent public release of the data (albeit at 50-m spatial resolution), the FNF maps, as well as the maps of Hansen et al. (2013) has made significant headway toward achieving this aim.

15.4.2 Comparison of Annual Area Estimates and Trends

The FAO indicated a general decrease of 27.903 million ha (−0.69%) between 2005 and 2010 (FAO 2010). The mapping from ALOS PALSAR data estimated that 1.620 million ha had been lost between 2007 and 2010, although losses and gains were variable in the intervening period. This may have been attributed to differences in HV γ^0 because of environmental effects but was also directly linked to forest growth and loss. Hansen et al. (2013) reported a loss of 230 million ha between 2000 and 2012 but also a gain of 80 million ha, with 20 million ha experiencing both a loss and a gain. These differences in estimated losses reflect the differences in the methods used for classifying the forests and detecting change as well as the variable timeframes of observation and hence are not directly comparable. However, the recommendation is that comparisons be undertaken between the Landsat-based estimates for 2007 through to 2010 and those generated using the ALOS PALSAR data.

The ALOS PALSAR comparison between years indicated a lower forest extent in 2010 compared to 2007 in Eurasia (by 1.23%), North America (2.36%), and Oceania (4.49%) but a larger extent in South America (0.41%), Asia (2.34%), and Africa (2.83%). While the total average estimate of forest area changes between 2007 and 2010 and FRA2005–2010 pointed in the same direction at a global level, there were some disagreements among regions. These differences in mapped forest area were considered to be due to the greater variation in the radar backscatter within the non-forested compared to the forested areas, which compromised FNF mapping, and also to prevailing environmental (e.g., surface moisture) effects (Lucas et al. 2010), even though efforts were made to minimize these by using data with minimal observed effects. The increases in Asia were attributed, in part, to the expansion of plantations, with the implementation of large reforestation projects in China (Christine et al. 2007) leading to an average annual percentage increase in the area of forest of 0.51% since the late 1970s (Shi et al. 2011). Over the forest areas, the decrease in HH and HV γ^0 was attributed to the replacement of forests by non-forested surfaces, a general smoothing of these surfaces (e.g., as a consequence of agricultural improvement), and an overall decrease in biomass within the vegetated component. These differences were not associated with degradation of the sensor performance over time.

The discrepancies in Eurasia and North and Central America between the FRA and PALSAR can be attributed as follows. For North and Central America, the FRA reported that the Canadian component was duplicated from the FRA2000 (FAO 2000) and hence decreases were not evident. However, the Canadian Forest Service (NRC 2012) reported a forest decrease with the ALOS PALSAR data for the four-year period also suggesting forest disturbance and decreases. Within Russia (Eurasia), the World Wildlife Fund (WWF) reported a decrease (WWF 2007), which concurs with that observed using the ALOS PALSAR data. Within Europe itself, a slight increase in the forest area was evident.

15.5   SUMMARY AND CONCLUSIONS

At a global level, mosaics of ALOS PALSAR data were generated for 2007, 2008, 2009, and 2010 at 25-m spatial resolution. Using thresholds of HV polarization data that varied regionally, maps of forest and non-forest were generated. The accuracy of the mapping was assessed against GEI, DCP, and FRA data.

The main conclusions from the study are:

a. The PALSAR remained stable (within 0.065 dB) over its lifetime (from 2006 to 2010), so changes in HV γ^0 over time could be attributed to changes in the land cover.

b. For forest areas, γ^0 remained stable at both HH and HV, with annual averages of the standard deviation being 0.21 ± 0.18 dB and 0.21 ± 0.19 dB, respectively.

c. The thresholds for HH and HV γ^0 for separating forest and non-forest were regionally variable, being -6.89 ± 0.95 dB in HH and -12.07 ± 1.52 dB in HV.

d. In comparison to the DCP, GEI, and FRA 2005/2010, accuracies of 84.86%, 91.25%, and 94.81% were obtained in the mapping of forest and non-forest at a global level with regional variations.

e. Based on these estimates, the decrease in forest cover between 2007 and 2010 was 1.620 million ha (−0.042%), with the FRA estimating a decrease of 27.903 million ha (−0.687%; based on FRA2010 and FRA2005).

f. γ^0 decreased by 0.040dB yr^{-1} in HH and 0.028dB yr^{-1} in HV globally and regionally, with this potentially related to decreases in forest area and AGB and a smoothing of the non-forest area (e.g., as a consequence of agricultural management leading to improvement of cleared areas).

Currently, the PALSAR FNF maps are derived from the PALSAR FBD mode, which is 10 m-slant range resolution. Refinement of the methods outlined here and the use of data from JAXA's forthcoming ALOS-2 PALSAR-2 is likely to better define the extent of forest but also changes in forest as a result of natural and human-induced events and processes. Estimates would also be improved by using combinations of the ALOS PALSAR data and the forest cover maps generated by Hansen et al. (2013). Hence, future effort should focus on understanding the synergies between these two data sets to reduce the uncertainties in forest losses and gains across the globe. Such assessments should be linked with those of the FAO, using both past and future data.

ALOS-2 was launched in 2014 carrying the PALSAR-2 and thrown to the operational monitoring of the global forest. The annual database for FNF has been produced since 2014 and is provided to the public from the JAXA/EORC (http://www.eorc.jaxa.jp/ALOS/en/palsar_fnf/fnf_index.htm). These data sets should be a part of the elemental monitoring of the global Earth and its environmental changes. (Shimada et al. 2014)

APPENDIX 15A-1: REQUIREMENT FOR THE NUMBER OF SAMPLES, N

Assume a data set follows the normal distribution of $n(\mu,\sigma)$, where μ is the mean and σ is its standard deviation, the possible error of the estimated mean using N samples is $\Delta_{X\%} \cdot \sigma / \sqrt{N}$, where $\Delta_{X\%}$ is the cumulative distribution function at the confidence level of X%.

Under the given error requirement, Δ_E, N can be obtained as follows:

$$N \geq \left(\frac{\Delta_{X\%} \cdot \sigma}{\Delta_E} \right)^2 . \tag{15A-1.1}$$

As an example, the $\Delta_{X\%}$ is 1.960 as 95%, σ is 2.15, and Δ_E is 0.01. N should be bigger than 178,000.

REFERENCES

Moderate Resolution Imaging Spectroradiometer

Achard, F., Eva, H., Mayaux, P., Stibig, H., and Belward, A., 2004, "Improved Estimates of Net Carbon Emissions from Land Cover Change in the Tropics for the 1990s," *Global Biogeochem. Cy.*, Vol. 18, No. 2, https://doi.org/10.1029/2003GB002142

Arino, O., Gross, D., Ranera, F., Bourg, L., Leroy, M., Bicheron, P., Brockman, C., et al., 2007, "GlobCover: ESA Service for Global Land Cover from MERIS," *Proc. International Geoscience and Remote Sensing Symposium, IGARSS 2007*, Barcelona, Spain, July 23–28, 2007, pp. 2412–2415.

Asner, G. P., Powell, G. V. N., Mascaro, J., Knapp, D. E., Clark, J. K., Jacobson, J., et al., 2010, "High-Resolution Forest Carbon Stocks and Emissions in the Amazon," *P. Natl. A. Sci USA*, Vol. 107, No. 38, pp. 16738–16742.

Bartholomé, E. and Belward, A. S., 2005, "A New Approach to Global Land Cover Mapping from Earth Observation Data," *Int. J. Remote Sens.*, Vol. 26, No. 9, pp. 1959–1977.

Bryant, D., Nielsen, D., and Tangley, L., 1997, *The Last Frontier Forests: Ecosystems and Economies on the Edge*, World Resources Institute, Washington, DC, p. 1.

Christine, J. T., Harrell, S., Hinckley, T. M., and Henck, A. C., 2007, "Reforestation Programs in Southwest China: Reported Success, Observed Failure, and the Reasons Why," *J. Mt. Sci.*, Vol. 4, No. 4, pp. 275–292.

Cloude, S. R. and Papathanassiou, K. P., 2003, "Three Stage Inversion Process for Polarimetric SAR Interferometry," *IEE P. Radar Son. Nav.*, Vol. 150, No. 3, pp. 125–134.

Dos Santos, R. J., Gonçalves, F. G., Dutra, L. V., Mura, J. C., and Paradella, W. R., 2007, "Analysis of Airborne SAR Data (L-Band) for Discrimination Land Use/Land Cover Types in the Brazilian Amazon Region," *Proc. International Geoscience and Remote Sensing Symposium, IGARSS 2007*, Barcelona, Spain, July 23–28, 2007, pp. 2342–2345.

Englhart, S., Keuck, V., and Siegert, F., 2011, "Aboveground Biomass Retrieval in Tropical Forests—The Potential of Combined X-and L-Band SAR Data Use," *Remote Sens. Environ.*, Vol. 115, No. 5, pp. 1260–1271.

Eva, H. D., Belward, A. S., De Miranda, E. E., Di Bella, C. M., Gond, V., Huber, O., Jones, S., Sgrenzaroli, M., and Fritz, S., 2004, "A Land Cover Map of South America," *Glob. Change Biol.*, Vol. 10, pp. 1–14.

Food and Agricultural Organization (FAO) of the UN, 2000, *Forest Resource Assessment (FRA) 2000*, Food and Agricultural Organization of the UN, Rome.

Food and Agriculture Organization (FAO) of the UN, 2010, *Global Forest Resources Assessment 2010, Country Report, CANADA*, FRA2010/036, Food and Agricultural Organization of the UN, Rome.

Food and Agricultural Organization (FAO) 2012a, *Global Forest Land-Use Change 1990–2005*, Food and Agriculture Organization of the United Nations, Rome.

Food and Agricultural Organization (FAO) 2012b, *Forest Resource Assessment (FRA) 2010*, Food and Agricultural Organization of the United Nations, Rome.

Friedl., M. A., McIver, D. K., Hodges, J. C. F., Zhang, X. Y., Muchoney, D., Strahler, A. H., Woodcock, C. E., et al. 2002, "Global Land Cover Mapping from MODIS: Algorithms and Early Results," *Remote Sens. Environ.*, Vol. 83, Nos. 1–2, pp. 287–302, 2002.

Giri, C., Ochieng, E. Tieszen, L.L., Zhu, Z., Singh, A., Loveland, T., Masek, J., and Duke, N., 2010, "Status and Distribution of Mangrove Forests of the World Using Earth Observation Satellite Data," *Global Ecol. Biogeogr.*, Vol. 20, No. 1, pp. 154–159.

Hansen, M. C. and DeFries 2004, Detecting long term forest change using continuous fields of tree cover maps from 8km AVHRR data for the years 1982–1999. *Ecosystems*, 7, 695–716.

Hansen, M. C., DeFries, R. S., Townshend, J. R. G., Carroll, M., Dimiceli, C., and Sohlberg, R. A., 2003, "Global Percent Tree Cover at a Spatial Resolution of 500 Metre: First Results of the MODIS Vegetation Continuous Fields Algorithm," *Earth Interact.*, Vol. 7, pp. 1–15.

Hansen, M. C., DeFries, R. S., Townshend, J. R. G., and Sohlberg, R., 2000, "Global Land Cover Classification at 1 km Spatial Resolution Using a Classification Tree Approach," *Int. J. Remote Sens.*, Vol. 21, pp. 1331–1364.

Hansen, M. C., Potapov, P. V., Moore, R., Hancher, M., Turubanova, S. A., Tyukavina, A., Thau, D., et al., 2013, "High-Resolution Global Maps of 21st-Century Forest Cover Change," *Science*, Vol. 342, No. 6160, pp. 850–853.

Hansen, M. C., Shimabukuro, Y., Potapov, P., and Pitman, K., 2008, "Comparing Annual MODIS and PRODES Forest Cover Change for Advancing Monitoring of Brazilian Forest Cover," *Remote Sens. Environ.*, Vol. 112, pp. 3784–3793.

Hansen, M. C., Stehman, S., Potapov, P., Arunarwati, B., Stolle, F., and Pittman, K., 2009, "Quantifying Changes in the Rates of Forest Clearing in Indonesia from 1990 to 2005 Using Remotely Sensed Data Sets," *Environ. Res. Lett.*, Vol. 4, No. 3, 034001.

Hawkins, R., Attema, E., Crapolicchio, R., Lecomte, P., Closa, J., Meadows, P. J., 2000, "Stability of Amazon Backscatter at C-Band: Spaceborne Results from ERS-1/2 and RADARSAT-1," ESA SP-450, *Proc. of the CEOS SAR Workshop*, Toulouse, France, October 26–29, 1999, pp.

Hoekman, D. H. and Quiriones, M. J. 2000, "Land Cover Type and Biomass Classification Using AirSAR Data for Evaluation of Monitoring Scenarios in the Colombian Amazon," *IEEE T. Geo. Sci. Remote*, Vol. 38, No. 2, pp. 685–696.

Iwao, K., Nishida, K., Kinoshita, T., and Yamagata, Y. 2006, "Validating Land Cover Maps with Degree Confluence Project Information," *Geophys. Res. Lett.*, Vol. 33, L23404.

Lefsky, M. A., 2010, "A Global Forest Canopy Height Map from the Moderate Resolution Imaging Spectroradiometer and the Geoscience Laser Altimeter System," *Geophys. Res. Lett.*, Vol. 37, L15401.

Longepe, N., Rakwatin, P., Isoguchi, O., Shimada, M., Uryu, Y., and Yulianto, K., 2011, "Assessment of ALOS PALSAR 50m Orthorectified FBD Data for Regional Land Cover Classification by using Support Vector Machines," *IEEE T. Geo. Sci. Remote*, Vol. 49, No. 6, pp. 2135–2150.

Loveland, T., Reed, B. C., Brown, J. F., Ohlen, D. O., Zhu, Z., Yang, L., and Merchant, J. W., 2000, "Development of a Global Land Cover Characteristics Database and IGBP DISCover from 1 km AVHRR Data," *Int. J. Remote Sens.*, Vol. 21, pp. 1303–1330.

Lucas, R. M., Armston, J., Fairfax, R., Fensham, R., Accad, A., Carreiras, J., Kelley, J., et al., 2010, "An Evaluation of the ALOS PALSAR L-Band Backscatter—Above Ground Biomass Relationship Queensland, Australia: Impacts of Surface Moisture Condition and Vegetation Structure," *IEEE J. Sel. Top. Appl.*, Vol. 3, pp. 576–593.

Luckman, A. J., Baker, J. R., Honzák, M. H., and Lucas, R. M., 1998, "Tropical Forest Biomass Density Estimation Using JERS-1 SAR: Seasonal Variation, Confidence Limits and Application to Image Mosaics," *Remote Sens. Environ.*, Vol. 62, No. 2, pp. 126–139.

Lukowski, T. I., Hawkins, R. K., Cloutier, C., Wolfe, J., Teany, L. D., Srivastava, S. K., Banik, B., Jha, R., and Adamovic, M., 2003, "RADARSAT Elevation Antenna Pattern Determination," *Proc. IGARSS 1997*, Singapore, August 3–8, 1997, pp. 1382–1384.

Mitchard, E. T. A., Saatchi, S. S., Lewis, S. L., Feldpausch, T. R., Woodhouse, I. H., Sonké, B., Rowland, C., and Meir, P., 2011, "Measuring Biomass Changes Due to Woody Encroachment and Deforestation/ Degradation in a Forest-Savanna Boundary Region of Central Africa Using Multi-Temporal L-Band Radar Backscatter," *Remote Sens. Environ.*, Vol. 115, pp. 2861–2873.

Motohka, T., Shimada, M., Uryu, Y., and Setiabudi, B., 2013, "Using Time Series PALSAR Gamma Naught Mosaics for Automatic Detection of Tropical Deforestation: A Test Study in Riau, Indonesia," *Remote Sens. Environ.*, Vol. 155, pp. 79–88.

Natural Resources Canada, 2012, *The State of Canada's Forests–Annual Report*, http://cfs.nrcan.gc.ca/pub-warehouse/pdfs/34055.pdf

Rakwatin, P., Longepe, N., Isoguchi, O., Shimada, M., Uryu, Y., and Takeuchi, W., 2012, "Using Multiscale Texture Information from ALOS PALSAR to Map Tropical Forest," *Int. J. Remote Sens.*, Vol. 33, No. 24, pp. 7727–7746.

Rosenqvist, A., Shimada, M., Chapman, B., Freeman, B., De Grandi, G., Saatchi, S., and Rauste, Y., 2000, "The Global Rain Forest Mapping Project: A Review," *Int. J. Remote Sens.*, Vol. 21, No. 6/7, pp. 1375–1387.

Rosenqvist, A., Shimada, M., Itoh, N., Shimada, M., and Watanabe, M., 2007, "ALOS PALSAR: A Pathfinder Mission for Global-Scale Monitoring of Environment," *IEEE T. Geo. Sci. Remote*, Vol. 45, No. 11, pp. 3307–3316.

Saatchi, S. S., Harris, N. L., Brown, S., Lefsky, M., Mitchard, E. T. A., Salas, W., Zutta, B., et al., 2011, "Benchmark Map of Forest Carbon Stocks in Tropical Regions across Three Continents," *P. Natl. A. Sci USA*, Vol. 108, No. 24, pp. 9899–9904.

Shi, L., Zhao, S., Tang, Z., and Fang, J., 2011, "The Changes in China's Forests: An Analysis Using the Forest Identity," *PLoS ONE*, Vol. 6, No. 6, e20778.

Shimabukuro, Y., dos Santos, J. R., Formaggio, A. R., Duarte, V., and Rudorff, B. F. T., 2013, "The Brazilian Amazon Monitoring Program: PRODES and DETER Projects," *Global Forest Monitoring from Earth Observation*, F. Archard and M. C. Hansen, Eds., CRC Press, Boca Raton, FL, pp. 167–184.

Shimada, M., 2005, "Long-Term Stability of L-Band Normalized Radar Cross Section of Amazon Rainforest Using the JERS-1 SAR," *Can. J. Remote Sensing*, Vol. 31, No. 1, pp. 132–137.

Shimada, M., 2010a, "On the ALOS/PALSAR Operational and Interferometric Aspects (in Japanese)," *J. Geodetic Society of Japan*, Vol. 56, No. 1, pp. 13–39.

Shimada, M., 2010b, "Ortho-Rectification and Slope Correction of SAR Data Using DEM and Its Accuracy Evaluation," *IEEE J-STARS Special Issue on Kyoto and Carbon Initiative*, Vol. 3, No. 4, pp. 657–671.

Shimada, M. and Freeman, A., 1995, "A Technique for Measurement of Spaceborne SAR Antenna Patterns Using Distributed Targets," *IEEE T. Geo. Sci. Remote*, 33, No. 1, pp. 100–114.

Shimada, M. and Isoguchi, O., 2002, "JERS-1 SAR Mosaics of Southeast Asia Using Calibrated Path Images," *Int. J. Remote Sens.*, Vol. 23, No. 7, pp. 1507–1526.

Shimada, M., Isoguchi, O., Longepe, N., Preesan, R., Motooka, T., Okumura, T., et al. 2011, "Generation of the 10 m Resolution L-Band SAR Global Mosaic and Forest/Non-Forest Map," *Proc. ISRSE*, Sydney, Australia, April 14, 2011.

Shimada, M., Isoguchi, O., Tadono, T., and Isono, K., 2009, "PALSAR Radiometric and Geometric Calibration," *IEEE T. Geo. Sci. Remote*, Vol. 47, No. 2, pp. 3915–3932.

Shimada, M., Itoh, T., Motooka, T., Watanabe, M., Shiraishi, T., Thapa, R., and Lucas, R., 2014, "New Global Forest/Non-forest Maps from ALOS PALSAR Data (2007-2010)," *Remote Sensing of Environment*, Vol. 155, pp. 13–31.

Shimada, M. and Ohtaki, T., 2010, "Generating Continent-scale High-quality SAR Mosaic Datasets: Application to PALSAR Data for Global Monitoring," *IEEE JSTARS Special Issue on Kyoto and Carbon Initiative*, Vol. 3, No. 4, pp. 637–656.

Shimada, M., Tadono, T., and Rosenqvist, A., 2010, "Advanced Land Observing Satellite (ALOS) and Monitoring Global Environmental Change," *Proc. IEEE*, Vol. 98, No. 5, pp. 780–799.

Shiraishi, T., Motooka, T., Thapa, R. B., Watanabe, M., and Shimada, M., 2014, "Comparative Assessment of Supervised Classifiers for Land Use–Land Cover Classification in a Tropical Region Using Time-Series PALSAR Mosaic Data," *IEEE JSATRS*, 2014, Vol. 7, No. 4, pp. 1186–1199.

Small, D., 2011, Flattening Gamma: Radiometric Terrain Correction for SAR Imagery, *Transaction on Geoscience and Remote Sensing*, Vol. 49, No. 8, pp. 3081.

Solomon, S., Qin, D., Manning, M., Chen, Z., Marquis, M., Averyt, K. B., Tignor, M., and Miller, H. L., 2007, *Contribution of Working Group I to the Fourth Assessment Report of the Intergovernmental Panel on Climate Change*, Cambridge University Press, Cambridge and New York.

Spalding, M., Blasco, F., and Field, C., 1997, *World Mangrove Atlas*, International Society for Mangrove Ecosystems, Okinawa, Japan.

Spalding, M., Kainuma, M. and Collins, L. 2010, *World Atlas of Mangroves*, Earthscan, London.

Thapa, R. B., Itoh, T., Shimada, M., Watanabe, M., Motohka T., and Shiraishi, T., 2013, "Evaluation of ALOS PALSAR Sensitivity for Characterizing Natural Forest Cover in Wider Tropics, *Remote Sens. Environ.*, Vol. 155, pp. 32–41.

Watanabe, M., Shimada, M., Rosenqvist, A., Tadono, T., Matsuoka, M., Romshoo, S. A., Ohta, K., Furuta, R., Nakamura, K., and Moriyama, T., 2006, "Forest Structure Dependency of the Relation Between L-Band $sigma^0$ and Biophysical Parameters," *IEEE T. Geo. Sci. Remote*, Vol. 44, pp. 3154–3165.

Way, J. and Smith, E. A., 1992, "The Evolution of Synthetic Aperture Radar Systems and Their Progression to the EOS SAR," *IEEE T. Geo. Sci. Remote*, Vol. 29, No. 6, pp. 962–985.

World Wildlife Fund (WWF) 2007, "Russia's Boreal Forests," WWF, Washington, DC.

Zink, M. and Rosich, B., 2002, "Antenna Elevation Pattern Estimation from Rain Forest Acquisitions," *ENVISAT/ASAR Calibration Review (ECR) of ESTEC*, European Space Agency (ESA), Noordwijk, Netherlands.

WWW REFERENCES

WWW1: http://www.eorc.jaxa.jp/ALOS/en/palsar_fnf/fnf_index.htm

WWW2: http://www.google.co.jp/intl/ja/earth/

WWW3: http://confluence.org/

WWW4: JAXA/EORC (http://www.eorc.jaxa.jp/ALOS/en/palsar_fnf/fnf_index.htm)

Index

Note: *f* indicate figure; *t* indicate table.

T - #0171 - 111024 - C392 - 254/178/18 - PB - 9780367570798 - Gloss Lamination